Water-Wise Cities and Sustainable Water Systems: Concepts, Technologies, and Applications

Water-Wise Cities and Sustainable Water Systems: Concepts, Technologies, and Applications

Edited by
Xiaochang C. Wang and Guangtao Fu

IWA PUBLISHING

Published by **IWA Publishing**
Unit 104-105, Export Building
1 Clove Crescent
London E14 2BA
Telephone: +44 (0)20 7654 5500
Fax: +44 (0)20 7654 5555
Email: publications@iwap.co.uk
Web: www.iwapublishing.com

First published 2021

British Library Cataloguing in Publication Data
A CIP catalogue record for this book is available from the British Library

ISBN: 9781789060751 (paperback)
ISBN: 9781789060768 (eBook)

Contents

Part II: New Paradigm of Systems Thinking and Technology Advances

Part III: Practices of Water-Wise Cities and Sustainable Water Systems

Preface

As urban areas expand worldwide and become densely populated, there is a growing interest in the building of water-wise cities in developed and developing countries alike. This growing interest is driven by the severe limitation of available water resources to meet the rapid urbanization and the need to improve urban liveability. There have been various approaches and conceptual developments for urban water management in various regions and countries, such as Low Impact Developments (LID) in North America, Sustainable Drainage Systems (SuDS) in Europe, Water Sensitive Urban Design (WSUD) in Australia, and more recently, Sponge Cities in China. In 2016, the International Water Association (IWA) launched the Principles for Water-Wise Cities at the Brisbane World Water Congress based on experiences from across the world. There is thus a need to compile a book providing comprehensive insights into the theoretical, systematic, and practical engineering aspects of water-wise cities with broad coverage of global issues.

In September 2018, Professor Xiaochang C. Wang (Xi'an University of Architecture and Technology, China) and Professor Guangtao Fu (The University of Exeter, UK) coordinated a three-day Workshop on Water-Wise Cities and Smart Water Systems in Xi'an, China jointly supported by the National Natural Science Foundation of China (NSFC) and British Council. This workshop covered topics on the concepts, technologies, systems analyses and case studies.

Editors: Xiaochang C. Wang and Guangtao Fu
doi: 10.2166/9781789060768_xv

It was well attended by senior researchers and early-career water professionals from both countries, where high-level presentations and thorough discussions were carried out on the latest developments and future research trends in China, UK, and beyond. Following the outputs from the Workshop and based on the common interests and motivations of the two research groups (The State International S&T Cooperation Centre for Urban Alternative Water Resources Development in Xi'an and The Centre for Water Systems in Exeter), it was agreed that editing a book on water-wise cities and sustainable water systems would be highly beneficial to the wider international research community.

This book aims to: (1) provide a critical review on theoretical developments in water-wise cities and the associated smart water systems, including the key concepts and principles, (2) provide new thinking on the design and management of sustainable urban water systems of various scales toward a paradigm shift under resource and environmental constraints, and (3) provide a technological package with successful examples of technology selection, integration, and optimization on a 'fit-for-purpose' basis. In addition to the outcomes from the Workshop, this book also includes contributions from senior researchers in the United States and Australia with their regional experiences and advances.

Per the abovementioned aims, this book consists of 15 chapters, which are categorized into three distinctive but interrelated parts as follows:

(1) Part I: Water Management Concepts and Principles (Chapters 1–5)
(2) Part II: New Paradigm of Systems Thinking and Technology Advances (Chapters 6–10)
(3) Part III: Practices of Water-Wise Cities and Sustainable Water Systems (Chapters 11–15)

Critical reviews of the existing literature (especially those published in the past 10 years) and/or provision of the latest cases and examples are the main features of each part and chapter. It is hoped to provide the latest information to a wide range of water professionals including researchers and practitioners across the world.

As co-editors of this book, Professor Xiaochang C. Wang is responsible for editing Chapters 2, 3, 4, 6, 11, 12, 13 and 14, whereas Professor Guangtao Fu is responsible for editing Chapters 1, 5, 7, 8, 9, 10 and 15.

The process of preparing this book, including the authors invitation, writing up, and editing, was severely hampered by the COVID19 pandemic. However, we are happy that most of the chapters in our book writing plan can be presented to readers in the present form, although slightly behind the schedule. Therefore, we are grateful to all the contributors for their hard work in this challenging time.

We sincerely thank the four reviewers of our book proposal for their valuable and insightful comments and suggestions, which helped to significantly improve the organization of this book more appropriately to meet the readers' expectations.

Gratitude is also given to Dr. Mawuli Dzakpasu, Associate Professor, School of Environmental and Municipal Engineering, Xi'an University of Architecture and Technology, for carefully proofreading all the chapters contributed by authors from China, and Ms. Fei Cong, Office Manager, The State International S&T Cooperation Centre for Urban Alternative Water Resources Development, for her valuable assistance in the whole process of editing the book.

Book editors

Xiaochang C. Wang
Professor
State International S&T Cooperation Centre for Urban
Alternative Water Resources Development
Xi'an University of Architecture and Technology, Xi'an, China

Guangtao Fu
Professor
Centre for Water Systems
College of Engineering, Mathematics and Physical
Sciences, University of Exeter, Exeter, UK

About the Editors

Xiaochang C. Wang is a Professor of Environmental Engineering at Xi'an University of Architecture and Technology, and Director of State International S&T Cooperation Centre for Urban Alternative Water Resources Development. His research focus is on wastewater treatment and reuse, urban water environmental protection and improvement, and urban water management. He is a Distinguished Fellow of the International Water Association (IWA). He has authored over 300 papers in international peer-reviewed journals, and received a number of awards including the 2012 IWA Project Innovation Award (Global Honour Award) and the 2014 Chinese National Science and Technology Progress Award.

Guangtao Fu, Professor of Water Intelligence at the University of Exeter, has a research focus on developing and applying new computer models, data analytics and artificial intelligence tools to tackle urban water challenges in water resources, urban flooding and urban wastewater management. Currently he is an IWA Fellow, a Royal Society Industry Fellow and a Turing Fellow at the Alan Turing Institute. He has authored over 130 papers in international peer-reviewed journals and conference papers, and received several awards including the 2014 'Quentin Martin Best Practice Oriented Paper' and the 2018 'Best Research-Oriented Paper' awards from the American Society of Civil Engineers.

Chapter 1

Pathways towards sustainable and resilient urban water systems

Guangtao Fu and David Butler

Centre for Water Systems, University of Exeter, Harrison Building, North Park Road, Exeter EX4 4QF, United Kingdom

1.1 INTRODUCTION

Urban Water Infrastructure (UWI) plays a central role in safeguarding water security and public health and welfare. Its main functions include abstracting, treating and delivering drinking water to communities and cities, collecting and treating wastewater to a standard before it can be safely discharged into a receiving water body, and collecting stormwater to prevent urban flooding. Traditionally, UWI consists of water supply systems, water distribution systems, water treatment works, urban drainage systems and wastewater treatment works; these systems were gradually built into a city and were generally designed, operated and managed in isolation without considering their interdependencies and wide impacts on the economy and society.

Nowadays, the function of UWI goes far beyond providing water and wastewater services in cities. The potential value of blue green infrastructure, which is part of the UWI, is recognised in climate change adaptation, reduction of heat island impacts, improvement of biodiversity, and community amenity. UWI also plays a key role in reduction, reuse and recovery of resources through optimisation of the water-energy-materials nexus, helping embedding a circular economy in our society. In the era of big data and artificial intelligence, UWI digitalisation is an

Editors: Xiaochang C. Wang and Guangtao Fu
doi: 10.2166/9781789060768_0003

essential component in the development of smart cities. However, operation and management of UWI systems are faced with huge challenges in population increase, urbanisation, climate change, stringent regulation and aging infrastructure (Larsen *et al.*, 2016; Yuan *et al.*, 2019). These challenges are explained in more detail in Chapter 2. The future UWI needs to meet the ever-evolving societal needs and challenges to achieve resilient and sustainable water management.

While it is difficult to define what the next generation UWI looks like, it is possible to identify the potential pathways that can lead to the future UWI. This chapter aims to provide such pathways from analysing the historical evolution of urban water systems. These pathways form a roadmap that provides a broad guide on the development of UWI. The Safe and SuRe framework is then introduced for intervention development that aims to transform existing water systems to sustainable and resilient ones.

1.2 THE EVOLUTION OF URBAN WATER SYSTEMS

Historically, UWI has evolved with the increasing needs of our society, as demonstrated in Figure 1.1, which shows the transitions framework for urban water management proposed by Brown *et al.* (2009). Though this framework was developed in the context of Australia, it represents a general transition pathway for cities moving towards sustainable urban water management.

From Roman times, water infrastructure was built to provide drinking water to growing cities such as Roman aquaducts. This is the stage of 'water supply city'

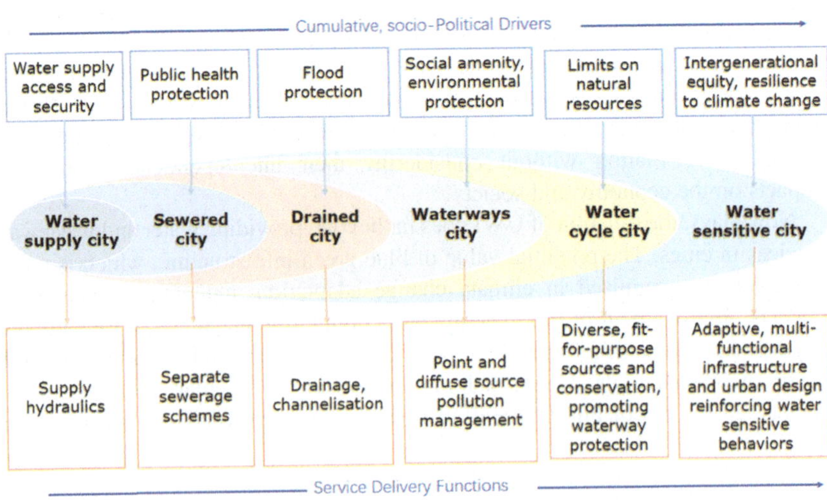

Figure 1.1 The transitions of urban water infrastructure (adapted from Brown *et al.*, 2009).

which aimed to provide water security for city residents. The Romans also built artificial drains, amongst which the most well-known is the cloaca maxima, built to drain the Roman Forum into a river (Butler & Davies, 2010). Only after the 19th century, however, did the extensive sewer system begin to be built in cities to tackle deteriorating public health problems such as outbreaks of water-borne diseases (e.g., cholera and typhoid) due to rapid population growth in cities (Burian *et al.*, 2000). This is the stage of 'sewered city' with a focus on the protection of public health. For example, in London, flush toilets discharging to cesspits became common around 1770–1780, however, connecting cesspits to the sewer system remained illegal until 1815 (Butler & Davies, 2010). This resulted in a serious health problem when the population of London expanded to more than one million. The only solution was to allow cesspit overflow to be connected to the sewer system. However, this moved the problem to the River Thames, which became heavily polluted by the 1850s. Following the great stink in 1858, an interceptor sewer system was built to take wastewater to the Thames Estuary, downstream of the main urban areas. Obviously it did not completely solve the problem but simply moved it downstream, polluting the estuary and its banks. This pollution problem started to improve only when biological wastewater treatment began to be built in the 1920s.

In the next stage of water system evolution, the separate sewer system was built to remove stormwater to tackle urban flooding which had become a high risk for many cities due to significantly increased impervious areas and population densities. As early as in the 1840s, Edwin Chadwick suggested the idea of separate systems, where wastewater is separated from surface runoff, to solve the River Thames pollution problem (De Feo *et al.*, 2014). However, at that time, it was impossible due to the complexity and capital costs of the dual system. It started to be adopted and implemented in practice in developed countries only after the 1940s. It was a period of rapid population growth and urbanisation following the Second World War, which significantly increased the risk of flooding in cities. This is referred to as 'drained city', which aimed to protect the city from flooding by quickly transporting excess stormwater downstream.

The 1970s saw a rapid environmental movement around the world, as many people worried about environmental catastrophe following substantial urban expansion of several decades. In urban water management, the focus was on reduction of pollutant discharges into water bodies and green measures such as wetlands, and bio-retention systems began to be developed and implemented in urban areas. The environmental movement eventually led to the development of the concept of sustainability in the 1980s. In this period, Sustainable Drainage Systems (SuDS) were developed in the UK as a new approach for urban stormwater management, aiming to achieve sustainable development. SuDS design generally considers achieving the benefits from four categories: water quantity, water quality, amenity and biodiversity. The concepts of waterways city and water cycle city reflect the concerns on environmental protection and limits of natural resources, respectively.

Water Sensitive Urban Design (WSUD) was a concept initially developed in Australia and accepted in many other countries. With WSUD, *water is given due prominence within the urban design process through the integration of urban design with the various disciplines of engineering and environmental sciences associated with the provision of water services including the protection of aquatic environments in urban areas*, according to the definition by Wong and Brown (2009). Building on waterways city and water cycle city, water sensitive city integrates urban water management with the natural and built environment and seeks to maximize opportunities for living with and exploiting the supply, use, reuse and management of water and stormwater to enhance and support human health and well-being by minimising the impacts of urbanisation on the natural environment and water cycle (Ashley *et al.*, 2013).

Sedlak (2015) presented a different revolution-based pathway for urban water management, largely driven by tackling human crises through technological advances. The first revolution was the development of the drinking water *supply* system, that is, the first generation of Urban Water Infrastructure (UWI 1.0) to meet the growing water demand in cities. The second revolution was the development of the drinking water *treatment* facility (UWI 2.0) using technologies such as filtration and chlorination to address the public health crisis in the late 19th and early 20th centuries. The third revolution was the invention and deployment of the wastewater treatment facility (UWI 3.0) as exemplified by activated sludge process to protect the environment. Chapter 2 provides more information on the historical development of water treatment and wastewater treatment technologies. The incoming fourth revolution is to achieve self-sufficiency through diversified water sources such as grey water, stormwater, and seawater. Again, this will be driven by developing new treatment technologies to meet stringent water quality standards.

1.3 PATHWAYS TOWARDS SUSTAINABLE WATER SYSTEMS

While it is difficult to be specific about the characteristics of future urban water systems, several general trends have become clear through research and practice in the last several decades: decentralisation, greening, circular economy, and digitalisation. These could be regarded as the pathways leading towards sustainable urban water systems (Figure 1.2) and are discussed below.

1.3.1 Decentralisation

Centralisation has been the key design principle of urban water systems from Romans water supply to modern urban water systems. In a centralised system,

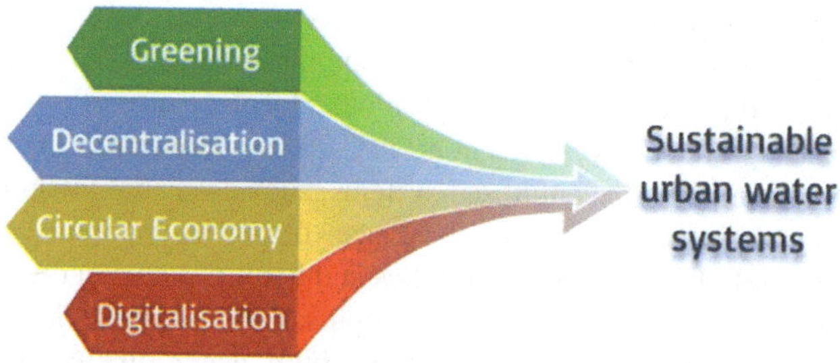

Figure 1.2 Pathways towards sustainable urban water systems.

water is often taken from a long distance and distributed from water treatment works (WTWs) at high pressure over a large area; and wastewater and stormwater are collected again for centralized treatment. Centralised systems, which are currently the normative and predominating paradigm in developed countries, can achieve cost savings in capital investments, operations and maintenance based on the theory of the economy of scale. In a centralised system, however, water, energy and materials may be unnecessarily lost, wasted and misused. For example, over 3 billion litres of water, which is about 20% of water consumption, is lost daily through leakage in the water distribution systems of England and Wales (Consumer Council for Water, 2019). Similarly in sewer systems, some pumping is unavoidable in many cases to divert wastewater to a centralised treatment plant, though collection and conveyance of wastewater mostly rely on gravity-based systems.

Decentralised systems have been regarded by many as a new paradigm for the future water system (Arora *et al.*, 2015). In a decentralised system, water is supplied from local and diversified sources and wastewater is treated within individual houses, buildings and communities. The key feature of decentralised systems is the use of regional or local facilities for water supply and wastewater treatment without long-distance transport and large-scale facilities (such as water treatment, wastewater treatment and pumping stations) so that they are more flexible than centralised systems to adapt to future uncertainties such as environmental change. Water supply self-sufficiency can be achieved with locally available water sources through improved water use efficiency, stormwater harvesting, greywater reuse, and recycled water. Decentralisation represents a pathway to the fourth urban water revolution envisioned by Sedlak (2015).

There has been much debate on centralised versus decentralised systems (Makropoulos & Butler, 2010). It was argued that decentralised wastewater systems would have the same capital, operating and maintenance costs as centralised systems to achieve similar levels of service (Ho & Anda, 2006), though decentralised systems are generally thought to have a greater energy and carbon footprint. Decentralised systems have many economic, social and environmental benefits, such as water saving through leakage reduction in large water distribution networks, flexibility in adapting to changing environments, lifestyle enhancement due to private green space and property value, and reduction of overflows. It is also suggested that decentralisation promotes local re-use of water and thus increase water productivity (Larsen *et al.*, 2016). However, there are many challenges in developing decentralised systems, such as spatial integration of such systems, energy intensity, social resistance from the public, lack of clear legislation, lack of clarity on responsibilities and liabilities (Arora *et al.*, 2015).

Technological advances will make decentralisation more feasible. On-site water and wastewater treatment can now achieve high water quality standards. For example, ecological wastewater treatment systems (e.g., wetlands and biofilters) and package biological plants have been applied in many countries in the world to treat residential, industrial, and municipal wastewater. Large-scale applications of these systems could potentially reduce the costs and improve system performance significantly.

1.3.2 Greening

The idea of green urban structures can be traced back to the concepts of urban farming and garden allotments in the 1870s (Pötz & Bleuzé, 2011). It is only since the 1980s, however, that concepts and terminologies, which highlight the importance of the use of nature-based solutions, have begun to emerge for urban water management, specifically in the area of urban stormwater management. The most common terminologies include low impact development (LID), Best Management Practices (BMPs), sustainable drainage systems (SuDS), water sensitive urban design, Urban Green Infrastructure (UGI) or Green Infrastructure, Blue-Green Infrastructure, the Blue-Green City (Thorne, 2020) and the Sponge City. These terminologies were developed in different countries and contexts and have different focuses and scopes, however they are all based on the same broad principle: mitigating the impact of urban developments through mimicking nature and achieving wider benefits than water quantity and quality (Fletcher *et al.*, 2015).

UGI has a very board scope far beyond water management. The concept of UGI originated from the field of landscape architecture and ecology in the USA in 1990s; it has been promoted as a network of near-natural and designed spaces and elements in cities, planned and maintained in such a way that the infrastructure as a whole offers high quality in terms of utility, biodiversity and aesthetic appeal while also delivering a broad range of ecosystem services. Indeed, the following strategic objectives are normally considered in developing UGI: improving health and quality of life, conserving biodiversity and ecological integrity, promoting social cohesion and inclusion, enhancing community resilience to environment change, boosting local economic development and attracting businesses. In a sense, it provides ecosystem services to urban residents through the value of green space.

In the field of urban stormwater management, UGI is defined as '*a network of decentralized stormwater management practices, such as green roofs, trees, rain gardens and permeable pavement, that can capture and infiltrate rain where it falls, thus reducing stormwater runoff and improving the health of surrounding waterways*' (Foster *et al.*, 2011). UGI is now widely accepted as part of UWI and implemented worldwide. It represents an approach that can adapt to local circumstances and address local concerns. In practice, it is common that the focus and strategic objective of UGI could vary significantly in different regions and countries. For example, it is mainly used for water scarcity management in Cape Town as discussed in Chapter 8.

In England, SuDS is primarily promoted to reduce flood risk. The current National Planning Policy Framework requires that priorities should be given to sustainable drainage and the impact of new development on flood risk should be considered. Non-statutory technical standards set out specific requirements on flood risk of both outside and inside the development as below (Defra, 2015).

For flood risk outside the development:

> 'For greenfield developments, the peak runoff rate from the development to any highway drain, sewer or surface water body for the 1 in 1 year rainfall event and the 1 in 100 year rainfall event should never exceed the peak greenfield runoff rate for the same event.
>
> Where reasonably practicable, for greenfield development, the runoff volume from the development to any highway drain, sewer or surface water body in the 1 in 100 year, 6 hour rainfall event should never exceed the greenfield runoff volume for the same event.'

For flood risk within the development:

> 'The drainage system must be designed so that, unless an area is designated to hold and/or convey water as part of the design, flooding does not occur on any part of the site for a 1 in 30 year rainfall event.
>
> The drainage system must be designed so that, unless an area is designated to hold and/or convey water as part of the design, flooding does not occur during a 1 in 100 year rainfall event in any part of: a building (including a basement); or in any utility plant susceptible to water (e.g., pumping station or electricity substation) within the development.
>
> The design of the site must ensure that, so far as is reasonably practicable, flows resulting from rainfall in excess of a 1 in 100 year rainfall event are managed in exceedance routes that minimise the risks to people and property.'

The narrow focus on flood risk in English legislation may be useful to encourage wide adoption of green infrastructure in flood-prone cities. However, its wider benefits in improving water security, urban pollution, ecosystem integrity, carbon reduction, public health and well-being should be recognised in the planning process (Thorne, 2020). Indeed, substantial evidence shows that it is an effective way to reduce the risks of climate extremes (Royal Society, 2014). The United Nations' Sustainable Development Goals and New Urban Agenda call for increased efforts in the development of UGI to tackle urban challenges and UGI should be put at the heart of governments' policies to achieve the long-term resilience and sustainability of cities.

The challenge of moving towards greening urban water infrastructure also lies in large-scale impact assessment and implementation, in addition to legislations, cost-effectiveness and community engagement. The delivery of expected services from UGI relies on spatial integration of individual green spaces and components, as implied in its definition as a network. This should go beyond urban catchments, where the focus is primarily on stormwater and wastewater management, to include the surrounding hinterlands and wider rural catchments. Nature-based solutions have been promoted for flood and water resources management at the catchment and river basin scales; they can improve the quality of water supply sources and reduce the fluvial flood risk in cities.

1.3.3 Circular economy

1.3.3.1 The linear model

The development of UWI has been following a linear 'take-make-use and dispose' model of growth in which resource recovery and reuse are not considered in the planning and design process, as illustrated in Figure 2.3 in Chapter 2. Historically, water resources were taken from the surrounding hinterlands or across river basins to meet growing water demands in cities; wastewater and storm water were regarded as waste products which need be removed from the site as soon as possible for treatment and disposal. Even today, linear thinking is still widely used to tackle water problems in many countries. When the local water sources run dry or become polluted, water is transferred from further away. This can be illustrated using the example of China where water is taken from an ever-longer distance via inter-basin water transfer projects to supply water scarce cities (Figure 1.3).

China has constructed a large number of water transfer projects to meet burgeoning water demands from rapidly urbanizing cities and expanding economies across its 10 first-order basins (Figure 1.3a). Water is transferred mainly from the Yangtze, Yellow and Southwest River basins to the Hai, Huai and Yellow River basins. It should be noted that the Yellow River Basin delivers water to other basins while receiving water mainly from the Yangtze River. The inter-basin water transfer capacity had been steadily increasing until 2000 when a building boom occurred to meet demands from rapidly expanding cities and economies (Figure 1.3b). One of the key projects constructed was the largest South-to-North water transfer project in the world; it includes eastern and central routes, each of which covers a distance of more than 1000 km and crosses four major river basins: the Yangtze River, Yellow River, Huai River, and Hai River, with a capacity of 25 billion m^3 yr^{-1} (Ding *et al.*, 2020). As a result, in 2016, ~48.3 billion m^3 yr^{-1} could be transferred via open channels or pipelines, many of which require pump stations. This is compared to the total demand of

(a)

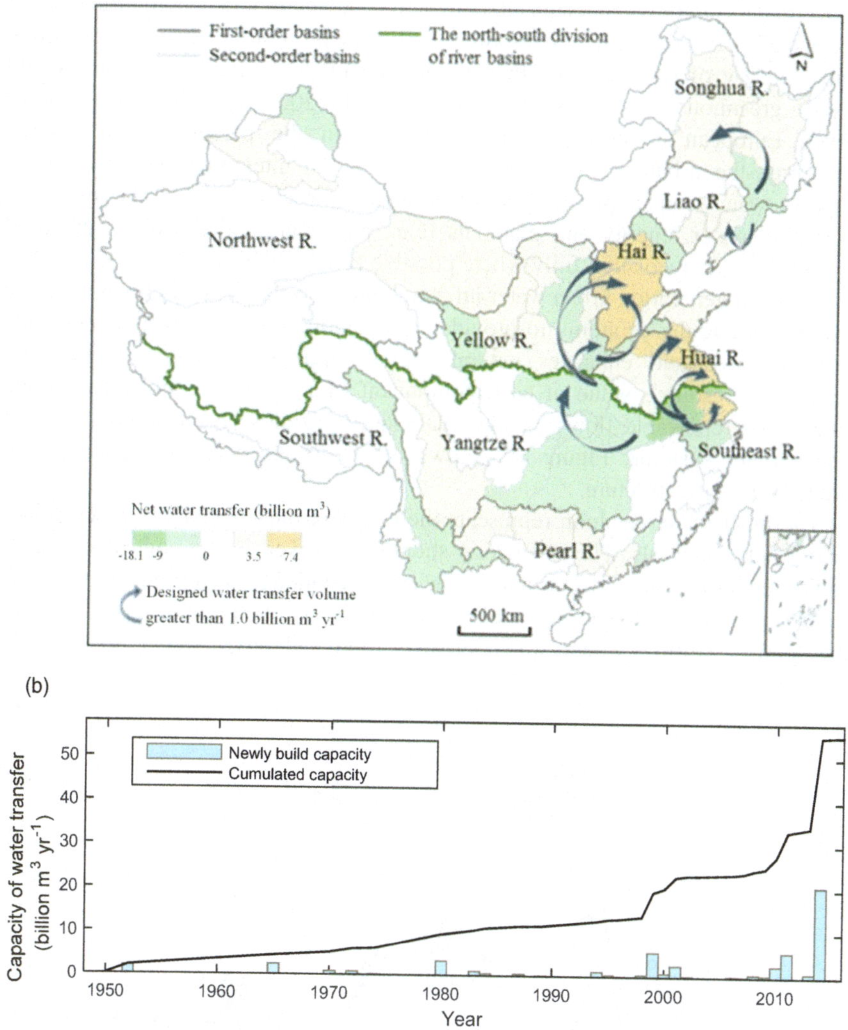

Figure 1.3 Inter-basin water transfer projects in China. (a) The water transfer volumes and directions. (b) Annual increase in water transfer capacity since 1950s.

~600 billion $m^3\ yr^{-1}$ in China, which includes residential, industrial, agricultural and ecological demands. The residential water demand was about 82.2 billion $m^3\ yr^{-1}$ in 2016. The large scale of water transfer in China demonstrates the dominance of linear thinking in the practical water management and the challenges in achieving water supply self-sufficiency in the region and river basin levels.

1.3.3.2 The circular economy model

The circular economy model, an alternative to the linear model, would turn goods at the end of their service life into resources for others, promoting reuse, recycle, repair and recovery of resources where possible. A shift to a circular economy could reduce greenhouse gas emissions by up to 70% by 2030, according to a study of seven European countries (Stahel, 2016). From the perspective of circular economy, the urban water system is defined as an integrated system of water supply, water consumption, wastewater and stormwater, where the resources including water, energy and materials (e.g., chemicals and biosolids) are used sustainably and recovered fully where possible (IWA, 2016).

The next generation urban water infrastructure, UWI 4.0, should be built on the concept of circular economy to provide a continuous positive development cycle that preserves and enhances natural capital, optimises resource yields, and maximizes resource value at each component of a system by managing finite stocks and renewable flows. A key challenge is to close the resources loop in linear water systems. Figure 1.4 shows the interlinks between resources and urban water infrastructure.

The inner resources loop represents the complex interrelations between water, energy and materials. The inner loop should be viewed in the context of water-energy-environment nexus, which is one grand challenge facing humanities. The water-energy-environment nexus has drawn increasing attention in recent years as the 'perfect storm', where water, energy, food and environment crises could

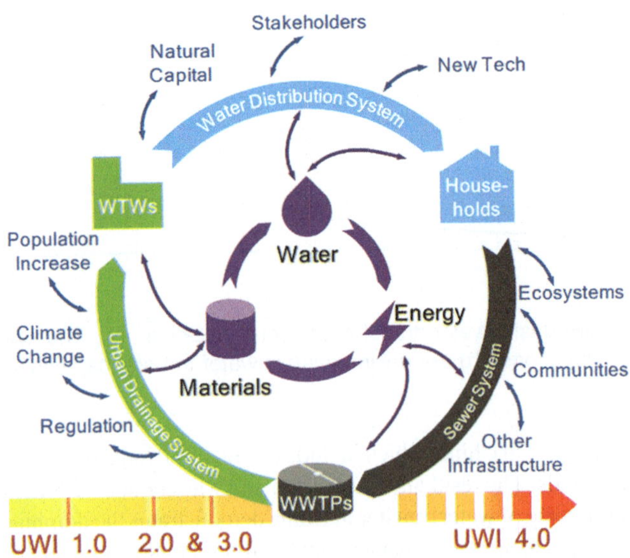

Figure 1.4 The water-energy-materials nexus with the urban water system.

occur simultaneously and thus significantly exaggerate the situation, became a concern for policy and decision makers (Olsson, 2015). The water sector is an energy-intensive sector, which consumes 4% of the energy globally. Understanding the nexus in the urban water systems is key to: (1) identify the fundamental and critical interrelations between factors; and (2) identify options to improve resource efficiency.

The outer water infrastructure loop integrates various water systems as an entire urban water infrastructure system. The outer loop builds on the integrated approach for urban water management (Butler & Schütze, 2005; Fu *et al.*, 2008). This requires a system of systems approach to consider interdependent natural, infrastructure and economic systems of different scales such as the hydrological system, land use, agricultural system, transportation system, building system, and social system (Bach *et al.*, 2014). Chapter 10 discusses the latest advances in integrated modelling and control of urban wastewater systems.

The two loops are intertwined. The architecture of water infrastructure will shape the flows of resources, while the nexus of water, energy and materials will determine the performance of water infrastructure. For example, the 'sulfide problem' in sewers, where sulfate (SO_4^{2-}) in wastewater is biologically converted into toxic hydrogen sulfide gas (H_2S) and further to corrosive sulfuric acid (H_2SO_4) under anaerobic conditions, leads to noxious odours and damage to sewer systems. Recent research reported that replacing a coagulant in the treatment of water, the SO_4^{2-} concentration in the wastewater can be reduced such that H_2S no longer affects sewer infrastructure (Pikaar *et al.*, 2014; Rauch and Kleidorfer, 2014). This example shows how the use of material in water treatment can affect the system performnace of the downstream sewers.

Understanding the two loops and their linkages is key to moving towards UWI 4.0. Research challenges remain in many areas, such as identifying key resources flows and their impacts on the nexus of water-energy-materials, characterising the UWI to maximize the reuse and recovery of resources, developing new approaches to upgrade and retrofit existing water systems. Such systems should be able to adapt to future uncertainties, while supporting both rural and urban communities and economies.

1.3.4 Digitalisation

The use of data analytics to support decisions on urban water management can be traced back to the work of John Snow in the 1850s when he analysed the spatial data of cholera victims in London to identify the source of the cholera outbreak (Eggimann *et al.*, 2017). This exemplar case illustrates the importance of data availability in tackling public health crises. Since then, data analytics has played an ever-increasing role in improving urban water and wastewater services. One of the milestones is the development and application of computer models for water management that emerged in the 1950s. This led to the establishment of a new research area: hydroinformatics (Abbott, 1991). In the early 2000s, new opportunities arose for water research communities and water utilities when deep learning technologies began to be developed and applied to a range of industries.

The Smart Water Networks Forum used a five-layer architecture to describe smart water networks (SWAN, 2020). This architecture is revised here to represent the key components of smart urban water infrastructure: physical infrastructure, sensing and control, collection and communication, data management and decision support systems, and data analytics and artificial intelligence. Figure 1.5 illustrates the five-layer architecture using the water supply system.

The physical layer consists of the grey and green structures that carry out hydrological, hydraulic, chemical and ecological functions (e.g., infiltration, storage, conveyance, treatment and purification of water, stormwater or wastewater) in the UWI system, such as swales, retention ponds, reservoirs, combined sewer overflows, storage tanks, pipes, water and wastewater treatment facilities. The evolution of the physical layer is described in Section 1.2.

The sensing and control layer consists of sensors, controllers, and actuators that carry out the remote control function. This layer is the hardware part that sits in or comes into contact with the physical water infrastructure, such as flow and pressure sensors, and remote-controlled devices. Those components (e.g., gates, valves), which cannot be remote-controlled, are part of the physical layer, as they do not have the data interfaces as part of the UWI smartness.

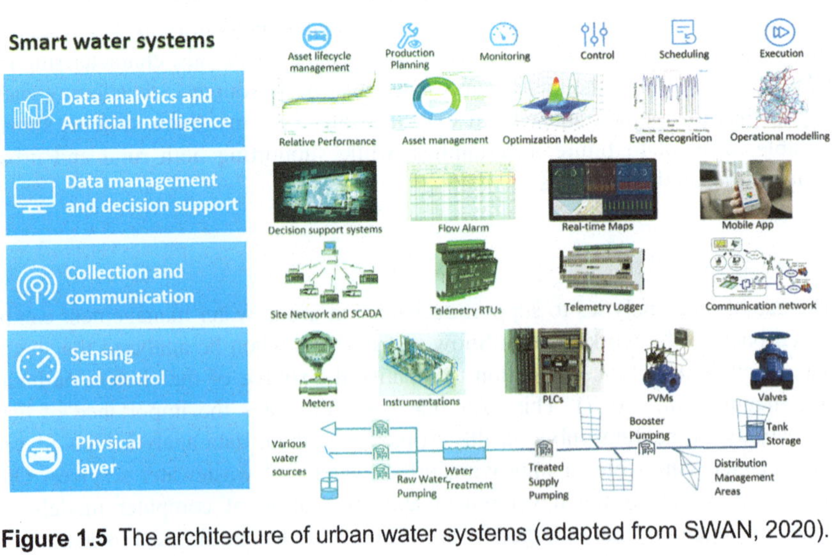

Figure 1.5 The architecture of urban water systems (adapted from SWAN, 2020).

The collection and communication layer carries out data collection, transmission, and storage functions. This layer relies on information and communication technologies such as mobile networks, satellites, and cloud data storage and thus is closely linked with cyber and communications infrastructure.

The next two layers – data management and decision support, and data analytics and AI – are the soft part of the architecture, which aims to provide optimal planning and management of UWI. It consists of databases, decision support systems, computer simulation models, machine learning and artificial intelligence algorithms.

The development of Instrumentation, Control and Automation (ICA) systems is a key part of the digitalisation pathway. ICA began to be applied in the water and wastewater treatment works as early as the 1970s (Yuan *et al.*, 2019). In the last decades, it has gained wide application in the water and wastewater industry, mainly due to the following reasons: (1) increasing pressure on the existing system capacity from urbanisation, climate change, stringent regulation, and aging infrastructure; (2) advances in the development of ICA technologies which enable real-time data collection and control while driving down the cost of ICA system installation; (3) increasing computing power that enables real-time, online optimisation; and (4) advances in the field of hydroinformatics with significantly improved modelling and predictive accuracy.

Digitalisation will not only transform how UWI is planned, operated and managed but will span into a wide range of issues including the nature of workforce operations, customer experience, and the role of the water sector in the development of smart and sustainable cities (IWA, 2019).

It should be noted that these pathways towards sustainable urban water systems are not mutually exclusive but inter-connected. For example, digitalisation can crossover all other pathways as digital solutions can be implemented to monitor and control green infrastructure, improve resource recovery efficiency through optimal control of the water-energy-material nexus, and accelerate decentralisation through autonomous operations of water systems locally and at a small scale rather than in a centralised control room.

1.4 A NEW PARADIGM TOWARDS SUSTAINABLE WATER MANAGEMENT

A fundamental question remains open: how do we know if an urban water system is sustainable? A prerequisite for this question is that the targets or ending points for sustainable water systems are known. It is been suggested that sustainability is not about the destination but better conceptualized as a journey (Butler & Davies, 2010). The four pathways discussed above provide a broad guide on the journey towards sustainable water systems. To ensure that our journey follows the pathways, performance measures are needed to assess the efficiency and effectiveness of the delivery of services from water infrastructure systems and provide an insight on where we are compared to past values, targets or other systems' performance;

frameworks are needed to develop intervention strategies to improve the system performance so that we do not move away from the pathways.

In this section, the intervention framework developed from the Safe & SuRe project is suggested as a new paradigm that enables the transformation of the existing water system towards a more sustainable and resilient one. More information can be found in the work by Butler *et al.* (2016).

1.4.1 Performance measures

It is argued that the performance of urban water systems can be broadly described within three categories: reliability, resilience and sustainability (Butler *et al.*, 2016), which are explained below.

Reliability has been at the heart of water system design and management for centuries. System reliability is defined as 'the degree to which the system minimises level of service failure frequency over its design life when subject to standard loading' (Butler *et al.*, 2016). It is generally interpreted as a probabilistic term considering uncertain design circumstances (e.g., stress and shock) and system responses which could occur within a specified design life. In the context of hydraulic reliability, the water distribution system design problem is normally to maximize the likelihood of water demand and/or pressure being met across all the nodes in the network and/or over an extended period where projected demands during normal or emergency water supply are considered. The sewer system design problem normally aims to minimize the likelihood of sewer flooding or surcharge across the network, given design rainfall events. Wastewater treatment plants are designed for the specified effluent quality being met. Reliability-based design and management aims to provide fail-safe performance. In other words, it simply aims to avoid failure as far as is (cost-effectively) possible but does not consider what happens when failure occurs.

The concept of resilience has been developed in the last decades to handle failures which occur when the system is subject to extreme shocks, exceeding the design conditions. In this context, resilience is defined as 'the degree to which the system minimises level of service failure magnitude and duration over its design life when subject to exceptional conditions' (Butler *et al.*, 2016). Essentially, resilience is a measure of system performance during or after a failure event, such as, can the system fail slowly, to what extend and magnitude the system can fail, how quickly and to what level can the system recover? There are various resilience measures for water distribution systems (Diao *et al.*, 2016; EPA, 2015; Meng *et al.*, 2018; Zhan *et al.*, 2020), urban drainage systems (Mugume *et al.*, 2015; Wang *et al.*, 2019) and wastewater treatment plants (e.g., Juan-García *et al.*, 2017; Meng *et al.*, 2017; Sweetapple *et al.*, 2017). All these measures aim to minimize the magnitude and duration of failure. Note, resilience has also been interpreted in a much broader context to include system properties and institutional capacity such as preparedness, recovery, robustness, redundancy, resourcefulness, and vulnerability. Resilience-based design and management aims to overcome failure and embraces a more 'safe to fail' concept.

Sustainability is typically represented by three pillars: social, economic and environmental. Thus, it is defined as 'the degree to which the system maintains levels of service in the long-term whilst maximising social, economic and environmental goals' (Butler *et al.*, 2016). This reflects the situation where many water infrastructure systems can provide services beyond their design life. For example, many sewer systems built in the Victorian era are still used today in the UK. Sustainability indicators vary greatly dependent on their purposes, contexts and users. For example, the International Water Association provided a comprehensive set of performance indicators for water supply services, which covers water resources, personnel, physical, operational, quality of service and economic and financial indicators (Alegre *et al.*, 2017). This is compared to a set of 18 indicators, grouped in the following four areas: customer experience, reliability and availability, environmental impact, and financial, which are used by the regulator Ofwat to measure the performance of water utilities in England. Sustainability aims to measure the performance at all levels, both above and below the required level of services.

Reliability, resilience and sustainability seem to cover different aspects of performance, yet are interlinked with a relationship: reliability is necessary but not sufficient for resilience, and resilience is necessary but not sufficient for sustainability. It is necessary to simultaneously consider multiple performance measures in the planning and management process (Casal-Campos *et al.*, 2015; Fu *et al.*, 2013).

1.4.2 Intervention framework

1.4.2.1 Four types of intervention

The Safe & SuRe framework for interventions development, as shown in Figure 1.6, describes the relationships between threats, UWI, impacts and consequences, and identifies four types of interventions that can be implemented to achieve long-term resilience and sustainability. A key feature of the framework is the distinction between impact and consequence. Impact describes the non-compliance degree of water services delivered by UWI, while consequences represent social, economic and environmental outcomes for a recipient from any impact.

Mitigation addresses the link between threat and water infrastructure and aims to reduce the likelihood, magnitude and duration of a threat through local or global actions. Examples include measures to reduce greenhouse gas emissions to mitigate climate change impacts which then reduce the likelihood of extreme weather, that is, an external threat to UWI.

Adaptation addresses the link between UWI and impact and is defined as any action to modify specific system properties aiming to enhance the system

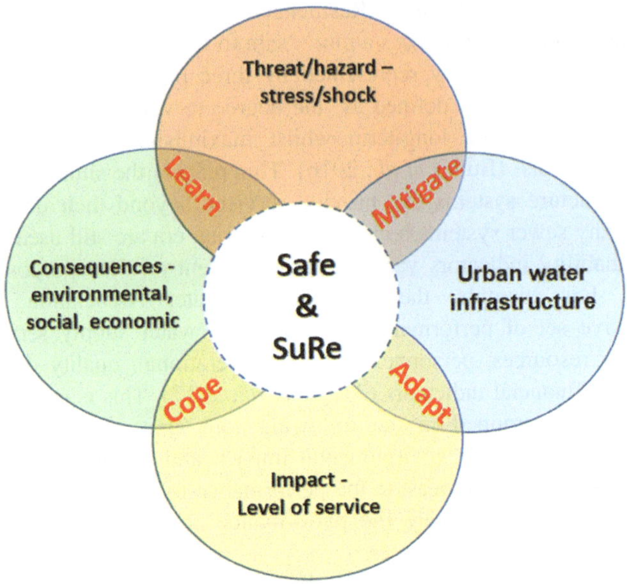

Figure 1.6 The new paradigm for sustainable urban water systems: Safe & SuRe.

capacity. It addresses the system failure that could result from any threats that cannot be mitigated. An example is use of green infrastructure to reduce urban flooding. The focus is on the UWI and its failure models, so it does not matter whether flooding results from climate change or urbanisation.

Coping addresses the link between impact and consequences and is defined as 'any preparation or action taken to reduce the frequency, magnitude or duration of the effects of an impact on a recipient' (Butler *et al.*, 2016). In an event of flooding, the consequences include damages to properties and traffic interruptions; the corresponding coping measures can be buying house insurance and taking a different route to avoid the flooded areas, respectively.

Learning addresses the link between consequences and threats, which closes the loop of the threat-system-impact-consequence chain. It aims to embed experiences and new knowledge for best practice. Examples of learning approaches include developing best practice from past events, and establishing pilot schemes to gain new knowledge or demonstrate best practice.

1.4.2.2 Analysis approaches

Distinguishing the four types of intervention maximizes the opportunity to develop effective intervention strategies. This is explicitly explored through four analysis directions: top-down, middle state-based, bottom-up and circular (Figure 1.7).

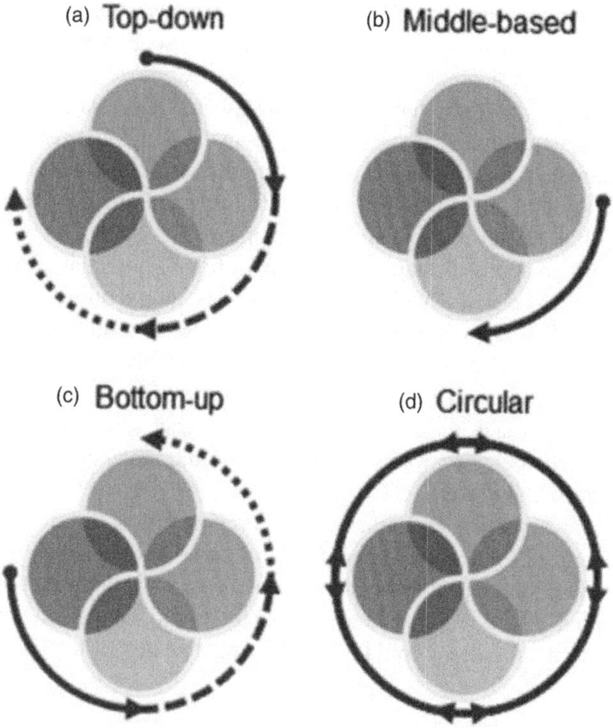

Figure 1.7 Four analysis approaches using the Safe & SuRe paradigm. Interlocking circles are interpreted in Figure 1.6. (a) Top-down. (b) Middle-based. (c) Bottom-up. (d) Circular. (Adapted from Butler *et al.*, 2016.)

The top-down approach starts from identifying and characterising the potential threats, and then propagates them through the water system to analyse the impact or consequences. This is the most commonly used approach in water management. For example, flood risk assessment is a typical top-down approach, and it assesses the consequences of hazards (e.g., rainfall events) on the water system. In practice, it is common to consider a single threat or consider different threats separately for assessment. Though there are numerous studies on characterising threats and their uncertainties, the key challenge with this approach lies in the identification of all threats, in particular, black swan (unknown) events, which are of most concern in achieving long-term resilience. Further, the top-down approach has difficulty in identifying effective interventions in the water system as their links to threats are often unclear.

The middle-based approach starts from identifying failure modes in the water system and then assesses their effects. It shifts the emphasis from identification of multiple threats to system failure modes (i.e., middle states), which are normally well-understood by domain experts. Most importantly, this makes it easier to identify interventions in response to system failure so that system performance could be improved. Compared to the top-down approach, this approach can effectively address multiple threats (including unknown events), which may result in the same system failure mode, in a single analysis. Global resilience analysis can be conducted on the basis of the middle-state approach for the water distribution system (e.g., Diao *et al.*, 2016), urban drainage (Mugume *et al.*, 2015) and urban wastewater system (Sweetapple *et al.*, 2019).

The bottom-up approach starts from identification of potential social, economic, or environmental consequences and then assesses interventions to reduce these consequences. This approach focuses on vulnerability reduction in the face of various threats and deep uncertainties and is increasingly used in water resource management in response to climate change (e.g., Poff *et al.*, 2016). It should be noted that the approach may start at either the impact stage or the system stage and proceed anticlockwise when the focus is on the water infrastructure and its service levels (i.e., impacts).

The circular approach encompasses all components of the framework with a focus on learning. When the three types of interventions, that is, mitigation, adaptation and coping are implemented in a strategy, it is important to understand their combined effects so that new knowledge could be gained and necessary adjustments to interventions could be implemented as part of the learning process.

1.5 CONCLUSIONS

This chapter analyses the potential pathways moving from the water systems of today to the next generation of systems. Four pathways are identified through the analysis of the historical evolution of urban water systems: decentralisation, greening, circular economy and digitalisation. These four pathways, which are not mutually exclusive but inter-connected, form a roadmap that provides a broad guide on the development of UWI towards sustainable water management. Three categories of performance measure, that is, reliability, resilience and sustainability, are suggested to assess the performance of water systems. The Safe and SuRe framework is then introduced for intervention development that can transform existing water systems to sustainable and resilient ones. This framework maximizes the opportunities to develop different types of interventions through four analysis approaches: top-down, middle state-based, bottom-up and circular.

Achieving smart and resilient water systems is a huge challenge. It requires an overhaul of institutional and regulatory systems, which explicitly facilitate the transformation through the Safe and SuRe framework to improve system performance. There are already many initiatives, such as Water Wise Cities promoted by the International Water Association, and many pilot cities practising along the pathways in the world, as discussed in the following chapters. This will provide valuable insight in the journey of water system transformation.

ACKNOWLEDGEMENTS

We would like to acknowledge the financial support from the UK EPSRC (Refs: EP/N010329/1 and EP/K006924/1) and Royal Society (Refs: IF160108 and IEC\NSFC\170249). Guangtao Fu is also supported by The Alan Turing Institute under the EPSRC (Ref EP/N510129/1).

REFERENCES

Abbott M. B. (1991). Hydroinformatics: Information Technology and the Aquatic Environment. Avebury, UK.

Alegre H., Baptista J. B., Cabrera E., Cubillo F., Duarte P., Hirner W., Merkel W. and Parena R. (2017). Performance Indicators for Water Supply Services. IWA Publishing, London, UK.

Arora M., Malano H., Davidson B., Nelson R. and George B. (2015). Interactions between centralized and decentralized water systems in urban context: a review. *Wiley Interdisciplinary Reviews: Water*, **2**(6), 623–634. doi: 10.1002/wat2.1099

Ashley R., Lundy L., Ward S., Shaffer P., Walker L., Morgan C., Saul A., Wong T. and Moore S. (2013). Water-sensitive urban design: opportunities for the UK. *Proceedings of the Institution of Civil Engineers: Municipal Engineer*, **166**(2), 65–76. doi: 10.1680/muen.12.00046

Bach P. M., Rauch W., Mikkelsen P. S., Mccarthy D. T. and Deletic A. (2014). A critical review of integrated urban water modelling – urban drainage and beyond. *Environmental Modelling and Software*, **54**, 88–107. doi: 10.1016/j.envsoft.2013.12.018

Brown R. R., Keath N. and Wong T. H. F. (2009) Urban water management in cities: historical, current and future regimes. *Water Science and Technology*, **59**(5), 847–855. doi: 10.2166/wst.2009.029

Burian S. J., Nix S. J., Pitt R. E. and Durans S. R. (2000). Urban wastewater management in the United States: past, present, and future. *Journal of Urban Technology*, **7**(3), 33–62. doi: 10.1080/713684134

Butler D. and Davies J. W. (2010). Urban Drainage, 3rd edn. Spon Press, London, UK.

Butler D. and Schütze M. (2005). Integrating simulation models with a view to optimal control of urban wastewater systems. *Environmental Modelling and Software*, **20**, 415–426.

Butler D., Ward S., Sweetapple C., Astaraie-Imani M., Diao K., Farmani R. and Fu G. (2016). Reliable, resilient and sustainable water management: the Safe and SuRe approach. *Global Challenges*, **1**(1), 63–77. doi: 10.1002/gch2.1010

Casal-Campos A., Fu G., Butler D. and Moore A. (2015). An integrated environmental assessment of green and gray infrastructure strategies for robust decision making. *Environmental Science and Technology*, **49**(14), 8307–8314. doi: 10.1021/es506144f

Consumer Council for Water. (2019). Water Water Anywhere. Available online: https://www.ccwater.org.uk/wp-content/uploads/2019/09/Water-water-everywhere-delivering-resilient-water-and-waste-water-services-2018-19.pdf, (accessed 28 August 2020).

Defra. (2015). Non-statutory Technical Standards for Sustainable Drainage Systems. Available online: https://assets.publishing.service.gov.uk/government/uploads/system/uploads/attachment_data/file/415773/sustainable-drainage-technical-standards.pdf, (accessed 28 August 2020).

De Feo G., Antoniou G., Fardin H. F., El-Gohary F., Zheng X., Reklaityte I., Butler D., Yannopoulos S. and Angelakis A. N. (2014). The historical development of sewers worldwide. *Sustainability*, **6**, 3936–3974. doi: 10.3390/su6063936

Diao K., Sweetapple C., Farmani R., Fu G., Ward S. and Butler D. (2016). Global resilience analysis of water distribution systems. *Water Research*, **106**, 383–393. doi: 10.1016/j.watres.2016.10.011

Ding W., Liu H., Li Y., Shang H., Zhang C. and Fu G. (2020). Unravelling the effects of long-distance water transfer for ecological recharge. *Journal of Water Resources Planning and Management*, **146**(9), 02520004. https://doi.org/10.1061/(ASCE)WR.1943-5452.0001272

Eggimann S., Mutzner L., Wani O., Schneider M. Y., Spuhler D., de Vitry M. M., Beutler P. and Maurer M. (2017). The potential of knowing more: a review of data-driven urban water management. *Environmental Science and Technology*, **51**(5), 2538–2553. doi: 10.1021/acs.est.6b04267

EPA. (2015). Systems Measures of Water Distribution System Resilience. U.S. Environmental Protection Agency, Washington, DC. EPA/600/R-14/383.

Fletcher T. D., Shuster W., Hunt W. F., Ashley R., Butler D., Arthur S., Trowsdale S., Barraud S., Semadeni-Davies A., Bertrand-Krajewski J. -L., Mikkelsen P. S., Rivard G., Uhl M., Dagenais D. and Viklander M. (2015). SUDS, LID, BMPs, WSUD and more – The evolution and application of terminology surrounding urban drainage. *Urban Water Journal*, **12**(7), 525–542. doi: 10.1080/1573062X.2014.916314

Foster J., Lowe A. and Winkelman S. (2011). The Value of Green Infrastructure for Urban Climate Adaptation. Centre for Clean Air Policy, Washington, DC.

Fu G., Butler D. and Khu S. T. (2008). Multiple objective optimal control of integrated urban wastewater systems. *Environmmental Modelling and Software*, **23**, 225–234.

Fu G., Kapelan Z., Kasprzyk J. and Reed P. (2013). Optimal design of water distribution systems using many objective visual analytics. *Journal of Water Resources Planning and Management*, **139**(6), 624–633.

Ho G. and Anda M. (2006). Centralised versus decentralized wastewater systems in an urban context: the sustainability dimension. In: Proceeding of the 2nd IWA Leading-Edge Conference on Sustainability, B. Beck and A. Speers (eds.), IWA Publishing, London.

IWA. (2016). Water Utility Pathways in a Circular Economy. IWA Publishing, London, UK.

IWA. (2019). Digital Water: Industry Leaders Chart the Transformation Journey. IWA Publishing, London, UK.

Juan-García P., Butler D., Comas J., Darch G., Sweetapple C., Thornton A. and Corominas L. (2017). Resilience theory incorporated into urban wastewater systems management. State of the art. *Water Research*, **115**, 149–161. doi: 10.1016/j.watres.2017.02.047

Larsen T. A., Hoffmann A., Lüthi C., Truffer B. and Maurer M. (2016). Emerging solutions to the water challenges of an urbanizing world. *Science*, **352**(6288), 928–933.

Makropoulos C. K. and Butler D. (2010). Distributed water infrastructure for sustainable communities. *Water Resources Management*, **24**(11), 2795–2816. doi: 10.1007/s11269-010-9580-5

Meng F., Fu G. and Butler D. (2017). Cost-effective river water quality management using integrated real-time control technology. *Environmental Science and Technology*, **51** (17), 9876–9886. doi: 10.1021/acs.est.7b01727

Meng F., Fu G., Farmani R., Sweetapple C. and Butler D. (2018). Topological attributes of network resilience: a study in water distribution systems. *Water Research*, **143**, 376–386. doi: 10.1016/j.watres.2018.06.048

Mugume S. N., Gomez D. E., Fu G., Farmani R. and Butler D. (2015). A global analysis approach for investigating structural resilience in urban drainage systems. *Water Research*, **81**, 15–26. doi: 10.1016/j.watres.2015.05.030

Olsson G. (2015). Water and Energy: Threats and Opportunities. IWA Publishing, London. https://doi.org/10.2166/9781780406947

Pikaar I., Sharma K. R., Hu S., Gernjak W., Keller J. and Yuan Z. (2014). Reducing sewer corrosion through integrated urban water management. *Science*, **345** (6198), 812–814. doi: 10.1126/science.1251418

Pötz H. and Bleuzé P. (2011). Urban Green-blue Grids for Sustainable and Dynamic Cities. Coop for Life, Delft. ISBN 978-90-818804-0-4.

Poff N. L., Brown C. M., Grantham T. E., Matthews J. H., Palmer M. A., Spence C. M., Wilby R. L., Haasnoot M., Mendoza G. F., Dominique K. C. and Baeza A. (2016). Sustainable water management under future uncertainty with eco-engineering decision scaling. *Nature Climate Change*, **6**(1), 25–34. doi: 10.1038/nclimate2765

Rauch W. and Kleidorfer M. (2014). Replace contamination, not the pipes. *Science*, **345** (6198), 734–735.

Royal Society. (2014). Resilience to Extreme Weather. Available online: https://royalsociety.org/topics-policy/projects/resilience-extreme-weather/, (accessed 28 August 2020).

Sedlak D. (2015). Water 4.0: The Past, Present, and Future of the World's Most Vital Resource. Yale University Press, Yale, USA.

Stahel W. R. (2016). The circular economy. *Nature*, **531**, 435–438. doi:10.1038/531435a

SWAN. (2020). A Layered View of Data Technologies for the Water Distribution Network. Available online: https://www.swan-forum.com/swan-tools/a-layered-view/, (accessed 26 August 2020).

Sweetapple C., Fu G. and Butler D. (2017). Reliable, robust, and resilient system design framework with application to wastewater-treatment plant control. *Journal of Environmental Engineering (United States)*, **143**(3), 04016086. doi: 10.1061/(ASCE)EE.1943-7870.0001171

Sweetapple C., Fu G., Farmani R. and Butler D. (2019). Exploring wastewater system performance under future threats: does enhancing resilience increase sustainability? *Water Research*, **149**. doi: 10.1016/j.watres.2018.11.025

Thorne C. R. (2020). Blue-green Cities: Integrating Urban Flood Risk Management with Green Infrastructure. ICE Publishing, London.

Wang Y., Meng F., Liu H., Zhang C. and Fu G. (2019). Assessing catchment scale flood resilience of urban areas using a grid cell based metric. *Water Research*, **163**, 114852. doi: 10.1016/j.watres.2019.114852

Wong T. H. F. and Brown R. R. (2009). The water sensitive city: principles for practice. *Water Science and Technology*, **60**(3), 673–682. http://dx.doi.org/10.2166/wst.2009.436

Yuan Z., Olsson G., Cardell-Oliver R., van Schagen K., Marchi A., Deletic A., Urich C., Rauch W., Liu Y. and Jiang G. (2019). Sweating the assets – the role of instrumentation, control and automation in urban water systems. *Water Research*, **155**, 381–402. doi: 10.1016/j.watres.2019.02.034

Zhan X., Meng F., Liu S. and Fu G. (2020). Comparing performance indicators for assessing and building resilient water distribution systems. *Journal of Water Resources Planning and Management*, **146**(12), 06020012. doi.org/10.1061/(ASCE)WR.1943-5452.0001303

Chapter 2

Water-wise cities and sustainable water systems: Current problems and challenges

Xiaochang C. Wang[1] *and Li Luo*[2]

[1]*State International S&T Cooperation Centre for Urban Alternative Water Resources Development, Xi'an University of Architecture and Technology, Xi'an, China*

[2]*School of Environmental and Municipal Engineering, Xi'an University of Architecture and Technology, Xi'an, China*

2.1 INTRODUCTION

Sustainable urban development is facing difficulties due to the limitation of available water resources to meet the increasing needs associated with fast urbanization in many countries and regions, especially the developing world. Following the discussion in Chapter 1 regarding the pathways towards sustainable and resilient urban water systems, this chapter is to provide a more detailed review of the current problems and needs for a paradigm shift in water system planning under the constraints of resources and environmental capacities.

Various resources, including water, are graces from the nature to human beings, but have their limitations due to the natural processes of their renewal and regeneration. For a very long time until the mid-1900s, natural resources had been considered unlimited, and if only technologically and economically feasible, exploitation of sufficient resources was the sole strategy to meet the requirement of urban development. The urban water system, covering the whole cycle of water supply, wastewater disposal and final discharge back to the nature, was

Editors: Xiaochang C. Wang and Guangtao Fu
doi: 10.2166/9781789060768_0025

traditionally a linear system characterized as 'end-of-the-pipe' (Novotny *et al.*, 2010) without worrying about its impact on water source availability and environmental quality. However, the explosive population growth between the mid-1950s and mid-1970s and the continuous growth in later years, associated with even faster urbanization, much wrecked the balance between water source availability and water demand. Continuous construction and/or expansion of the urban water systems following the traditional manner inevitably brought about overexploitation of water resources which seriously damaged the source water system, deteriorated the aquatic environment, and degraded the urban liveability.

Under such a circumstance, needs were growing worldwide for taking brand-new strategies and measures to change our traditional manner of water system design, operation, and management. The turn of the century thus became the turning point for water professionals to strive for building the so-called water-wise cities with sustainable water systems to solve the current problems and adapt to the ever-changing situations in the future.

2.2 FACTS OF OUR LIVING CONDITIONS ON THE EARTH

2.2.1 Population and cities

The latest estimation of the world's population in August 2019 is about 7.723 billion (World Population Review, 2019), a significant increase over that of 7.2 billion in 2015. About 53.2% of these people currently live in urban areas (Worldometers, 2019) and the percentage is projected to increase to 68% by 2050 (UN Department of Economic and Social Affairs, 2019).

Urbanization usually results from industrialization and economic development. Taking China as an example (Figure 2.1(a)), as a typical agricultural country from the mid-1950s through to the early 1980s, its urban population remained less than 20%. However, increased urbanization was seen in the recent 30 years along with its rapid economic growth. The urban population of China reached 50% in 2010, and of the current population of 1.42 billion, 60.4% are living in cities. Urbanization has also been speeded up in India, the country with the second largest population in the world (Figure 2.1(b)). The urban population was less than 20% before the early 1970s when its total population was about 550 million but increased to over 33% recently along with the remarkable total population growth to about 1.37 billion in 2019. Compared to developing countries, urbanization in developed countries proceeded much earlier. In the USA for example (Figure 2.1(c)), the urban population in 1955 was already as high as 66.8% but only gradual increases were seen afterward to reach the current level of 83.9%.

According to the United Nations' data booklet (UN Department of Economic and Social Affairs, 2019), by 2018, there were 33 megacities with inhabitants of more than 10 million in the world, and 48 cities with inhabitants between five and 10 million. An overwhelming majority of the world's cities have fewer than five

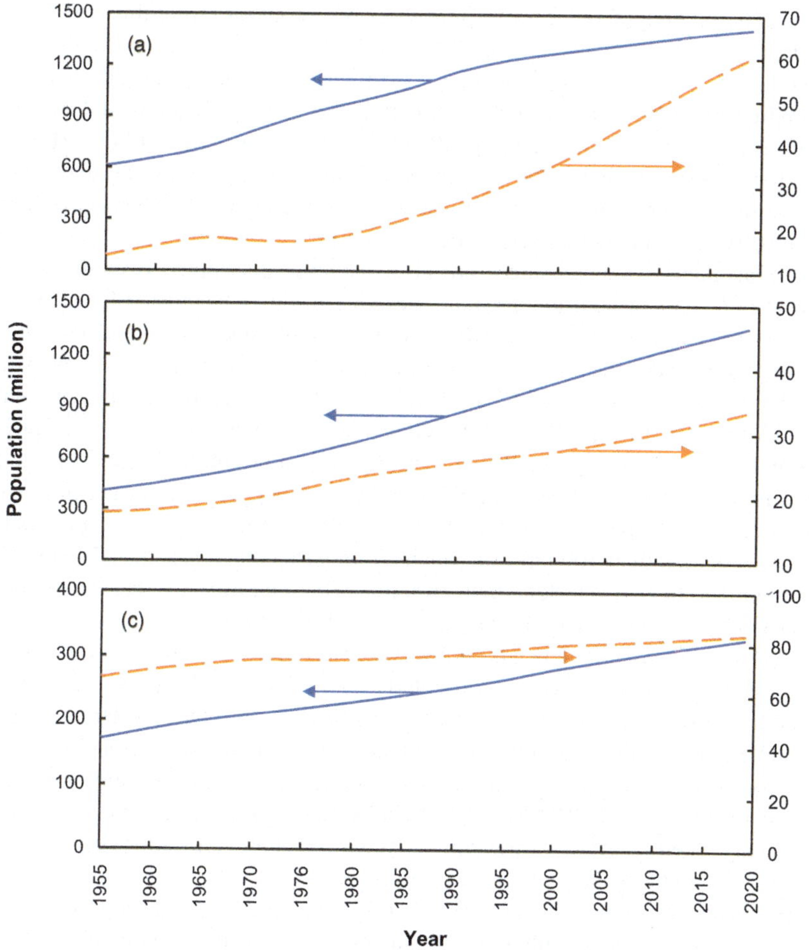

Figure 2.1 Total population and urban population percent in China (a), India (b), and USA (c) from 1955 to 2019. (Data source: Worldometers, 2019.)

million inhabitants. In 2018, there were 467 cities with between one and five million inhabitants and an additional 598 cities with between 500,000 and one million inhabitants. The concentration of inhabitants in cities, especially the large cities, is the most prominent feature of our modern society.

2.2.2 Available water resources

Water resources are natural resources that are potentially useful to human beings. About 97.4% of the water on earth is saline (96.5% in oceans and 0.9% in other saline waters). Only about 2.5% is freshwater but about 68.7% of this is frozen in glaciers and polar ice caps. About 30.1% of the freshwater is groundwater, while about 1.2% is surface and/or other freshwater, including ground ice and permafrost (69.0%), lakes (20.9%), soil moisture (3.8%), swamps and marshes (2.6%), rivers (0.49%), living things (0.26%) and the atmosphere (3.0%) (USGS, 2019). The water available for human use is limited to the freshwater extractable from groundwater aquifers and surface water (mostly rivers), which takes only about 0.003% of the global water (Gleick, 2014).

The available water resources are also unevenly distributed over the world. Taking into account both the internal renewable resource calculated based on the internal river flows and groundwater from rainfall and the population in a country, the renewable internal freshwater resources per capita varies between about 0 m^3 in Kuwait to 10.66 million m^3 in Greenland based on 2014 data (IndexMundi, 2019). In addition to Greenland, there are five other countries (Iceland, Guyana, Suriname, Papua New Guinea, and Bhutan) with per capita water resources of more than 1.0 million m^3 due to their rich freshwater resources and small population. In contrast, there are 44 countries with per capita water resources of less than 1000 m^3, of which 27 countries have less than 500 m^3 and 10 countries have even less than 100 m^3. The majority of these extremely water-deficient countries are in Africa and the Middle East.

Almost every country depends on freshwater resources to sustain domestic water supply and economic activities, particularly agriculture and industry. Agriculture (including irrigation, livestock, and aquaculture) is by far the largest water consumer, accounting for 69% of annual water withdrawals globally, while industry (including power generation) accounts for 19% and domestic (household) for 12% (WWAP, 2019). Recreational and environmental water uses have long been paid less attention in water resource allocation but recently growing percentages are seen in many countries.

The Water Resources Institute (WRI) has developed the updated Aqueduct™ water risk framework combining 13 water risk indicators, including quantity, quality, and reputational risks, into a composite overall water risk score (Hofste *et al.*, 2019). One important indicator is the Baseline Water Stress (BWS) that measures the physical water shortage as a ratio of total water withdrawals, including domestic, industrial, irrigation, and livestock consumptive and non-consumptive uses, to the available renewable surface and groundwater supplies, including the impact of upstream consumptive water users and large dams on downstream water availability. The BWS score ranges between 1 and 5, with higher values indicating more competition among users. An assessment based on current situations in 189 countries (WRI, 2019) indicated that 17 countries (Qatar, Israel, Lebanon, Iran, Jordan, Libya, Kuwait, Saudi Arabia, Eritrea, United Arab Emirates, San Marino, Bahrain, India, Pakistan, Turkmenistan, Oman, and Botswana) have BWS scores ranging from 4.97 to 4.02 and belong to extremely high-risk countries. Another 27 countries (Chile,

Cyprus, Yemen, Andorra, Morocco, Belgium, Mexico, Uzbekistan, Greece, Afghanistan, Spain, Algeria, Tunisia, Syria, Turkey, Albania, Armenia, Burkina Faso, Djibouti, Namibia, Kyrgyzstan, Niger, Nepal, Portugal, Iraq, Egypt, and Italy) have BWS scores ranging from 3.98 to 3.01 and belong to high-risk countries.

Within a country with a large territorial area, the situation may differ largely between different regions. For example, the 2019 BWS scores for China, USA, Australia and United Kingdom are 2.40 (medium–high risk), 1.85 (low–medium risk), 2.67 (medium–high risk), and 1.40 (low–medium risk), respectively. However, at the province/state level, there are areas with much higher BWS scores such as Beijing, Hebei, and Shandong with extremely high risks (4.55–4.11), and Shanxi, Tianjin, Nei Mongol, Xinjiang Uygur, Henan, and Ningxia Hui with high risks (3.72–3.05) in China. Similarly, New Mexico has an extremely high risk (4.26), and California, Arizona, Colorado, and Nebraska have high risks (3.72–3.16) in the USA, whereas Victoria has a high risk (3.53) in Australia (WRI, 2019).

Another indicator is the Drought Risk (DR), which measures where droughts are likely to occur, the population and assets exposed, and the vulnerability of the population and assets to adverse effects (Hofste *et al.* 2019). An assessment based on current worldwide situations (WRI, 2019) indicated that Moldova and Ukraine have the highest DR of 0.8162 and 0.8050, respectively, and are high draught risk countries. With DR scores of between 0.7909 and 0.6019, 43 countries (Bangladesh, India, Serbia, Syria, Morocco, Haiti, Romania, Indonesia, Cambodia, Togo, Uzbekistan, Poland, Lithuania, Hungary, Pakistan, Tunisia, Vietnam, Belarus, Cuba, Dominican Republic, Rwanda, Macedonia, Lebanon, Slovakia, Bosnia and Herzegovina, North Korea, Denmark, Bulgaria, Azerbaijan, Italy, Tajikistan, Burundi, Croatia, Nicaragua, Sri Lanka, Czech Republic, Ghana, China, Kenya, UK, France, Thailand, and Turkey) are medium to high drought risk countries.

In contrast to drought, usually as an event of prolonged shortages of natural precipitation, some countries and regions may suffer in areal flooding resulting from an overflow of water from water bodies, or an accumulation of rainwater on saturated ground. One indicator of the flooding risk is the Riverine Flood Risk (RFR) which measures the percentage of the population expected to be affected by riverine flooding in an average year, accounting for existing flood-protection standards (Hofste *et al.*, 2019). The RFR scores correspond to the risk of populations to be affected and are categorized into extremely high risk (more than one in 100), high risk (six in 1000 to one in 100), medium–high risk (2–6 in 1000), low–medium risk (1–2 in 1000), and low risk (0–1 in 1000). There are 32 countries assessed as extremely high RFR (Somalia, Mauritania, Liberia, Bangladesh, Cambodia, Myanmar, South Sudan, Kyrgyzstan, Afghanistan, Tajikistan, Djibouti, Vietnam, Laos, Indonesia, Guyana, Madagascar, Rwanda, Chad, Lebanon, Mozambique, Egypt, Mali, North Korea, Sierra Leone, Malawi, Guinea-Bissau, Papua New Guinea, Kenya, Sri Lanka, Yemen, Suriname, and Georgia), and 34 countries as high RFR (Sudan, Senegal, Eritrea, Liechtenstein,

Tanzania, Nepal, Niger, Mongolia, Libya, Morocco, Honduras, Guatemala, United Arab Emirates, Iraq, Central African Republic, Ghana, Syria, Pakistan, Guinea, Timor-Leste, Gambia, Belize, India, Benin, Democratic Republic of the Congo, Burundi, Bhutan, Philippines, Thailand, Gabon, Ethiopia, Cuba, Panama, and Togo) (WRI, 2019).

Notably, several Asian countries, such as Bangladesh, Cambodia, Afghanistan, Vietnam, and Indonesia, have both extremely high riverine flood risk and relatively high drought risk or baseline water stress. Extremely uneven rainfall and poor infrastructure for flood protection may be the main reason.

2.2.3 Imbalanced resource provision and consumption – biocapacity and ecological footprint as indicators

Population growth and urbanization inevitably result in increased consumption of various resources from the earth. Since the 1990s, biocapacity has been used to evaluate the capacity of the natural ecosystem to produce biological materials for human consumption and to absorb waste materials generated by humans, under current management schemes and extraction technology (Global Footprint Network, 2019a). To measure how much biocapacity, in terms of area of biologically productive land and water, an individual, population, or activity requires, Ecological Footprint (EF) has been introduced as an indicator, which can indicate the human impact on Earth's ecosystem (Global Footprint Network, 2019a).

According to the latest data on EF evaluation (Global Footprint Network, 2019b), of the 187 countries evaluated, 136 have a biocapacity deficit, namely ecological footprint exceeding biocapacity. Only 51 countries have a biocapacity reserve, namely biocapacity exceeding ecological footprint. For countries with large populations (>100 million), only Russia and Brazil are still with a biocapacity reserve. Others (China, India, USA, Indonesia, Pakistan, Nigeria, Bangladesh, Mexico, Japan, Ethiopia, Philippines, and Egypt) have biocapacity deficits exceeding 600%. Figure 2.2 shows the variation of world population, global biocapacity, and ecological footprint from 1961 to 2016. Before 1970, the global EF was lower than the global biocapacity, indicating that our world was with a biocapacity reserve until the world population reached 3.7 billion in 1970. However, with population growth, the global EF continued to increase, and our world turned to be biocapacity-deficit globally – an indication that our earth can no longer produce biological materials for human consumption and to absorb waste materials generated by humans under current management schemes and extraction technologies. In 2016, the global biocapacity was 12.17 billion global hectares (gha) but the global EF was as high as 20.51 billion gha. This means that to sustain human consumption of biological materials at the current level, we may need 1.7 planets like our earth to provide sufficient biocapacity, not accounting for the additional demand from the further increases in world

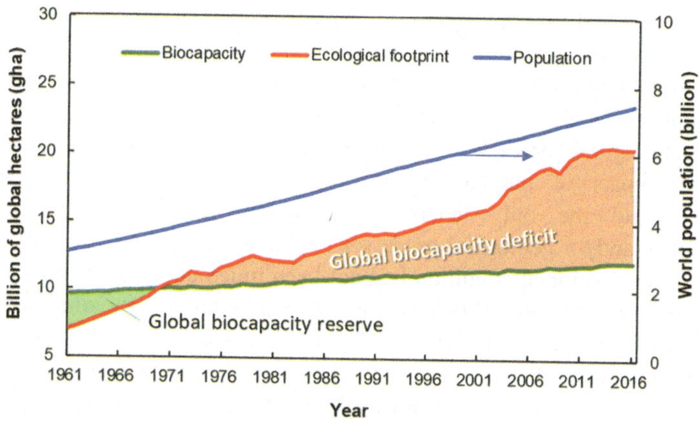

Figure 2.2 World population, biocapacity, and ecological footprint from 1961 to 2016. (Population data source: Worldometers, 2019; Biocapacity and ecological footprint data source: Global Footprint Network, 2019b.)

population, which was projected by the United Nation (UN, 2019) to be 8.55 billion in 2030, 9.74 billion in 2050 and 10.87 billion in 2100. Otherwise, the ecological condition of our earth will be considerably deteriorated due to the destructive exploitation of biological materials.

2.3 URBAN WATER SYSTEMS: HISTORY AND DEVELOPMENT

2.3.1 Water and human settlements

Water is indispensable for human life. For this simple reason, from the beginning of civilization, human beings located their settlements near waters (Carrasco & Dangol, 2019). The primary demand for water was for drinking and dining, so from the taste and appearance of the water, people understood which area was more suitable for living. This explains why human civilization originated and developed mostly on river banks, such as the close relationship of the Mesopotamia Civilization with the Tigris and Euphrates river system, the Egyptian Civilization with the Nile River, the Harappan Civilization with the Indus River, and the Chinese Civilization with the Yellow River (Isachenko, 2011). Along with population growth and enlargement of agricultural and commercial activities, the settlement scales expanded from scattered houses to villages, and then to towns and cities.

The primitive manner of water-use and used-water disposal was very simple. People just fetched water from a nearby water body and poured out the used water arbitrarily to a soiled ground surface or into natural ditches, and/or even back into the water body. This was, in fact, not a 'poor' manner but a 'luxury' one because at that time the ecological capacity was sufficient enough for either a water body or soil system to accommodate and assimilate the very low pollutant loading from human life. As long as the source water was fresh and clean, the water quality was always good and directly potable. The only thing people would do was select good source water for their consumption. On the other hand, at that time the impact of human activities on natural aquatic systems was also negligible and the only determinative factor was the action of natural hydrological and/or geological processes (Valhondo & Carrera, 2019).

The abovementioned condition had not been much changed in most regions of the world until the early 1700s before the first Industrial Revolution when the world population was far below one billion (Strong, 1990). The model of human's utilization of water in this long period can be characterized as 'nature dependent', namely, negligible artificial modification of natural water systems and full dependence on nature's action to secure water quality.

2.3.2 Pre-modern urban water systems

Even before the 1700s, a number of densely populated cities appeared in some regions of the world, such as Alexandria in Egypt with 300,000–600,000 inhabitants in 200BC and about one million inhabitants in 100BC (Goerke-Shrode, 1998), Rome, in Italy, with 1–1.2 million inhabitants in 1–300AD (Pena & Morley, 1998), Chang'an in China with up to one million inhabitants in 700–900AD (Zhang, 2008), and Baghdad in Iraq with 1.2 million inhabitants in 1000–1200AD (Bosworth, 1995). For these ancient large cities, engineered systems were required for both water supply and sanitation. Taking Rome as an example, a system of eleven aqueducts were built to supply the city with water from as far away as the river Anio. A complex system of sewers covered with stones was also provided for collecting waste flushed from the latrines which then flowed into a nearby river or stream. The Roman aqueducts provided the inhabitants of Rome with water of varying quality, the best being reserved for potable supplies, and poorer quality water being used in public baths and latrines. Another example is Chang'an, the ancient capital of China in 618–907AD (Wang & Chen, 2014). Located on the right bank of the Wei River, the largest tributary of the Yellow River, a series of open channels flowing through the city were built and a water network formed through connecting these channels with the Wei River and several branch rivers. The water network was used for both water supply and discharging used water according to the flowing direction of each channel, namely fetching potable water in the upstream side and discharging wastes in the downstream side, so the used water finally flowed back to the river. The design of such a water network for water supply and sanitation may have followed the belief that 'flowing water would not rot'. Although people at that time did not know the principle that a sufficient volume of flowing water could dilute the incoming pollutants and provide a good self-purification condition, they certainly understood from their experiences that drinking water should be collected from the upstream side of the channel and used water should be dumped to the downstream side. Good availability of natural flow much favored such an urban water network.

The engineered water system for ancient Rome and the semi-engineered water system for ancient Chang'an were also 'nature dependent' to a great extent for water quantity and quality in urban water supply and sanitation. The impact of artificial modification of natural water systems was very small due to the limited population and rich biocapacity.

2.3.3 Modern urban water systems

The so-called 'modern urban water system' is depicted conceptually in Figure 2.3.

The system consists of four subsystems: (1) source water subsystem for provision of source water for water supply, (2) water supply subsystem for purification of the source water to the quality meeting drinking water requirement (usually in a water treatment plant, WTP) and distributed to users in the urban area (usually through a distribution pipe network), (3) wastewater subsystem for collection of the used water from users (usually through a collection pipe network) and treatment of the collected wastewater to meet discharge requirement (usually in a wastewater treatment plant, WWTP), and (4) waste discharge subsystem for final disposal of the treated wastewater (usually to a receiving water body). It is notable that comparing with the pre-modern urban water systems discussed in section 2.3.2, WTP and WWTP are added to the system for artificial water quality conversion before supplying to the city and after being used.

2.3.3.1 Needs for drinking water purification

Until the early 1800s, water from natural sources was mostly used directly for drinking purposes without special treatment. However, people at that time knew

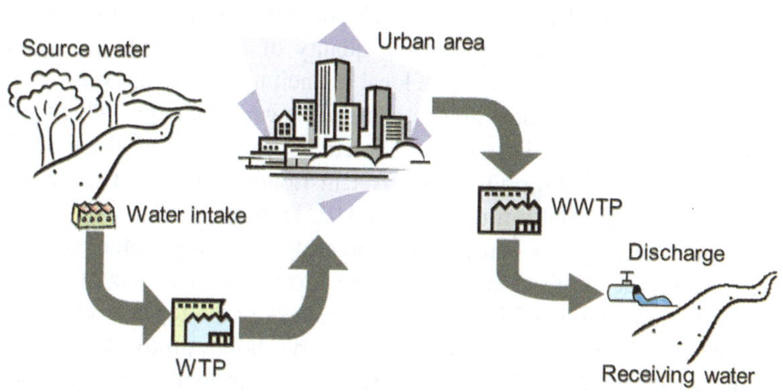

Figure 2.3 Conceptual depiction of the modern urban water system. (Figure by authors; WTP: water treatment plant; WWTP: wastewater treatment plant.)

that clarity (appearance) and palatability (taste) were important features of the water for drinking and they tried to obtain drinking water from streams and/or groundwater wells that appeared clean and tasted good. In case clean source water was unavailable, people knew from their primary experiences that after placing water in a vessel for a while, visible particulates (suspended matter) could settle to the bottom and the supernatant became clear. This was the early practice of 'sedimentation' for removing particulates from water. There were even recorded experiences of using natural potassium aluminum sulfate (crystalized $KAl(SO_4)_2 \cdot 12H_2O$) to reduce the visible cloudiness (turbidity), which was the early practice of 'coagulation' for more effective removal of suspended and colloidal particles from water (Faust & Aly, 1999).

The systematic drinking water purification practice in the world started in the early 1800s as symbolized by the use of slow sand filters in Paisley, Scotland in 1804, and Chelsea, London in 1829. The installation of the filtration facilities provided filtered water for every resident in the service area, and the network design was widely copied throughout the UK in the ensuing decades (Logston, 2008).

The need for drinking water purification increased with fast urbanization in the UK. Taking London as an example, its population grew rapidly from approximately 750,000 in 1760 to over 1.4 million by 1815 and then to about 3.2 million by 1860 (Emsley *et al.*, 2019), bringing about increased demand for a centralized water supply with an enlarged scale of water withdrawal from available sources which may not be sufficiently clean. The primary objective of filtration by slow sand filters was to remove turbid and colored substances from the apparently polluted surface waters, e.g. from the River Thames in London. The practice of water purification by slow sand filters soon became mainstream, and the true virtues of the system were made starkly apparent after investigations by the physician John Snow, who demonstrated the role of the water supply in spreading of the cholera epidemic during the 1854 Broad Street cholera outbreak (William *et al.*, 2008). He made use of a dot distribution map and statistical proof to illustrate the connection between the quality of the groundwater source and cholera cases. His data convinced the local council to disable the water pump that withdrew sallow groundwater downstream of sewer discharge, which promptly ended the outbreak. The discovery of cholera as a waterborne disease led to the promulgation of the Metropolis Water Act in London, in which it was required that all water be 'effectually filtered' from 31 December 1855.

The virtues of the slow sand filter systems for drinking water purification to prevent waterborne diseases are due to the formation of a gelatinous layer (or biofilm) called the hypogeal layer or *Schmutzdecke* in the top few mm of the fine sand layer. The *Schmutzdecke* consists of bacteria, fungi, protozoa, rotifers, and a range of aquatic insect larvae. As water passes through the hypogeal layer, particles of foreign matter are trapped in the mucilaginous matrix and soluble organic material is adsorbed. The contaminants are metabolized by bacteria, fungi, and protozoa. The water produced from an exemplary slow sand filter is of excellent quality with 90–99% bacterial cell count reduction (National Drinking Water Clearinghouse, 2000).

Water treatment came to the USA in 1872 when Poughkeepsie, NY, opened the first slow sand filtration plant, and its design criteria were used throughout the country in the following years (Fuller, 1903). However, due to the low filtration rate (as slow as 0.1–0.3 m/h), slow sand filters unavoidably require an extensive land area for a large municipal system. On the other hand, slow sand filters are generally with no pretreatment applied to the influent water, so it is difficult to use such a system to cope with high-turbidity source water. From the late 1890s to early 1900s, many municipal systems in the USA that initially used slow sand filters gradually switched to the rapid filtration systems which were usually a train of coagulation (with the commercialization of aluminum sulfate, or alum, as coagulant), sedimentation, and rapid sand filters (with a filtration rate up to 10 m/h or more), followed by chlorination (with the commercialization of liquid chlorine as a disinfectant). Such a water treatment process, with its capability to cope with various source waters, soon became the mainstream of drinking water purification in the world, and are still the basic processes adopted in most countries (Sikosana *et al.*, 2019).

2.3.3.2 Needs for wastewater treatment

As described in section 2.3.2, many ancient cities had drainage systems, but these were primarily used to carry rainwater away from roofs and pavements. When privy vaults and cesspools were used, most wastes were simply dumped into gutters to be flushed through the drains by floods. Water closets were installed in houses in some cities, such as London, in the early 1800s, but they were usually connected to cesspools, not to sewers. In densely populated areas, local conditions soon became intolerable because the cesspools were seldom emptied and frequently overflowed. The threat to public health became apparent, so as to cause outbreaks of cholera. It soon became necessary for all water closets in the larger towns to be connected directly to the stormwater sewers. This transferred sewage from the ground near houses to nearby water bodies. Thus, a new problem emerged: surface water pollution due to excess discharge of sewage to the receiving water (Laconte & Haimes, 1982).

To treat the sewage to some degree before disposal, the construction of centralized wastewater treatment plants began in the late 1800s and early 1900s, almost the same period as the centralized drinking water purification described in section 2.4.1, principally in the UK and the USA. Instead of discharging sewage directly into a nearby water body, the collected sewage was first passed through a combination of physical, biological, and chemical processes that removed some or most of the pollutants. Also, beginning in the 1900s, new sewage-collection systems were designed to separate stormwater from domestic wastewater, so that treatment plants did not become overloaded during periods of wet weather (Szule, 2014).

The wastewater treatment process also underwent development from simple primary treatment by lime-precipitation (Mennell *et al.*, 1974) which could raise the pH of sewage water and precipitate the coagulable organic and inorganic pollutants, thus, preventing offensive odor from the sewage and reducing pollutant loads to the receiving water, to secondary treatment with the discovery of the activated sludge process at Manchester in 1913 and its full-scale application at Worcester in 1916 (Eddy & Fales, 1916). The secondary wastewater treatment process, as is still used worldwide, is basically a train of treatment units including grid and screen chambers to remove the coarse floating and suspended matter, sand chambers to remove the heavy inorganic particulates, primary settlers to remove the settleable substances, followed by bioreactors with aeration to sustain an aerobic condition for biomass growth and the degradation of organic substances through the action of aerobic microorganisms. Secondary settlers are installed for separating the treated effluent from the mixed liquor from the bioreactor and part of the settled sludge returns to the bioreactors for sustaining sufficient biomass for continuous biodegradation. In most cases the treated effluent is disinfected with chlorine or other disinfectants for killing pathogens. Such a typical wastewater treatment process has been widely known as the activated sludge process since then, and it has become the mainstream of domestic wastewater treatment in the world (Kaur *et al.*, 2017).

2.3.3.3 Needs for urban watershed management and aquatic system conservation

Cities are located within watersheds. The natural waters, such as rivers, streams, lakes, and groundwater aquifers, in a city-related watershed provide source waters for domestic and industrial supplies and, on the other hand, receive urban drainage and sewer discharges. Principally, natural waters have the capability to assimilate pollutants through a series of natural physical (dilution, sedimentation, entrapment), physicochemical (natural coagulation, complexation/precipitation, filtration, adsorption, ion-exchange, etc.), chemical (oxidation, etc.), and biological (decomposition, degradation, etc.) actions (Mikhailovskii & Fisenko, 2000; Obst, 2003). The capability of a water body to accommodate pollutants without substantial change in its background quality is called the 'carrying capacity', and the associated natural process within the carrying capacity is called 'self-purification' (Chen *et al.*, 2016).

In early times, as human settlements were sparsely scattered in various watersheds or different parts of a watershed, their arbitrary water use and used water discharge did not bring about the apparent change of water quality because the pollutant loading never exceeded the self-purification capacity of the receiving water. This may also be the condition in some sparsely populated remote villages adjacent to waters nowadays. As discussed in section 2.3.3.2, wastewater treatment in cities can substantially reduce the pollutant loading from sewage discharge to receiving waters, but the discharged treated effluent still carries residual pollutants with concentrations usually much higher than the background level to result in water pollution near the point of discharge and extend the pollution to a larger water area or even the whole water body. In fact, the pollutant loading to a water body is not limited to the so-called point sources of treated and/or untreated wastewater discharge, but also the nonpoint or diffuse sources such as those from surface runoff during rainy days (Loague & Corwin, 2006).

The consequence of water pollution in cities and densely populated areas mainly includes the following:

1. Degradation of source water quality for drinking water supply: Many rivers flowing through cities are important source waters for drinking water supply, such as the Thames for London, the Seine for Paris, the Yangtze for Wuhan, and Nanjing in China, and so on. These rivers, as well as other types of source waters, if receiving excessive pollutant loading, will no longer be suitable to provide source water for a safe drinking water supply.
2. Damage to water ecological environment: Surface waters, including rivers, streams, and lakes, are habitats of aquatic organisms. The polluted surface water may not only be with excessively high concentrations of organic substances, in terms of COD and/or BOD, which may result in depletion of dissolved oxygen, and high concentration of nutrients, which may bring about water eutrophication, but also potentially hazardous inorganic and organic chemicals, which, even at trace concentrations, may cause damage to aquatic flora and fauna to destroy the water ecological health (Walker *et al.*, 2019).
3. Deterioration of urban water landscape: Urban waters provide important landscapes for recreation and entertainment. As a consequence of water pollution, the sensory property of the waters will be considerably degraded.

Under such conditions, there have been growing needs since the 1980s for urban watershed management and aquatic ecosystem conservation (Zhang & Luo, 2013).

2.4 INTERNATIONAL ACTIONS FOR BUILDING WATER WISE CITIES

2.4.1 Cities of the future program implemented by the International Water Association

The International Water Association (IWA) launched the Cities of the Future (CoF) program since 2009 in response to the need for a different approach to urban water management. Such needs are growing due to dramatic changes in our world in the past decades along with climate change, population growth, growing resource constraints, and rapidly increasing global urbanization. Therefore, the city of the future must integrate water management planning and operations with other city services to meet the needs of humans and the environment in a dramatically superior manner. The CoF program is the first research program of its kind within IWA which covers a wide stretch of issues related but not limited to world water. The overall objective of the IWA CoF program is to recognize that water, and its interactions with other urban sectors (e.g. energy, transportation, etc.), is a central focus in the development and redevelopment of urban areas in the developed and less developed world. The more specific objective is to encourage urban water managers to systematically collaborate with other professionals and the local community to redesign water management systems, integral with other city services to deliver sustainable water services, and at the same time to enhance life both within and beyond the urban environment (Binney *et al.*, 2010).

In the past decade, IWA and its CoF Steering Committee held many Cities of the Future Conferences with world and/or region-oriented topics, and a number of special workshops were organized during IWA World Water Congresses, IWA Development Congresses, IWA regional conferences, and other international conferences such as the Singapore International Water Week, Stockholm International Water Week, and so on. In such a way, the IWA is leading in this field by convening and stimulating discussion and actions around these important topics in the area of water and sanitation services, and with other planning agencies.

2.4.2 The IWA principles for water-wise cities

The most important and comprehensive output from the IWA CoF Program is the *IWA Principles for Water-Wise Cities* (IWA, 2016) launched during the IWA World Water Congress, Brisbane, October 2016. As has been stated in the introduction section of the document, the Principles are to assist leaders to develop and implement their vision for sustainable urban water, beyond equitable universal access to safe drinking water and sanitation. The Principles underlie resilient planning and design in cities. The ultimate goal of these Principles is to encourage collaborative action, underpinned by a shared vision, so that local governments, urban professionals, and individuals actively engage in addressing and finding solutions for managing all waters of the city.

Figure 2.4 shows a framework of the Principles, which consists of five building blocks and four levels of actions. The call for building water-wise cities implies a paradigm shift from the highly engineering-dependent conventional urban water system as depicted in Figure 2.3 (current paradigm), to a brand-new urban water system characterized by engineering in nature (future paradigm). Such a paradigm shift needs a new set of instruments including the new *Vision*, more adaptive *Governance*, enriched *Knowledge and Capacities*, novel *Planning Tools*, and effective *Implementation Tools* – these are *The Five Building Blocks* to deliver sustainable urban water. With the application of the five building blocks, *The Four Levels of Actions*, which are not individual ones but interrelated with each other within one framework, have to be taken for building water-wise cities.

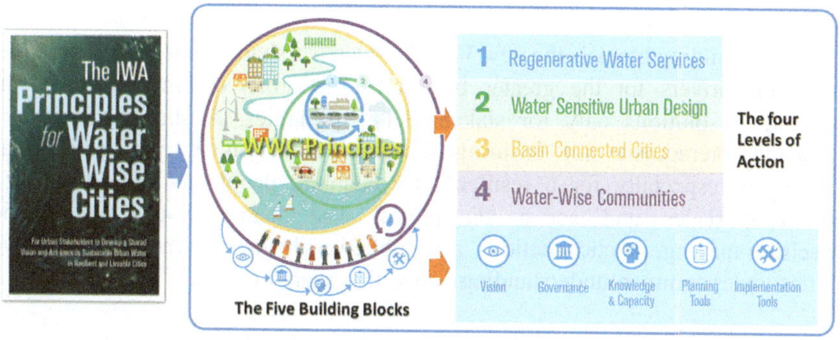

Figure 2.4 The 'IWA Principles for Water Wise Cities' framework (adapted from IWA, 2016).

Regenerative Water Services, at the first level, are the basic objectives, and in order to achieve such a goal, *Water Sensitive Urban Design* is the action at the second level as the precondition for providing regenerative water services. *Basin Connected Cities*, the action at the third level, is to support water sensitive urban design because of the close linkage of a city with the watershed where the city is located and water resources and aquatic environment provided. All these actions from the first level to the third level need the mobilization and collaboration of people of all walks, which leads to the action at the fourth level – *Water-Wise Communities*.

2.4.2.1 The five building blocks

Figure 2.5 outlines the five building blocks to deliver sustainable urban water. The essence of *sustainable urban water* is that the water from available sources, which is limited in its total inflow to a city, can be wisely utilized to meet various demands for urban water supply, not only for the present time but also for the future.

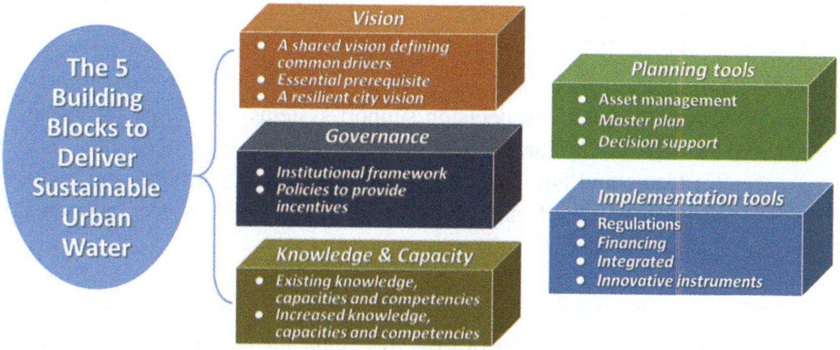

Figure 2.5 The 'Five Building Blocks' in the IWA Principles for Water-Wise Cities (adapted from IWA, 2016).

2.4.2.1.1 Vision

The first building block is the new *Vision*, which is a shared vision defining a set of common drivers for the greater benefit of the urban community rather than defending solutions only for stakeholders' specialties. A shared vision is an essential prerequisite for ensuring the implementation of new policies and strategies, especially for a resilient city enabling people to work together at different scales and across disciplines. The shared vision can support decision-making, practical actions, and public involvement in building water-wise cities under common understanding and with a shared will.

2.4.2.1.2 Governance

The second building block is *Governance* that is more adaptive to building water-wise cities. The primary objective of governance is to provide an institutional framework for urban stakeholders to work together for integrating water in all related urban services at various scales ranging from building, neighborhood, to city, and catchment. Better governance also depends on apt policies that provide incentives for urban stakeholders to unlock the synergies across sectors so that the benefits of water to cities can be maximized.

2.4.2.1.3 Knowledge and capacities

The third building block is enriched *Knowledge and Capacities* which can sufficiently support implementing the sustainable urban water vision. This starts with the existing knowledge, capacities, and competencies of the different urban stakeholders who have accumulated rich experiences in long-term urban development and realized the need for a paradigm shift. However, towards the new paradigm of water-wise cities, we also need increased knowledge, capacities, and competencies to fully realize the new vision, through sharing success stories from other cities, increasing knowledge from 'learning by doing', and opening to other sectors' approaches and methods.

2.4.2.1.4 Planning tools

The fourth building block is a set of novel *Planning Tools* to support asset management, master plans, or decision support systems, which are the means for urban stakeholders to initiate action. These tools are developed and used by cross-sectoral teams and are useful for assessing risks, identifying and analyzing benefits and co-benefits of projects, defining levels of service, and ensuring ownership by stakeholders.

2.4.2.1.5 Implementation tools

The fifth building block is a set of effective *Implementation Tools* for delivering sustainable urban water. Regulations are, in any sense, the top instruments because they create incentives, and provide a solid frame for stakeholders to invest in sustainable urban water based on quality assurance, social equity, transparency, accountability, and sound financing. Financing tools are extremely important as they are linked to rigorous asset management plans, and enable long-lasting improved service levels with a well-maintained infrastructure. Financing tools, which value the ability of solutions to adapt to changes or recover from disasters, allow cities to adopt more efficient solutions and transition towards systems requiring smaller and more frequent investments. Integrated services, combined with shorter investment cycles and co-benefit values, are absolutely necessary for bringing new funding opportunities and providing options to overcome the lack of financial capacity for cities. The augmentation of traditional financing and contracting models further needs the provision of innovative instruments involving private and public financing, including circular economy mechanisms, to open new funding opportunities that promote regenerative water services.

2.4.2.2 *The four levels of actions*

Figure 2.6 outlines the four levels of actions to build water-wise cities. They are based on the principle that all city-dwellers have access to safe drinking water and sanitation services, which requires planning, prioritization, monitoring, and reporting of the human rights to water and sanitation.

2.4.2.2.1 Level 1 – Regenerative water services

The main goal of regenerative water services is to ensure public health and satisfy all current needs while protecting the quality and quantity of water resources for future generations, which accords with the principle of sustainable development (Blewitt, 2015). Regenerative water services require efficient production and use of water, energy, and materials, and are underpinned by the principle of 3Rs, namely *Replenish*, *Reduce* and *Reuse*.

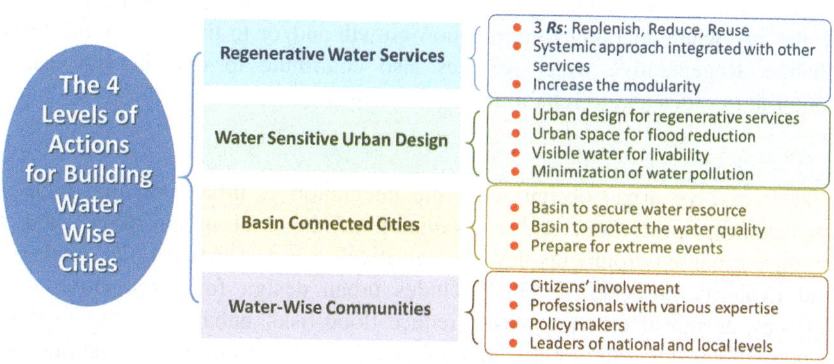

Figure 2.6 The 'Four Levels of Actions' in the IWA Principles for Water-Wise Cities (adapted from IWA, 2016).

Replenish is to renew the aquatic environment of water bodies and their ecosystems by taking from or discharging to them only what can be given or absorbed by the natural environment so that the quality of water resources can be protected. The actions include the reduction of water intakes from the water bodies to match their bearing capacity and use with minimal treatment requirements for wastewater and urban runoff entering the water bodies.

Reduce is to minimize the amount of water used in accordance with storage capacities, and to minimize the energy used in moving and treating urban waters, including rainwater.

Reuse is to utilize water from all available sources, especially alternative ones including the treated wastewater that matches the purpose of use, applying the 'fit for purpose' water quality approach, so as to decrease the urban water demand for freshwater resources. Integrated water resources management becomes important for optimizing the reuse scheme. Reuse is usually associated with the recovery of energy from water whether through heat, organic energy, or hydraulic energy, and the recycling of 'upcycled' materials, such as nutrients or organic matter.

Regenerative water services can only be realized through a *Systemic Approach* integrated with other urban services. The different parts of a water system and other closely related services such as waste or energy should be considered as a whole, so as to enable apt solutions for replenish, reduce and reuse while improving services costs efficiently.

Methodologically, we have to *Increase the Modularity* and ensure that multiple resources, treatment, storage, and conveyance options are available throughout the system for ensuring service levels and resilience of urban water systems in the face of either gradual or sudden changes.

By embedding the abovementioned principles in water and wastewater systems rehabilitation, extension, or new development, it will be ensured that the resource is protected and not overused. New revenue is to be generated from the required financing while delivering broader economic, social, and environmental benefits to the city when adapting to population growth and/or to the impacts of climate change. Regenerative water services also contribute to reducing the carbon footprint of cities and to rehabilitating their basins.

2.4.2.2.2 Level 2 – Water sensitive urban design

Water sensitive urban design seeks the integration of urban planning with the management, protection, and conservation of the total urban water cycle to produce urban environments that are 'sensitive' to water sustainability, resilience, and livability co-benefits. This includes urban design for regenerative water services, design of urban spaces to reduce flood risks, enhancement of livability with visible water, and minimization of water pollution from urban materials.

Domestic and industrial precincts and buildings are elements of a city. Each of them should be planned and implemented in a way to enable regenerative water services which benefit both citizens and their living environment. The reduction of water, energy, and carbon footprints of housing, through various measures, contributes to its affordability to citizens through lower monthly bills. It also leads to cleaner waterways, benefits ecosystems, and people, while also improving social and urban amenities.

A well-designed city should provide ample spaces to reduce flood risks. This needs the development of apt urban drainage solutions integrated with urban infrastructure design in a novel manner so that the city can act as a 'sponge' which absorbs and stores stormwater, limits surges, and releases rainwater as a resource. Building green infrastructures at various scales is one of the most important measures. Vital infrastructure should also be planned to enable quick disaster recovery.

Making water visible is important for enhancing a city's livability. This needs the creation of water-related spaces and infrastructures, such as protection and/or rehabilitation of urban lakes/ponds and streams, provision of an urban waterfront area by artificial impoundment with water from alternative sources, building roadside green infrastructure and blue-green corridors, and so on. Urban water services are essential for ensuring sustainable irrigation of parks and gardens, providing shade and mitigation of heat islands.

Measures for water pollution control should also be incorporated into water sensitive urban design, especially an effective reduction of pollutants possibly released from roofs, walls, surfaces, roads, and urban furniture when exposed to sun and rain. The selection and modification of urban materials to minimize their impact on water quality are thus required.

2.4.2.2.3 Level 3 – Basin connected cities

A city is built either within a basin or in the neighborhood of multiple basins. In any case, the city depends on the related basin(s) for obtaining source water and accommodating urban drainage. Hydrologically, a basin depends on the natural water cycle to sustain its ability to serve the city, as well as other economic activities, especially agriculture and industries, in its service area. Overexploitation of water resources, excessive discharge of pollutants into water bodies, and improper utilization of the basin area will definitely damage the water cycle and degrade the basin's service level. Therefore, integrated basin management is the key point of this third level action.

The first principle of integrated basin management is to secure the water resource for water supply to the city. As the total renewable source water is limited and may also fluctuate annually and seasonally in a basin, the water resource management plan should include two aspects, namely allocation of water resources among all users and water storage for seasonal regulation.

The allocation of available water resources is to share the quantity of freshwater among various sectors that depend on the basin for water provision. Any increase in water demand for the city may result in a need for increasing its share of water resources from the related basin. If the total amount of water resource is fixed, the increasing share for the city may only be compensated by decreasing the share for other sectors, such as industrial and agricultural users, which usually needs negotiation among various users, and/or measures for water-saving from certain users.

Increasing the basin-wide water storage capacity for accommodating larger quantities of water can efficiently mitigate the imbalance between water supply and demand so as to increase the availability of water resources. This is also an extremely important strategy for drought mitigation.

The second principle of integrated basin management is to protect the quality of the water resource so as to ensure high-quality drinking water achieved with minimal treatment and energy requirements. This needs the implementation of a basin-wide water quality protection strategical plan. However, because the city is usually the largest polluter, in terms of its pollutant load exposed to the water environment of the basin, the reduction of pollution from the city should always be prioritized. Water quality protection can also upgrade ecosystem services for both the basin and the city through the provision of forest catchment areas and wetlands.

The third principle of integrated basin management is to prepare for extreme events, such as storms and heavy rains. This requires the implementation of a basin-wide hydraulic and hydrological management plan to manage flow regimes in rivers and streams and to maintain adequate porous soft surfaces and vegetated areas in the basin so that flash floods can be minimized. It is extremely important to invest in coastal storm risks mitigation and flood warning systems.

2.4.2.2.4 Level 4 – Water-wise communities

The implementation of the previous three levels of actions requires a holistic approach and strong partnerships. Therefore, there comes the fourth level of action for mobilizing people to become 'water-wise' and devoted to building water-wise cities.

The full involvement of all citizens in building water-wise cities is the fundamental requirement. With their understanding of the risks related to flooding and water scarcity, and opportunities (resource recovery, reducing dependency on uncertain future resources, increasing well-being) to overcome the current and envisaged difficulties, water-wise citizens can drive urban planning and design in a new direction. By sharing the sustainable urban water vision, water-wise citizens will adapt their behavior to new situations, develop their acceptance to various alternative solutions that enable regenerative water services, and increase their confidence toward future changes along with their willingness to bare certain payment within affordability for building a water-wise city.

Professionals will be more water-wise within their areas of expertise to perform dominant roles in water-wise city planning and design. With their professional understanding of the market and non-market value of the co-benefits associated with an integrated urban agenda, they may enable innovative sustainable solutions. Building a water-wise city requires the collaboration of professionals with various expertise covering finance, technical, and social aspects. There exist synergies and dependencies between water and various urban services, such as urban planning, architecture, landscaping, energy, waste management, and transportation. The integration of all these interrelated sectors within the framework of a water-wise city relies on collaborative efforts of transdisciplinary planning and operation teams through a coordinated approach. A city planning organization recognizing these inter-relations and bridging over existing individual departments is needed to enable urban professionals to implement sustainable urban water.

The implementation of the aforementioned principles for regenerative water services, water sensitive urban design, and basin-connected cities also needs water-wise policymakers who establish policies and financing mechanisms (tariffs and partnerships that are responsive and adaptive to future changes) to drive and enable sustainable urban water through incentivizing and rewarding innovative solutions. Water-wise policymakers will phase out the existing subsidies and tax advantages that are environmentally harmful, and monitor, evaluate, and adjust the policies based on future needs as they change over time.

We further need water-wise leaders who govern at the national and local levels for enabling sustainable urban water through coordination and integration, leveraging effective and efficient governance enhancing thrust and engagement. Water-wise leaders provide the progressive vision and a governance structure to coordinate work at catchment, metro, neighborhood, and household scales for implementing sustainable urban water.

The abovementioned people at four levels, namely 'citizens-professionals-policymakers-leaders', form the water-wise communities to implement the principles for water-wise cities. By working together, they will use the building blocks to put the principles into action.

2.4.3 Envisaged solutions

The implementation of water-wise cities needs new solutions for solving current problems and ensuring sustainable urban development toward the new paradigm. The envisaged solutions implicit in the IWA Principles for Water-Wise Cities are summarized in Figure 2.7.

2.4.3.1 Systematic solutions

Systematic thinking is the key to urban planning and design toward the future. In such a way, all the systems to serve the daily lives of urban dwellers can be viewed as elements of an integrated system. For example, the layout of an urban plan determines the scheme of water supply and sewerage services, not only for the residents' daily life, but also for urban architecture and landscapes which need water to decorate, provision of urban green belt to assist nonpoint pollution

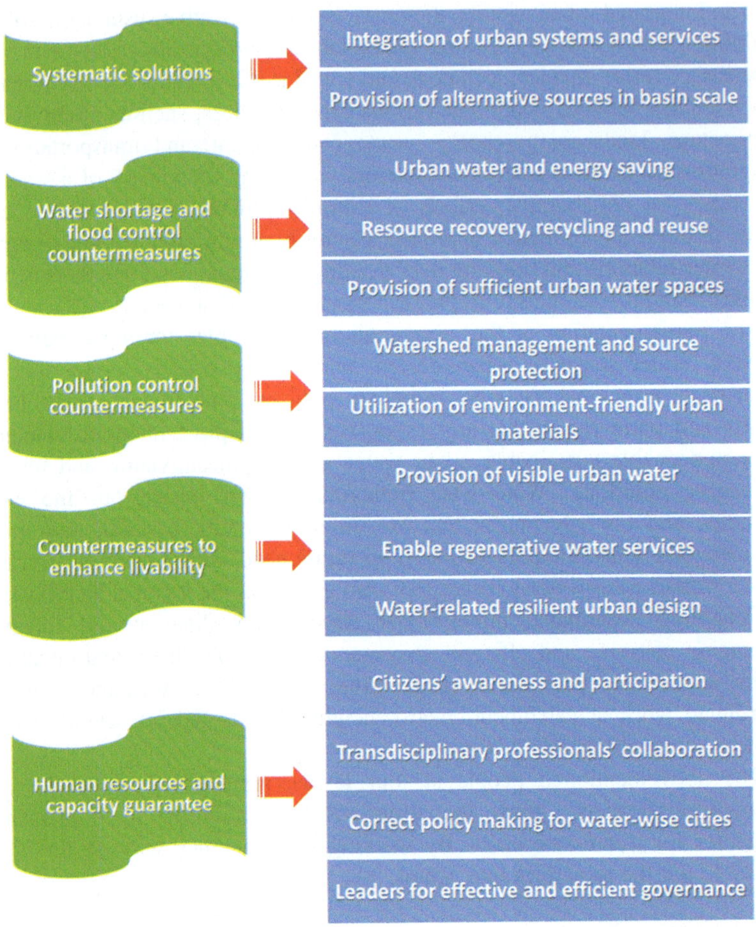

Figure 2.7 Envisaged solutions for building water-wise cities (adapted from IWA, 2016).

control, and even the use of urban water to produce energy. All these are interrelated with each other and integration of them in a systematic scheme is extremely important for gaining benefits, reducing water and energy consumption, and upgrading urban service levels. Following the principle of basin connected cities, we understand that urban services depend much on the related basins not only for water resources, but also for energy, food materials, and ecological service. Therefore, the water-energy-food nexus is always a hot topic closely related to urban system design and optimization (Heard *et al.*, 2017; Zhang *et al.*, 2019a).

As will be discussed in Chapter 6 of this book, a basin connected with cities provides an aquatic environmental buffer zone through the basin-scale hydrological cycle (St. George Freeman *et al.*, 2020). Due to the limited capacity of the basin to provide freshwater resources, there are growing needs for alternative water resource development including water reclamation and reuse (Maryam & Büyükgüngör, 2019). Such an element should also be integrated into the source water system at the basin scale.

2.4.3.2 Water shortage and flood control countermeasures

Many cities in the world face both the problems of water shortage and urban flooding. Without effective measures to combat these two extreme conditions (too little water and too much water), a water-wise city will be a meaningless term.

The key measures to combat water shortages are always the reduction of water demand and the increase of available water resources. Urban water saving is the most realistic way of using a limited amount of water to serve more users and can simultaneously result in energy savings (Lam *et al.*, 2016). As the process of water use in most circumstances does not really bring about substantial water loss but a significant change in the water quality, the used water may be used again only if the quality can be adjusted to a level that meets the requirement of subsequent water use (Rizzo *et al.*, 2018). Wastewater treatment and reclamation have already become common practices in many cities. Because the wastewater contains organic substances, nutrients, and other reclaimable materials, by proper selection of treatment processes, resource and energy recovery can also be made possible (Sgroi *et al.*, 2018).

The occurrence of urban flooding is, in most cases, due to the insufficient capacity of the urban watershed to accommodate and/or smoothly transport rainwater runoff (Sarmah & Das, 2018). For this reason, urban flood control principally relies on two categories of measures, namely reduction of surface runoff and provision of sufficient urban water space. Low impact development (LID) and provision of green infrastructures, as will be discussed intensively in other chapters of this book, can keep the underlying urban surface as permeable as possible and substantially delay and reduce the peak flow of stormwater runoff. Lakes, ponds, rivers, streams, and wetlands are important water elements for a city. In addition to their landscape and eco-environmental values, they provide spaces to accommodate rainwater and efficiently reduce the risk of urban flooding. The sufficient storage volume of these water bodies also benefits the utilization of rainwater as an alternative water source (Lee *et al.*, 2016; Nguyen *et al.*, 2019).

2.4.3.3 Pollution control countermeasures

Successful control of water pollution for the city and its related basin needs the adoption of various measures for point and nonpoint source control, and water quality improvement in water bodies by increasing their self-purification capacities (Wang *et al.*, 2012).

Although measures for source control should be taken at each location where water pollution may occur, the comprehensive improvement of urban aquatic environmental quality requires systematic measures at the basin-scale within the watershed management scheme. As discussed in section 2.3.2.3 regarding water quality, the principal target of watershed management is to ensure high-quality drinking water achieved with minimal treatment and energy requirements. This needs an integrated plan to reduce pollutant loading from all possible sources and increase the carrying capacity of the source water body to withstand external impacts.

The release of pollutants from building roofs, walls, surfaces, roads, and urban furniture may also result in water pollution. Therefore, careful selection of urban materials and development and utilization of environment-friendly materials are also important countermeasures for pollution control.

2.4.3.4 Countermeasures to enhance liveability

As explained in section 2.3.1 of this chapter, because water is indispensable for human life, from the beginning of civilization human beings located their settlements near waters. When the most basic requirement for water, such as drinking and sanitation, is no longer the sole target as it was in the past, human beings seek more livable environments through proper utilization of waters (Maftuhah *et al.*, 2018). This is the reason to put the enhancement of livability into our targets of water-wise cities.

In the IWA Principles, the first solution related to this is the provision of visible urban water, including the protection and rehabilitation of existing water bodies (lakes and streams, etc.) and the enlargement of the waterfront area along with the construction of infrastructure for rainwater storage and urban flood control. Public parks, gardens, green corridors, and wetlands are also water-related urban landscapes which increase visible water and greenness in a city.

Water services for all citizens should be regenerative within the scheme of water-wise cities. Regenerative means the ability to heal or become active again after being damaged or inactive. So, the provision of regenerative water services could make citizens lead more comfortable lives sustainably. As discussed in section 2.4.2.2.1, the three Rs, namely Replenish, Reduce and Reuse, are the core principles of regenerative water services.

With the close relationship between water and livability, water-related resilient urban design is indispensable. Drought and flooding are usually extreme events occurring occasionally in many cities and causing damages (Zhang *et al.*, 2019b). Therefore, a city needs to be redesigned so as to increase its resilience to such extreme events and recover easily and quickly from any damage.

2.4.3.5 Human resources and capacity guarantee

In Section 2.4.2.2.4, we stressed the importance of mobilizing water-wise communities, which involve the wide participation of human resources at four levels, namely citizens, professionals, policymakers, and leaders. Without collaborative efforts from all of them, it is impossible for urban stakeholders to act towards sustainable urban water in resilient and livable cities.

Citizens can drive urban planning and design toward the new paradigm because they fully understand the risks related to flooding and water scarcity and opportunities for better well-being brought to them by building water-wise cities. Professionals with various expertise who understand the co-benefits across urban sectors can find and implement the best solutions for meeting citizens' requirements through transdisciplinary collaboration. Policymakers establish policies and financing mechanisms to drive and enable sustainable urban water, and leaders provide the governance structure to coordinate work at various scales. All these, as a whole, can guarantee the capacity of cities to implement their plans and reach their goals for building water-wise cities.

REFERENCES

Binney P., Donald A., Elmer V., Ewert J., Phillis O., Skinner R. and Yong R. (2010). IWA Cities of the Future Program Spatial Planning and Institutional Reform Conclusions from the World Water Congress. IWA World Water Congress, September 2010, Montreal, Canada.

Blewitt J. (2015). Understanding Sustainable Development, 2nd edn. Routledge, New York.

Bosworth A. (1995). World cities and world economic cycles. In: Civilizations and World Systems: Studying World-Historical Change, S. K. Sanderson (ed.), AltaMira Press, New York, pp. 206–227.

Carrasco S. and Dangol N. (2019). Citizen-government negotiation: cases of in riverside informal settlements at flood risk. *International Journal of Disaster Risk Reduction*, **38**, 101195.

Chen H., Qu F. and Chen L. (2016). The Right to Use Environmental Capacity: Legislation for Energy Conservation and Emissions Reductions. Springer, Singapore.

Eddy H. P. and Fales A. L. (1916). The activated-sludge method of sewage purification: the activated-sludge process in treatment of tannery wastes. *Industrial and Engineering Chemistry*, **8**(7), 648–651.

Emsley C., Hitchcock T. and Shoemaker R. (2019). London History – A Population History of London, Old Bailey Proceedings Online, Ver. 7.0. Available from: www.oldbaileyonline.org

Faust S. D. and Aly O. M. (1999). Chemistry of Water Treatment, 2nd edn. Ann Arbor Press, Chelsea, MI.

Fuller G. W. (1903). Present status of the purification of public water supplies. *Journal of the American Medical Association*, **41**(18), 1084–1090.

Gleick P. H. (2014). Water, drought, climate change, and conflict in Syria. Weather, Climate, and Society. **6**(3), 331–340.

Global Footprint Network (2019a). About the Data. Available from: https://data.footprintnetwork.org/#/abouttheData

Global Footprint Network (2019b). Country Trends. Available from: https://data.footprintnetwork.org/#/countryTrends?cn=5001&type=BCtot,EFCtot

Goerke-Shrode S. (1998). Alexandria in Egypt – yesterday and today. *Calliope*, **9**(4), 34.

Heard B. R., Miller S. A., Liang S. L. and Xu M. (2017). Emerging challenges and opportunities for the food–energy–water nexus in urban systems. *Current Opinion in Chemical Engineering*, **17**, 48–53.

Hofste R., Kuzma S., Walker E. H., Sutanudjaja E. H., Bierkens M. F. P., Kuijper M. J. M., Sanchez M. F., Beek R. V., Wada R., Rodriguez S. G. and Reig P. (2019). Aqueduct 3.0: Updated Decision-Relevant Global Water Risk Indicators, Technical Note. World Resources Institute, Washington, DC.

Indexmundi (2019). Renewable Internal Freshwater Resources, Total. Available from: www.indexmundi.com/facts/indicators/ER.H2O.INTR.K3

Isachenko A. G. (2011). Geographical roots of ancient civilizations (on the 120th anniversary of L.I. Mechnikov's Civilization and Great Historic Rivers): Part II. *Regional Research of Russia*, **1**(2), 177–194.

IWA (2016). The IWA Principles for Water Wise Cites. International Water Association, London. Available from: https://iwa-network.org/publications/the-iwa-principles-for-water-wise-cities/

Kaur J., Punia S. and Kumar K. (2017). Need for the Advanced Technologies for Wastewater Treatment. In: Advances in Environmental Biotechnology, R. Kumar, A. Sharma and S. Ahluwalia (eds.), Springer, Singapore, pp. 39–52.

Laconte P. and Haimes Y. Y. (1982). Water Resources and Land-Use Planning: A Systems Approach. NATO Advanced Study Institutes Series, Louvain-la-Neuve, Belgium.

Lam K. L., Lant P. A., O'Brien K. R. and Kenway S. J. (2016). Comparison of water-energy trajectories of two major regions experiencing water shortage. *Journal of Environmental Management*, **181**, 403–412.

Lee K. K., Mokhtar M., Hanafiah M. M., Halim A. A. and Badusah J. (2016). Rainwater harvesting as an alternative water resource in Malaysia: potential, policies and development. *Journal of Cleaner Production*, **126**, 218–222.

Loague K. and Corwin D. L. (2006). Point and nonpoint source pollution. In: Encyclopedia of Hydrological Sciences, M. G. Anderson, J. J. McDonnell and T. Gale (eds.), John Wiley & Sons, Hoboken, NJ.

Logston G. S. (2008). Water Filtration Practices: Including Slow Sand Filters and Precoat Filtration American Water Works Association, Denvor, CO.

Maftuhah D. I., Anityasari M. and Sholihah M. (2018). Model of urban water management towards water sensitive city: a literature review. *IOP Conference Series: Materials Science and Engineering*, **337**, 012047.

Maryam B. and Büyükgüngör H. (2019). Wastewater reclamation and reuse trends in Turkey: Opportunities and challenges. *Journal of Water Process Engineering*, **30**, 100501.

Mennell M., Merril D. T. and Jorden R. M. (1974). Treatment of primary effluent by lime precipitation and dissolved air flotation. *Journal – Water Pollution Control Federation*, **46**(11), 2471–2485.

Mikhailovskii V. and Fisenko A. I. (2000). The Physical-Chemical Mechanism of Water Stream Self-Purification Working paper, Los Alamos National Laboratory.

National Drinking Water Clearinghouse (2000). Slow Sand Filtration. Available from: www.nesc.wvu.edu/pdf/DW/publications/ontap/tech_brief/TB15_SlowSand.pdf

Nguyen T. T., Ngo H. H., Guo W. S., Wang X. C., Ren N. Q., Li G. B., Ding J. and Liang H. (2019). Implementation of a specific urban water management – Sponge City. *Science of the Total Environment*, **652**, 147–162.

Novotny V., Ahern J. and Brown P. (2010). Water Centric Sustainable Communities: Planning, Retrofitting, and Building the Next Urban Environment. John Wiley & Sons Inc., Hoboken, New Jersey, USA.

Obst U. (2003). Strategies of maintaining the natural purification potential of rivers and lakes. *Environmenal Science and Pollution Research International*, **10**(4), 251–255.

Pena J. T. and Morley N. (1998). Metropolis and Hinterland: The City of Rome and the Italian Economy 200 B. C.-A. D. 200. *American Journal of Archaeology*, **102**(2), 451–452.

Rizzo L., Krätke R., Linders J., Scott M., Vighi M. and de Voogt P. (2018). Proposed EU minimum quality requirements for water reuse in agricultural irrigation and aquifer recharge: SCHEER scientific advice. *Current Opinion in Environmental Science and Health*, **2**, 7–11.

Sarmah T. and Das S. (2018). Urban flood mitigation planning for Guwahati: A case of Bharalu basin. *Journal of Environmental Management*, **206**, 1155–1165.

Sgroi M., Vagliasindi F. G. A. and Roccaro P. (2018). Feasibility, sustainability and circular economy concepts in water reuse. *Current Opinion in Environmental Science & Health*, **2**, 20–25.

Sikosana M. L., Sikhwivhilu K., Moutloali R. and Madyira D. M. (2019). Municipal wastewater treatment technologies: A review. *Procedia Manufacturing*, **35**, 1018–1024.

St. George Freeman S., Brown C., Cañada H., Martinez V., Nava A. P., Ray P., Rodriguez D., Romo A., Tracy J., Vázquez E., Wi S. and Boltz F. (2020). Resilience by design in Mexico City: A participatory human-hydrologic systems approach. *Water Security*, **9**, 100053.

Strong M. (1990). The Environment: Pioneers and Scepticism. In: The United Kingdom – The United Nations, J. Erik and F. Tomas (eds.), Palgrave Macmillan, UK.

Szule B. (2014). Wastewater Treatment in the United Kingdom. Faculty of Agricultural and Food Sciences, University of West Hungary, London.

UN (2019). The Sustainable Development Goals Report 2019. New York, United Nations. Available from: https://unstats.un.org/sdgs/report/2019/

UN Department of Economic and Social Affairs (2019). 68% of the World Population Projected to Live in Urban Areas by 2050, says UN. Available from: www.un.org/development/desa/en/news/population/2018-revision-of-world-urbanization-prospects.html

USGS (2019). Water Basics. U.S. Geological Survey, Washington DC, USA. Available from: www.usgs.gov/special-topic/water-science-school/science/water-basics

Valhondo C. and Carrera J. (2019). Chapter 1 – Water as a finite resource: From historical accomplishments to emerging challenges and artificial recharge. In: Sustainable Water Wastewater Process, C. M. Galanakis and W. Agrafioti (eds.), Elsevier, Cambridge, pp. 1–17.

Walker D. B., Baumgartner D. J., Gerba C. P. and Fitzsimmons K. (2019). Chapter 16 – Surface Water Pollution. In: Environmental and Pollution Science, 3rd edn., M. L. Brusseau, L. L. Pepper and C. P. Gerba (eds.)., Academic Press, London, pp. 261–292.

Wang J., Liu X. D. and Lu J. (2012). Urban river pollution control and remediation. *Procedia Environmental Sciences*, **13**, 1856–1862.

Wang X. C. and Chen R. (2014). Xi'an: Water Management and the Development of City Planning in History. In: A History of Water, Series III, Vol. **1**: Water and Urbanization, T. Tvedt and T. Oestigaard (eds.), I. B. Tauris & Co. Ltd., New York, pp. 71–88.

William S., Gunn A. and Masellis M. (2008). Concepts and Practice of Humanitarian Medicine. Springer, New York, NY.

Worldometers (2019) Population. Available from: www.worldometers.info/population/

World Population Review (2019). Available from: https://worldpopulationreview.com/

WRI (Water Resources Institute) (2019). Aqueduct 3.0 Country Rankings. Available from: www.wri.org/resources/data-sets/aqueduct-30-country-rankings

WWAP (UNESCO World Water Assessment Programme) (2019). The United Nations World Water Development Report 2019: Leaving No One Behind. UNESCO, Paris.

Zhang T. H. (2008). On the Population of Chang'an in the Tang Dynasty-Concurrently Commenting on Studies about Population of Chang'an in Tang Dynasty during the Latest 15 Years. *Tangdu J.*, **24**(3), 11–14 (Chinese).

Zhang H. and Luo Y. M. (2013). Assessment system for watershed ecological health in the United States: Development and application. *Ying Yong Sheng Tai Xue Bao*, **24**(7), 2063–2072 (in Chinese).

Zhang P. P., Zhang L. X., Chang Y., Xu M., Hao Y., Liang S., Liu G. Y., Yang Z. F. and Wang C. (2019a). Food-energy-water (FEW) nexus for urban sustainability: A comprehensive review. *Resources. Conservation, and Recycling*, **142**, 215–224.

Zhang X., Chen N. C., Sheng H., Ip C., Yang L., Chen Y. Q., Sang Z. Q., Tadesse T., Lim T. P. Y., Rajabifard A., Bueti C., Zeng L. L., Wardlow B., Wang S. Q., Tang S. Y., Xiong Z., Li D. R. and Niyogi D. (2019b). Urban drought challenge to 2030 sustainable development goals. *Science of the Total Environment*, **693**, 133536.

Chapter 3

Chinese version of water-wise cities: Sponge City initiative

Nanqi Ren, Xiuheng Wang and Shunwen Bai

School of Environment, Harbin Institute of Technology, Harbin, China

3.1 INTRODUCTION

By the end of 2018, China had 669 cities of which 297 were national or local central cities (prefecture-level and above) of large sizes and population. Altogether, these cities occupy an urban area of 200,897 km^2, with an average population density of 2546 people/km^2. According to the National Bureau of Statistics of the People's Republic of China, the urbanization rate of the permanent population has increased from 10.64% in 1949 to 59.58% in 2018. During this period, China experienced the largest and fastest urbanization process in the history of the world. Along with the rapid urbanization and social economy, the original hydrological and ecological environments in China have been severely damaged, resulting in obvious urban water environmental problems (Cheng & Hu, 2011; Jia *et al.*, 2013). For instance, two-thirds of China's cities are suffering from water shortage. Poor water quality was identified in 61.3% of the groundwater monitoring sites in 202 cities. According to the China Flood and Drought Disaster Bulletin, 83 cities were flooded in 2018, with a direct economic loss of 16 billion Yuan, equivalent to 0.18% of the GDP that year. With the manifold pressures from population growth, resource constraint, and economic development, novel concepts and principles are needed to guide urban water systems planning towards sustainable development.

Editors: Xiaochang C. Wang and Guangtao Fu
doi: 10.2166/9781789060751_0053

Consequently, the Sponge City strategy has been set by the central government. On 12 December 2013 at the Central Urbanization Work Conference, President Xi Jinping emphasized the need to prioritize stormwater reservation in the building of urban drainage systems, the utilization of natural forces for water discharge, and the construction of 'Sponge City' with natural storage, infiltration, and purification as sole connotations. A national program was subsequently initiated for implementing pilot projects of Sponge City Construction (SCC) in selected cities with financial support from the central government. Efforts were also made in the standardization of SCC technologies and methods for project evaluation, such as the Evaluation Standard for Sponge City Construction (GB/T 51345-2018) approved in 2018. All these marked the beginning of a new era to transform the urban water environment from grey-based systems into more flexible, resilient, and sustainable systems.

Consequently, research on SCC technologies and applications is burgeoning in recent years. As SCC aims at developing a systematic strategy to deal with complicated urban water problems, the first effort was to interpret and advance technological guidelines at the national level, associated with the identification of opportunities and challenges to promoting an overall SCC progress (Chan *et al.*, 2018; Griffiths *et al.*, 2020; Ren *et al.*, 2017; Thu *et al.*, 2020; Xia *et al.*, 2017). Another effort was to solve urban water problems through the development of sponge-based technologies such as permeable materials (Shen *et al.*, 2020) or greening plants (Li *et al.*, 2019). In addition, as SCC often involves large-scale public projects that are subject to public financial support, efforts were also made to improve the financial sufficiency, including investigations on public willingness to pay for SCC projects (Wang *et al.*, 2020), and identifications of risks in cooperative arrangements such as public-private partnerships (Zhang *et al.*, 2019). Besides, there were studies focusing on the establishment of methodological frameworks to improve performance evaluation and decision support (Chang & Su, 2020; Meng & Li, 2020; Thu *et al.*, 2020).

3.2 PROBLEMS TO SOLVE

Before the proposal of the Sponge City strategy, many countries in the developed world proposed various strategies for urban rainwater management, which have greatly changed the traditional way of thinking and gradually shifted focus from individual technical development to a comprehensive governance strategy advancement.

In the United States, a number of strategies have been proposed and widely adopted since the 1970s, such as the proposal of Best Management Practice (BMP) in 1972 for non-point source pollution control (Karr & Schlosser, 1978), Low Impact Development (LID) in 1990 (Carlson *et al.*, 2015), and Green Infrastructure (GI) in the early 1990s (Hiltrud & Pierre, 2011). BMP focuses on measures for flood peak flow and pollutant control, groundwater recharge and storage, and ecological sensitivity management. The core of LID is to control the development area as close to the natural hydrological cycle as possible through decentralized and small-scale source control and ultimately achieve stormwater runoff and pollution control. GI aims to form an interconnected and unified green network system composed of natural regional elements.

Also, in the 1990s in Australia, Water Sensitive Urban Design (WSUD) was proposed with primary concerns on rainwater utilization to combat water shortage under arid climates (Sharma *et al.*, 2016). In the UK, Sustainable Urban Drainage System (SUDS) has been adopted to solve the problems of frequent flooding and ecological pollution through optimized regional drainage system design (Ellis & Lundy, 2016). In 2006, the Active, Beautiful, and Clean Waters Programme (ABC) was launched in Singapore (Lim & Lu, 2016) as a comprehensive urban environment improvement measure. The purpose of ABC is to transform drainage channels and reservoirs into clean and beautiful rivers and lakes, integrate them into the entire urban environment and provide new urban public space for citizens.

The abovementioned strategies and measures have, so far, been widely applied worldwide for comprehensive urban water environmental planning. These experiences are no doubt beneficial in China for Sponge City Construction. However, for solving the current problems associated with rapid urbanization in Chinese cities, what we need to do is not just an adaptation of the existing experiences to local situations but the development of novel Chinese models based on similar concepts.

Most Chinese cities are facing the common problems of uncontrolled urban runoff, lack of water resources, and pollution of the water environment, which have considerably restricted the sustainable development of the economy and society.

Urban runoff is a problem that may not be easily coped with by the current urban water systems in China. On one hand, uncontrolled urban runoff results in frequent waterlogging, causing huge economic losses and casualties. On the other hand, urban runoff is the cause of urban nonpoint source pollution and the destruction of the ecological environment. Surface runoff in rainy days usually carries pollutants into urban water bodies through mixed rainwater and sewage overflows. Due to insufficient baseflow in urban river channels and shortage of source water to replenish urban lakes, as well as the high pollutant loading from point and nonpoint sources, some urban water bodies have become black and odorous, bringing about a serious urban water environmental problem of public concern.

Statistical data show that the per capita water resource in China is far below the world average level. With the rapid industrialization and urbanization, urban water demand continues to increase, and there is an unbalanced supply–demand relationship in many cities. The water shortage problem will be further deepened in the coming years. Meanwhile, urban sewage discharge is also increasing year by year, but the existing sewage treatment facilities and their treatment capacities cannot meet the increasing needs. Although investments in water and wastewater related infrastructure has been increasing, the development speed of sewage treatment plants and auxiliary facilities (including sludge disposal systems) are still lagging behind the speed of economic growth and urban and industrial development. The overall quality of the water environment has not been improved as expected, and even continues to decline in some cities. Water shortage also threatens urban water safety and public health.

Table 3.1 summarizes the statistical data from 2014 to 2018 on urban water-related conditions in China.

It should also be pointed out that, as a large country, there are great differences between cities in different regions in terms of their natural conditions and development level. Facing the complicated problems of the water environment,

Table 3.1 Urban water-related data from 2014 to 2018.

Year	2014	2015	2016	2017	2018
Urban population density (person/km^2)	2419	2399	2408	2477	2546
Number of cities with waterlogging	125	168	192	104	87
Direct economic losses caused by waterlogging (100 million CHY)	1573	1661	3643	2142	1615
Percent of centralized drinking water supply in cities of prefecture-level and above (%)	96	90	90	91	91
Water resource per capita (m^3/person)	1999	2039	2355	2074	1972
Total urban water supply (100 million m^3)	547	560	581	594	615
Urban sewage discharge (100 million m^3)	445	467	480	492	521
Percent of urban sewage treatment (%)	90	92	93	95	95
Investment in fixed assets of urban public water supply facilities (100 million CHY)	475	620	546	580	543
Investment in fixed assets for the construction of urban public drainage facilities (100 million CHY)	900	983	1223	1344	1530
Investment in fixed assets for the construction of urban public sewage treatment and recycling facilities (100 million CHY)	404	513	490	451	803
Investment in fixed assets for the construction of urban public utility tunnel facilities (100 million CHY)			295	673	619

Data source: summarized by authors based on original data from https://data.stats.gov.cn (National Bureau of Statistics of China).

the governance of urban water requires systematic solutions to be worked out under governmental leadership and coordination, and the participation of all related sectors. Sponge City is a new form of city that conforms to the requirement of development in China, and it is also a new concept to lead the formulation of a theoretical and methodological framework for aquatic ecological environment governance, with stress on the interlinkage between the social water cycle and the natural water cycle. The construction of a Sponge City aims to solve the problem of urban waterlogging, to improve the urban water environment with the elimination of urban black and odorous waters as the core objective, and more importantly, to make cities more sustainable.

As a governmental oriented action, Sponge City Construction stresses an overall improvement of the urban water environment. Related to the national development goal and institutional context, Sponge City distinguishes from other approaches proposed so far in other countries. In addition to water quantity and quality, it stresses the social and natural attributes of the entire water cycle to balance environmental and economic interests.

3.3 CONVENTIONAL SOLUTIONS: GRAY ENGINEERING MEASURES

The Sponge City initiative in China is deeply committed to solving multiple urban water problems. To better understand the context and significance of Sponge City, this section will introduce the status quo of China's current urban water systems along with the systematic problems.

3.3.1 Urban water system 1.0

In ancient times, transport aqueducts were developed to meet water supply and discharge demands for centralized urban residents, forming the primary urban water system 1.0 as shown in Figure 3.1 (De Feo *et al.*, 2014). With a small population size (several thousand to tens of thousands) and agriculture as the

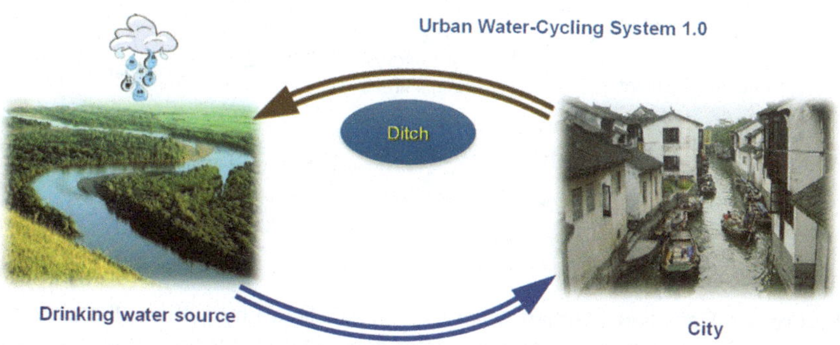

Figure 3.1 Depiction of Urban water system 1.0 (figure by authors).

main production activity, such a type of system model could ensure human access to adequate and safe water and sanitation through natural water circulation and purification.

3.3.2 Urban water system 2.0

Unexpected challenges, i.e. water-related diseases spreading and industrialization, stimulated the revolution of urban water systems. From the late 1800s to the early 1900s, drinking water and wastewater treatment technologies and infrastructures were emerging in succession and brought about the formulation of urban water system 2.0. This is a common system model applied over the world until now, even in many industrialized countries including China. As depicted in Figure 3.2, urban water system 2.0 is a conspicuously linear model consisting of a series of large-scale centralized infrastructure, including water supply and wastewater treatment plants.

China's urban development has long relied on urban water system 2.0 to meet the needs for water supply, sanitation, and drainage. However, as the rate of urbanization often exceeds the rate of construction of new water infrastructure, a series of urban water problems have emerged (Figure 3.3). From 1978 to 2014, the urbanization rate in China increased from 18 to 55% and socio-economic development proceeded at a very high speed, as indicated by the GDP increase of 174 times. In the meantime, urban domestic water consumption also increased by several times, resulting in an imbalance between the growing demands and available freshwater resources. In many cities, pollutant loading has exceeded water environmental capacity, albeit most of the wastewater treatment plants have made efforts to meet the effluent discharge standards. It is usually a challenge for

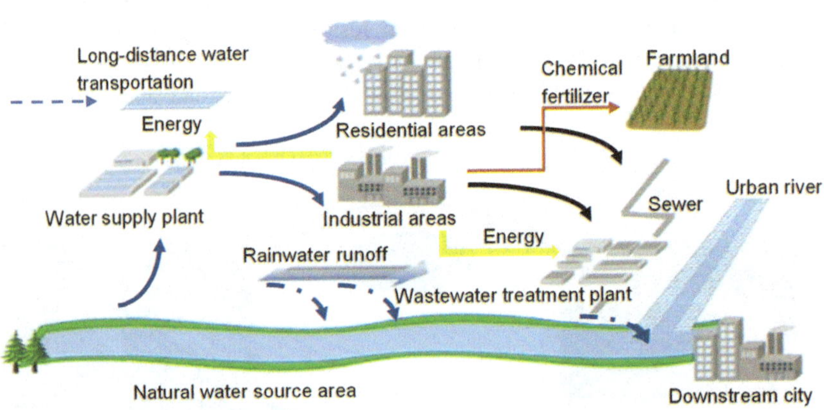

Figure 3.2 Depiction of urban water system 2.0 (figure by authors; the blue, grey, brown and yellow arrows represent water flow, wastewater flow, resource supply, and energy supply, respectively).

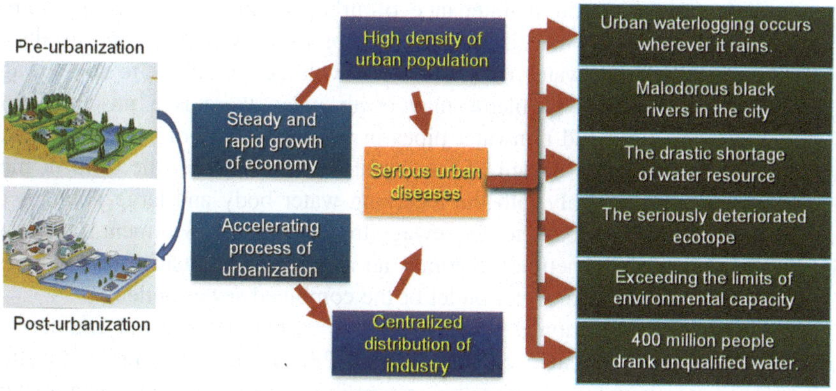

Figure 3.3 Major urban water problems generated along with rapid urbanization and social economy growth (figure by authors).

water and wastewater systems to be designed in a way to accommodate the ever-increasing demands and pollutant loading by following the water system 2.0 model. Other factors, such as climate change, the extension of impervious covering, and shrinkage of water areas aggravate the urban water problems as well, resulting in insufficient groundwater recharge and increasing surface runoff (Jia *et al.*, 2013). Water system 2.0 is also incommensurate to the desire of citizens for urban liveability and water landscape.

Under the water system 2.0, either combined or separate sewer systems are adopted for wastewater and stormwater runoff transport. For a combined system, domestic sewage, industrial wastewater, and rainwater runoff are collected and mixed in the same pipe system, and then transported to a sewage treatment plant for combined treatment before final discharge to a receiving water body. During periods of heavy rainfall, however, the water volume may far exceed the capacity of the combined sewer system or the treatment plant, so overflow occurs occasionally and the excess volume of the mixed wastewater flows directly to nearby streams, rivers, or other water bodies. The so-called combined sewer overflows (CSOs) contain not only stormwater but also untreated human and industrial wastes, toxic materials, and debris, which may cause serious pollution of the receiving waters. In contrast, a separate sewer system consists of separate pipelines to collect municipal wastewater and surface runoff for transport to different destinations (sewage treatment plant and receiving water, respectively). So, sewer overflow can be prevented and the sewage treatment plant can be operated as usual during rainy periods. Due to rapid urbanization, the drainage systems in many cities are mixtures of both combined and separate sewers. There is always the problem of malfunctioning of different sewer systems to result in point source pollution of water bodies, even though the coverage of sewer collection and treatment is not low.

Improper operation and maintenance of urban drainage systems is another problem in some cities where groundwater seeps into sewer pipes and results in the dilution of sewage water to increase the volume of inflow to the sewage treatment plants. Similar problems may occur when there is a misconnection between sewage pipes and rainwater pipes in some locations of a separate sewer system. Due to improper design of the outfalls in some cities, the effluent may not be discharged smoothly into the receiving water body and large volumes of water can be discharged back to the sewage treatment plant. Sediment deposition in drainage pipelines sometimes becomes an additional pollutant source when it is scoured out from the rainwater outlet or the combined sewer outlet.

Facing the abovementioned water problems, various fragmented engineering measures have been taken to enhance system 2.0 (Table 3.2). Similar to other countries, water issues are under the responsibilities of different governmental agencies and authorities.

The provision of engineering infrastructure is necessary and effective but needs large investment for their construction, which may pile pressures on the economy for sustainable development. For example, the investment cost of the South-to-North Water Transfer Project is as high as 500 billion CHY, and the unit cost of water transfer is 8–10 CHY per cubic meter. Another example is the upgrading of wastewater treatment facilities for meeting the increasingly stringent effluent discharge standard, which has caused the unit treatment cost to climb to 1 CHY per cubic meter while gaining little net benefit in environmental improvement (Wang *et al.*, 2015).

The system 2.0 may be enhanced by these engineering measures, but the system's inherent problems are inevitably amplified, especially those related to hydrological and ecological issues. For instance, from the hydrological standpoint, channelization is a measure widely used for flooding control while environmentally, it may result in the loss of self-purification ability, ecological function, and landscape values for urban rivers.

For these reasons, the enhancement of system 2.0 by fragmented engineering measures may not be adequate to overcome the shortcomings of a linear system due to its strong dependence on resource consumption and unsustainability. There is a growing requirement for China to take new measures to build urban water systems to be more sustainable, resilient, and multi-functional.

3.4 TOWARDS A MULTI-PURPOSE WATER-WISE SYSTEM: SPONGE CITY

3.4.1 Urban water system 3.0 as a new approach

The establishment and advancement of Sponge City marks the beginning of a new era to transform China's urban water system into a multiple-purpose water-wise system. In this regard, an innovative urban water system 3.0 is proposed based on technology advancement and a specific domestic context. Figure 3.4 shows the

Table 3.2 Fragmented engineering measures to enhance system 2.0 and outlook for future water systems (summarized by authors).

Goals	Elements in System 2.0	Responsible Authorities and Their Actions to Enhance System 2.0	Endeavors in Future
Water supply	River and groundwater	Ministry of Water Resource	Develop new water source
Drinking water quality	Water supply plant	Long-distance water transportation and reservoir construction Minister of Health of the People's Republic of China releasing new drinking water sanitary standard in 2006	Non-potable water supply Water conservation
Water environment quality Eliminating black and malodorous waters	Centralized sewage treatment plant (STP) sewer system	Ministry of Environmental protection releasing action plant for water in 2015 Upgrading STP discharge standards in 1996, 2002 and 2015 Ministry of Housing and Urban-Rural Development STP construction	Decentralized system for in situ reuse Balance environment and economic benefits
Flood/waterlogging prevention	Urban river stormwater system	Ministry of Water Resource channelized river for flood discharging Ministry of Housing and Urban-Rural Development/Local Authorities sponge city construction,16 cities started LID in 2015	Natural hydrological cycle protection Urban resilience
Recreation aesthetics	Urban river	Local authorities Inland rivers replenished by long-distance transported water revetment in garden/artificial landscape	Natural landscape Ecology recovery habitability

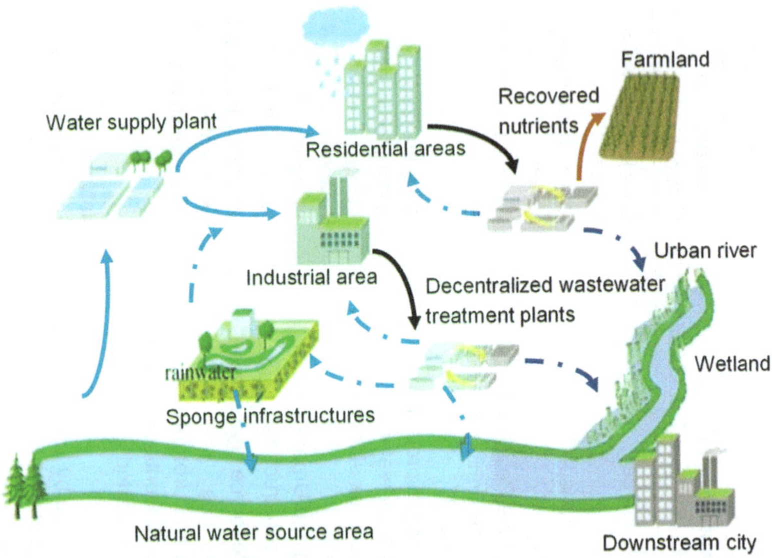

Figure 3.4 Depiction of urban water system 3.0 (figure by authors; blue, grey, brown and yellow arrows represent water flow, wastewater flow, resource supply, and energy supply, respectively).

structural and functional characteristics of urban water system 3.0, which is expected to be a flexible, resilient, and sustainable system that can provide efficient solutions to current problems.

In this new system model, sponge infrastructures are introduced to facilitate unconventional water resources development, such as rainwater harvesting and water reclamation from treated wastewater, not only for source enlargement to meet the increasing demand but also for reducing pollutant loading to receiving waters. A number of water cycles can be formed through linkages between wastewater treatment facilities, urban water bodies, wetlands, and other water elements so that better water circulation conditions can be ensured. Toward the future, the wastewater treatment facilities can be energy-neutral and provide fertilizers to farmlands through resource recovery. Decentralization of the wastewater treatment facilities may be more appropriate for such a purpose.

On the basis of system 3.0, the authors further propose a newer version of urban water system 4.0, whereby large-scale drainage facilities are added to the system, including large tunnels, culverts, deep trenches, and ponds of sufficient storage volumes for accommodating stormwater runoff. System 4.0 is more suitable for cities in areas with annual precipitation higher than 1000 mm where the drainage facilities are of smaller scales, as shown in Figure 3.4, and may not be capable of coping with extremely heavy rainfall.

The proposal of system 3.0 coincides with the International Water Association (IWA) Principles for Water-Wise Cities (IWA, 2016), which calls for actions at four levels, namely regenerative water services, water sensitive urban design, basin-connected cities, and water-wise communities. However, faced with the current problems and special situation in China, the main themes related to system 3.0 can be set as below.

3.4.1.1 Sustainable water services

The main measures for realizing sustainable water services in all cities in China include in situ amplification of water resources by unconventional water utilization and local circulation and protection of natural water resources by the restoration of water bodies and their aquatic ecosystems.

3.4.1.2 Improvement of overall environmental quality, resilience, and liveability in urban areas

For the improvement of the overall urban environmental quality, measures should be taken mainly for onsite wastewater treatment and reclamation, the introduction of diversified treatment processes, and enhancement of ecological purification in receiving water bodies, along with holistic process control measures for effective pollutant reduction. A resilient mechanism should be established for mitigating urban waterlogging and hazards from other catastrophic and extreme events. The building of sponge infrastructures and implementing decentralized sewage systems are the main measures. To make cities more liveable, a series of ecological measures should be implemented, such as the provision of wildlife habitats, an increase of green entertainment spaces, improvement of liveable microclimate, and upgrading of aesthetic values of urban water landscapes.

3.4.1.3 Water-wise communities

Figure 3.5 shows a framework for the establishment of water-wise communities. This framework depends much on public awareness and acceptance, the participation of professionals with various expertise, cooperation between different governmental agencies, and integrated urban planning. Various advanced communication tools and methods should also be used, such as wireless devices and effective data management (Chung & Yoo, 2015; O'Donovan *et al.*, 2015), and the introduction of the a public–private partnership (PPP) model to facilitate implementation and practice.

3.4.1.4 Reviving water culture

Figure 3.6 is a depiction of the unique Chinese water culture to be revived via Sponge City construction. Water culture is a crucial element of Chinese culture from ancient times. As many cities have been built alongside waters, the water

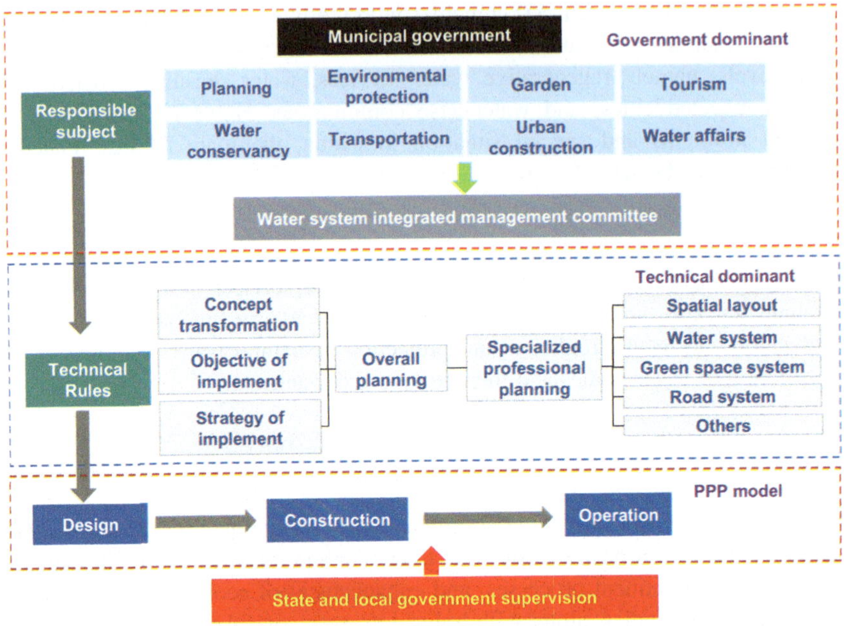

Figure 3.5 Water-wise communities: organization and implementation of sponge city construction and integrated water system regulation (figure by authors).

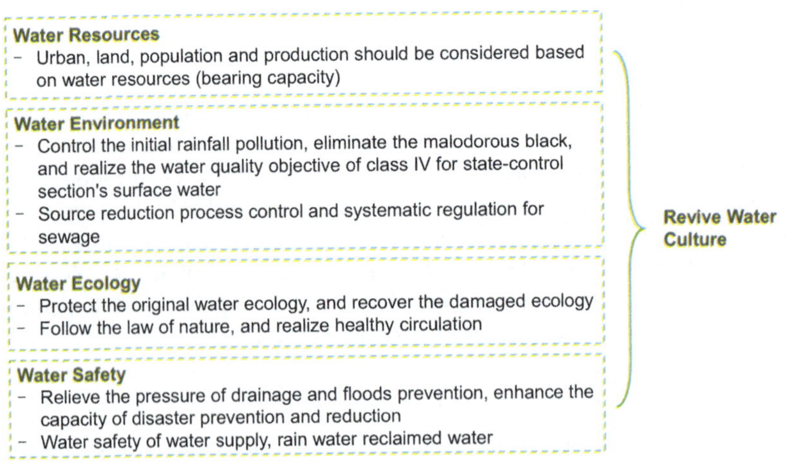

Figure 3.6 Formation of Chinese unique water culture via sponge city construction (figure by authors).

culture of a city is recognized as the soul of that city. The water culture to be revived usually satisfies the requirement of multiple objectives, including water resources, water environment, water ecology, and water safety.

3.4.2 Main functional elements of the water system 3.0

3.4.2.1 Sponge infrastructure

Figure 3.7 shows the sponge infrastructures at different utilization levels. By definition, sponge infrastructures are green infrastructures (Breuste *et al.*, 2015). Green roofs, permeable pavements, low elevation greenbelts, bioretention facilities, just to mention a few, are typical green infrastructures designed to provide the functions of permeation, retention, storage, purification, utilization, and drainage of rainwater. With the implementation of these green infrastructures, a Sponge City is not limited to its capability of runoff control for waterlogging prevention and rainwater utilization, but also non-point source pollution control, natural hydrological protection, and ecological recovery. Under the current situations of high population density, severe pollution of first flush runoff, and occurrence of intense rainfall, it is necessary to combine a series of green infrastructure with functional grey facilities, such as integrative initial rain intercepting chambers and underground reservoirs, to provide more robust and efficient solutions. Compared with the large-scale deep tunnels widely built in European cities, sponge infrastructures are more flexible and easier to be integrated into a city and implemented without disruption of traffic and other public utilities. Furthermore, the future Sponge City should not be limited to these scattered infrastructures or merely serve for rainwater management. It can

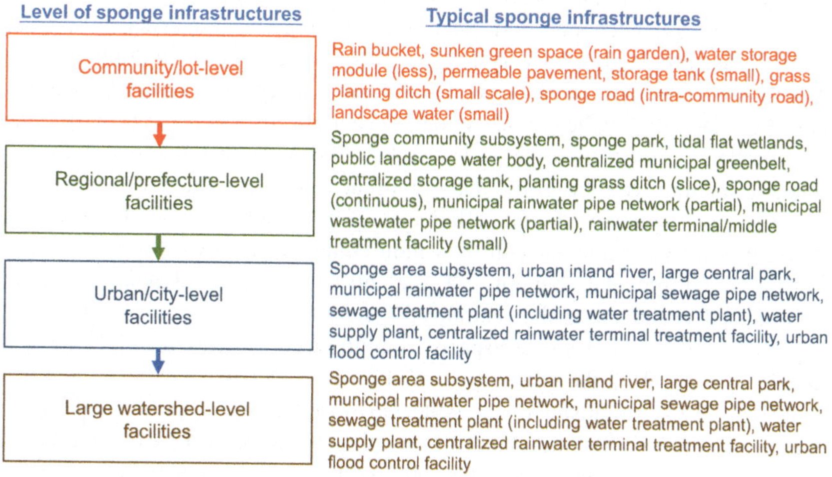

Figure 3.7 Sponge infrastructure at different utilization levels (figure by authors).

play roles in either synergistic wastewater disposal in dry seasons in coordination with decentralized sewage systems or in establishing an interconnected urban water body network and a healthy ecosystem by integrated planning.

3.4.2.2 Decentralized sewage system

Decentralized sewage systems are widely recognized as a potentially suitable approach to relieve environmental stress and mitigate water shortage. It is referred to as an approach related to wastewater treatment in a non-centralized way, with more than one treatment plant or sites for varying conditions in the whole water system, rather than merely a small scale pattern (Libralato *et al.*, 2012). Decentralized systems are usually flexible in their scale and can be classified into several levels with various service populations, such as:

- small scale to serve an individual house or a dozen of typical houses;
- middle scale to serve communities or larger blocks like school and hospital;
- larger scale to serve districts with larger populations, also referred to as semi-centralized systems.

In China, the middle and larger scaled systems are more suitable for densely populated urban areas. Aimed at onsite wastewater treatment and reclamation, an assortment of treatment technologies are available for meeting the requirements of various system scales, such as simple sanitation systems for communities, including septic tanks, membrane bioreactors, etc. (Fane & Fane, 2005), sewer mining in collection processes by forward osmosis membrane distillation, etc. (Xie *et al.*, 2013), mechanical–biological methods including sand filter, aerobic lagoons, and constructed wetlands, etc. (Massoud *et al.*, 2009), and treatment associated with materials recovery, including anaerobic reactors to produce methane, and ion exchange and electrodialysis to recover nutrients (Li *et al.*, 2015). A decentralized sewage system may also be applied for multiple purposes of pollution reduction, water reclamation, and energy and materials recovery. With the relatively smaller scale system, decentralized systems can be designed and installed in more flexible ways (Bakir, 2001) so that the distribution of pollutant loading can be fully taken into account to reduce the burden on enterprises for achieving higher efficiency of pollutant removal. Decentralized treatment is also suitable for eutrophication control in receiving water bodies to increase environmental benefits. These are the main advantages of the decentralized systems over the centralized systems.

In the economic and social aspects, decentralized systems are drawing wide interests as well. With the characteristics of onsite collection, treatment, and even reuse, long-distance transfer of the collected wastewater can be avoided and the investment is mainly for the treatment facilities. It has been reported that for the centralized sewage system, more than 60% of the capital costs are for collection and transfer pipelines (Massoud *et al.*, 2009). Moreover, resources and energy recovery in wastewater treatment can be realized in much easier ways in decentralized systems than centralized sewage systems. Thus, if resources and energy recovery are accounted for, the benefits will be even larger (Li *et al.*, 2015; Wilderer & Schreff, 2000).

3.4.2.3 Fit-for-purpose water supply system

In China, it is already required that all cities in water-deficient regions (mostly in north China) should promote the use of alternative or unconventional water resources, such as rainwater and reclaimed wastewater (seawater desalination is site-specific and not included in system 3.0). The potential for using unconventional water resources to mitigate urban water shortages is great. For example, the northeast city of Harbin has a population of about five million. It is estimated that the potential of rainwater harvesting amounts to 42 million m^3 per year, equivalent to 9% of the total annual water supply to the city. The potential of water reclamation from domestic wastewater is even larger. At present, only 7% of the treated wastewater is reused. If the percentage can be increased to 30%, the total amount will reach 14 billion m^3 per year, which is sufficient to cover the water consumption in 20 megacities.

Water use in a city is for various purposes. In addition to potable water supply for households and other domestic and municipal uses, environmental water uses for urban irrigation, landscape water replenishment, and so on usually consume large amounts of water. As different water uses have different quality requirements, a fit-for-purpose water supply system should be configured when unconventional water is added to the available water sources. An envisaged strategy is to combine unconventional water use with the implementation of sponge facilities and formulate a diversified water supply system framework where the harvested rainwater and reclaimed wastewater, after proper treatment to meet non-potable quality requirements, are directly supplied for environmental water use. By contrast, the potable water from the existing urban water supply network is solely for domestic water supply with a much-reduced volume.

3.4.2.4 Near-natural ecological zones

Nowadays, there are increasing efforts to switch hard engineering solutions for near-natural measures for the improvement of degraded urban water ecology. Examples of such efforts include those of Santa Ana River in Southern California, USA, and Cheonggyecheon River in Seoul, Korea (Gret-Regamey *et al.*, 2016) via the introduction of wetland systems and near-natural waterways. The near-natural and ecological approaches may not be introduced for short-term economic benefits but for the long-term ecosystem and social services such as to improve the quality of the water environment, facilitate groundwater recharge, provide wildlife habitat and promote urban liveability, resilience, and aesthetic satisfaction.

Near-natural ecological zones are important components of the framework for the urban water system 3.0. Ecological slope-maintenance, native species selection and configuration, artificial floating islands, and wetlands are major green elements to be introduced to replace or supplement grey engineering elements. The provision of crucial passages for flooding flow should be combined with the provision of purification capacity and aquatic landscapes. Reclaimed water and harvested rainwater can be used for supplementing ecological baseflow for urban rivers and streams so that long-distance transfer of clean water for water replenishment can be avoided. Wetlands can be constructed in association with rivers and streams to provide buffer zones for runoff pollution reduction and nonpoint source control.

3.4.2.5 *Intelligent water management system*

Sponge City construction not only needs the introduction of sponge facilities and various sponge measures but also an advanced managerial scheme to ensure that the urban water system is operated smoothly to achieve the prescribed goals. Figure 3.8 shows an example of the intelligent water management system (IWMS) for a Sponge City. The IWMS is a monitoring network with the application of various advanced tools and methodologies for performance evaluation and assessment. It is generally composed of five parts, including

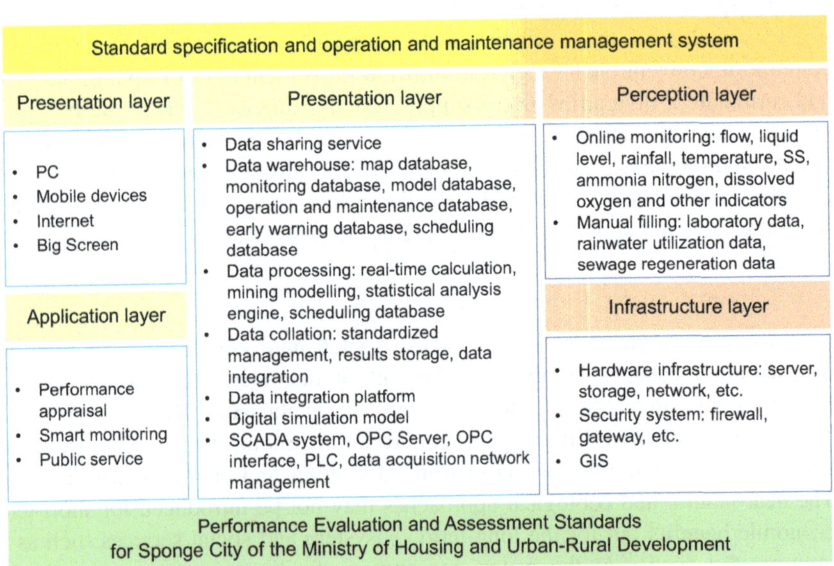

Figure 3.8 Intelligent water management system (IWMS) for Sponge City construction (figure by authors).

information structure, monitoring, data processing, application, and visualization. The core function of IWMS is the monitoring, collection, processing, integration, and sharing of large amounts of data related to the performance of the urban water system. The use of big data and machine learning can assist the establishment of relationships between governance objectives and construction measures, and further contribute to the identification of the key features of Sponge City construction in various regions with different economic and natural conditions.

The IWMS can also assist the synthesis and normalization of data in varied dimensionalities, and thus provide apt guidance to the development and management of practical sponge projects by the formulation of integrated management schemes, and generalization and specialization of technologies.

As shown in Figure 3.8, the IWMS has a presentation layer for the visualization and all-round display of Sponge City construction effects. This needs the implementation of hardware equipment and software platforms to ensure effective and sustainable online data generation and transmission.

Following the above discussion, we can see that the urgent need for solutions to the current water problems in Chinese cities provides the driving force for the implementation of Sponge Cities based on an innovative concept. The proposed urban water system 3.0 has provided an integrated multi-purpose strategy for cities in China toward sustainable development. A systematic approach with full consideration of regional situations is the core of the planning and design of such an urban water system.

3.5 FUTURE PERSPECTIVES

3.5.1 Enhancing system monitoring and evaluation and promoting multi-channel cooperation management

Sponge City projects require a systematic process of monitoring, evaluation, and management (Hakimdavar *et al.*, 2016). Prior to the construction, an integrated data system should be established for collecting information on (1) suitable technologies, (2) investment sources and their allocation, (3) aspects related to system planning and project design, and (4) potential outcomes of the projects, including the long-term effects and life cycle benefits (Jia *et al.*, 2015). A long-term monitoring program should be implemented so that the Sponge City's performance on urban water improvement can be well evaluated. The accumulation of data by long-term monitoring and proper evaluation will surely assist the development of Sponge Cities in China and even other countries of the world. This needs multi-channel cooperation among various government agencies and different sectors so that data and information can be well shared toward the common aims and policies for maximization of benefits.

3.5.2 Developing decision support tools for sustainable implementation of sponge city

The development of decision support tools is extremely important for Sponge City construction. Model simulation is, in any sense, indispensable for supporting policymakers, designers, and practitioners in building water systems at both urban and watershed scales (Bach *et al.*, 2014; Golden & Hoghooghi, 2018; Stanchev & Ribarova, 2016; Zhang & Chui, 2019). There are many urban stormwater models developed and applied globally, so far, such as SWMM, MIKE URBAN, and Info-works CS models, which can provide useful tools. However, as Sponge City construction is for solving complex problems, more comprehensive models have to be developed to meet the new needs. The basic requirements for the new models may include the following. First, they can well predict urban surface runoff in accordance with the complex underlying characteristics of urban regions. Second, they can well simulate the physical and chemical processes related to pollutants transport and reduction in green infrastructures, and third, they can forecast the ability of a Sponge City to prevent or minimize urban flooding and waterlogging.

Effective utilization of these models toward sustainable urban water management should also be ensured by incorporating the following:

- integration of modelling with available online and physical tools (Butler & Schutze, 2005);
- full attention paid to tackle uncertainties in model simulation (Liu *et al.*, 2008);
- modelling for obtaining longer time series results at high integration levels (Urich *et al.*, 2013).

3.5.3 Valuing Sponge City ecosystem services

Proper valuation of ecosystem services is important for raising the perceptions of Sponge Cities, so that good public–private partnerships can be promoted. The implementation of a Sponge City involves a range of biophysical, economic, cultural, and health values. Cost-benefit assessment is needed for all the projects implemented by the public and/or private sectors (Toran, 2016). Cost-effectiveness analysis for sponge projects may be more complicated due to unknown factors, such as cost calculation for project maintenance and monitoring, and evaluation of the life cycle benefits related to social and ecological amenities (Liang, 2018; Mao *et al.*, 2017). The benefits related to biodiversity, recreational spaces, urban heat reduction are important factors. The impact of Sponge City construction on the value of properties located in and/or around the project areas (Zhang *et al.*, 2018) should also be assessed.

To answer the question of the cost-effectiveness of Sponge City construction requires further analysis of the economic benefits in a large-scale context (Chui *et al.*, 2016). Regional differences in geographic location, hydrological conditions, social status, and urban infrastructure level may result in different outcomes and levels of success. Therefore, pilot projects and simulation scenarios may help identify the real costs involved and a more precise understanding of the benefits to all stakeholders, as well as public involvement and people's willingness to pay for Sponge City initiatives.

In the USA, CIRIA (the Construction Industry Research Information Association) has developed W045 BeST (Benefit of SuDS Tool) for the evaluation of the benefits from ecological services and economic benefits of SuDS (Sustainable Drainage Systems). This tool can be utilized to estimate the overall advantage of SuDS practice versus existing urban water management practices (CIRIA, 2015). Also, this tool can be partially applied for analysing Sponge City projects with the incorporation of other methods to account for future spatial and temporal changes that may affect the performance of Sponge City infrastructure.

3.5.4 Developing local guidelines and standards for Sponge City implementation

Every urban area has its own hydrological and climatic features, as well as its history and characteristics of urban development. Although national laws and regulations have provided general guidance of Sponge City construction for the whole nation, they may not be completely applicable to the local context. Therefore, understanding local conditions is key to the successful implementation of the Sponge City concept, and additional local guidelines and standards should be developed based on local needs. In many cases, blindly copying the experiences from other cities or just following instructions from the central government may be inappropriate in the planning, design, construction, operation, and evaluation of Sponge City projects. At the provincial, municipal, or even project levels, research should be conducted on the feasible engineering measures and available technologies that best fit the regional or local situations.

3.5.5 Promoting Sponge City construction in watershed-scales based on data and information sharing

Sponge City construction is a new concept that can be applied for projects of various scales. However, as the goal of Sponge City construction is to build a sustainable urban water system in the form of system 3.0 shown in Figure 3.4, it is not limited to the implementation of individual sponge facilities or a series of such facilities. Rather, it encompasses a reformation of the whole framework of the water-related infrastructure of a city closely related to the local watershed. Therefore, it is important to consider beyond individual project sites and pay attention to an integrated and watershed-scale approach, aiming to solve all the interrelated problems and creating a sustainable water environmental system covering broader areas. Sponge City construction at the watershed-scale can potentially avoid segmentation and isolation of the sponge facilities, maximize the overall benefits of ecosystem conservation, water quality improvement, flooding and waterlogging control, and create a healthy environment in the entire watershed.

As an innovative and revolutionary approach, Sponge City is drawing wide attention from the world, and in the process of Sponge City construction, China has learnt a lot from other countries with similar concepts and technologies such as BMPs, LID, SUDs, WSUD, etc. Within China, different cities are learning from each other to obtain the latest successful experiences, especially through the central government-oriented national pilot program (refer to Chapter 11 of this book). In the long term, data and information sharing will become more and more important for the advancement of Sponge City technologies. It is also important to improve coordination across governmental agencies by the establishment of a Sponge City database and an experiences-sharing network.

A proper social and economic evaluation of Sponge City practices is also indispensable for highlighting the whole life cycle benefits and risk of failure to deliver useful information to the public and increase their knowledge and perceptions, to enhance public willingness to support the implementation of Sponge Cities. In an era of climate change and rapid urbanization, the formulation of strategies and policies focusing on the promotion of the Sponge City concept will play an extremely important role in developing healthy, resilient, and sustainable cities.

REFERENCES

Bach P. M., Rauch W., Mikkelsen P. S., McCarthy D. T. and Deletic A. (2014). A critical review of integrated urban water modelling Urban drainage and beyond. *Environmental Modelling and Software*, **54**, 88–107.

Bakir H. A. (2001). Sustainable wastewater management for small communities in the Middle East and North Africa. *Journal of Environmental Management*, **61**(4), 319–328.

Breuste J., Artmann M., Li J. X. and Xie M. M. (2015). Special issue on green infrastructure for urban sustainability. *Journal of Urban Planning and Development*, **141**(3), 1–5.

Butler D. and Schutze M. (2005). Integrating simulation models with a view to optimal control of urban wastewater systems. *Environmental Modelling and Software*, **20**(4), 415–426.

Carlson C., Barreteau O., Kirshen P. and Foltz K. (2015). Storm water management as a public good provision problem: survey to understand perspectives of low-impact development for urban storm water management practices under climate change. *Journal of Water Resources Planning and Management*, **141**(6), 1–13.

Chan F. K. S., Griffiths J. A., Higgitt D., Xu S., Zhu F., Tang Y.-T. and Thorne C. R. (2018). 'Sponge City' in China – A breakthrough of planning and flood risk management in the urban context. *Land Use Policy*, **76**, 772–778.

Chang H.-S. and Su Q. (2020). Research on constructing Sponge City indicator and decision evaluation model with fuzzy multiple criteria method. *Water Environment Research*, **92**, 1910–1921.

Cheng H. F. and Hu Y. A. (2011). Economic transformation, technological innovation, and policy and institutional reforms hold keys to relieving China's water shortages. *Environmental Science and Technology*, **45**(2), 360–361.

Chui T. F. M., Liu X. and Zhan W. T. (2016). Assessing cost-effectiveness of specific LID practice designs in response to large storm events. *Journal of Hydrology*, **533**, 353–364.

Chung W. Y. and Yoo J. H. (2015). Remote water quality monitoring in wide area. *Sensors and Actuators B-Chemical*, **217**, 51–57.

CIRIA (2015). Evaluating the Benefits of SuDS Using CIRIS's BeST, the Construction Industry Research Information Association. Available from: www.susdrain.org/resources/presentations.html

De Feo G., Antoniou G., Fardin H. F., El-Gohary F., Zheng X. Y., Reklaityte I. and Angelakis A. N. (2014). The historical development of sewers worldwide. *Sustainability*, **6**(6), 3936–3974.

Ellis J. B. and Lundy L. (2016). Implementing sustainable drainage systems for urban surface water management within the regulatory framework in England and Wales. *Journal of Environmental Management*, **183**, 630–636.

Fane A. G. and Fane S. A. (2005). The role of membrane technology in sustainable decentralized wastewater systems. *Water Science and Technology*, **51**(10), 317–325.

Golden H. E. and Hoghooghi N. (2018). Green infrastructure and its catchment-scale effects: an emerging science. *Wiley Interdisciplinary Reviews – Water*, **5**(1), 1254.

Gret-Regamey A., Weibel B., Vollmer D., Burlando P. and Girot C. (2016). River rehabilitation as an opportunity for ecological landscape design. *Sustainable Cities and Society*, **20**, 142–146.

Griffiths J., Chan F. K. S., Shao M., Zhu F. and Higgitt D. L. (2020). Interpretation and application of Sponge City guidelines in China. *Philosophical Transactions of the Royal Society A – Mathematical Physical and Engineering Sciences*, **378**(2168), 1–20.

Hakimdavar R., Culligan P. J., Guido A. and McGillis W. R. (2016). The Soil Water Apportioning Method (SWAM): an approach for long-term, low-cost monitoring of green roof hydrologic performance. *Ecological Engineering*, **93**, 207–220.

Hiltrud P. and Pierre B. (2011). Urban Green-blue Grids for Sustainable and Dynamic Cities. Coop for Life, Delft.

Jia H. F., Yao H. R. and Yu S. L. (2013). Advances in LID BMPs research and practice for urban runoff control in China. *Frontiers of Environmental Science and Engineering*, **7** (5), 709–720.

IWA (2016). IWA Principles for Water-Wise Cities. International Water Association, London, UK.

Jia H. F., Yao H. R., Tang Y., Yu S. L., Field R. and Tafuri A. N. (2015). LID-BMPs planning for urban runoff control and the case study in China. *Journal of Environmental Management*, **149**, 65–76.

Karr J. R. and Schlosser I. J. (1978). Water resources and the land-water interface. *Science*, **201**(4352), 229–234.

Li W. W., Yu H. Q. and Rittmann B. E. (2015). Chemistry: Reuse water pollutants. *Nature*, **528**(7580), 29–31.

Li C., Huang M., Liu J., Ji S., Zhao R., Zhao D. and Sun R. (2019). Isotope-based water-use efficiency of major greening plants in a sponge city in northern China. *Plos One*, **14**(7), e0220083.

Liang X. (2018). Integrated economic and financial analysis of China's sponge city program for water-resilient urban development. *Sustainability*, **10**(3), 1–12.

Libralato G., Ghirardini A. V. and Avezzu F. (2012). To centralise or to decentralise: an overview of the most recent trends in wastewater treatment management. *Journal of Environmental Management*, **94**(1), 61–68.

Lim H. S. and Lu X. X. (2016). Sustainable urban stormwater management in the tropics: an evaluation of Singapore's ABC Waters Program. *Journal of Hydrology*, **538**, 842–862.

Liu Y. Q., Gupta H., Springer E. and Wagener T. (2008). Linking science with environmental decision making: experiences from an integrated modeling approach to supporting sustainable water resources management. *Environmental Modelling and Software*, **23** (7), 846–858.

Mao X. H., Jia H. F. and Yu S. L. (2017). Assessing the ecological benefits of aggregate LID-BMPs through modelling. *Ecological Modelling*, **353**, 139–149.

Massoud M. A., Tarhini A. and Nasr J. A. (2009). Decentralized approaches to wastewater treatment and management: Applicability in developing countries. *Journal of Environmental Management*, **90**(1), 652–659.

Meng F. and Li S. (2020). A new multiple attribute decision making method for selecting design schemes in sponge city construction with trapezoidal interval type-2 fuzzy information. *Applied Intelligence*, **50**(7), 2252–2279.

O'Donovan P., Coburn D., Jones E., Hannon L., Glavin M., Mullins D. and Clifford E. (2015). A cloud-based distributed data collection system for decentralised wastewater treatment plants. *Procedia Engineering*, **119**, 464–469.

Ren N., Wang Q., Wang Q., Huang H. and Wang X. (2017). Upgrading to urban water system 3.0 through sponge city construction. *Frontiers of Environmental Science and Engineering*, **11**(4), 1–8.

Sharma A. K., Pezzaniti D., Myers B., Cook S., Tjandraatmadja G., Chacko P. and Walton A. (2016). Water sensitive urban design: an investigation of current systems, implementation drivers, community perceptions and potential to supplement urban water services. *Water*, **8**(7), 272.

Shen W., Liu Y., Wu M., Zhang D., Du X., Zhao D., Xu G., Zhang B. and Xiong X. (2020). Ecological carbonated steel slag pervious concrete prepared as a key material of sponge city. *Journal of Cleaner Production*, **256**, 120244.

Stanchev P. and Ribarova I. (2016). Complexity, assumptions and solutions for eco-efficiency assessment of urban water systems. *Journal of Cleaner Production*, **138**, 229–236.

Thu T. N., Huu Hao N., Guo W. and Wang X. C. (2020). A new model framework for sponge city implementation: emerging challenges and future developments. *Journal of Environmental Management*, **253**, 109689.

Toran L. (2016). Water level loggers as a low-cost tool for monitoring of stormwater control measures. *Water*, **8**(8), 346.

Urich C., Bach P. M., Sitzenfrei R., Kleidorfer M., McCarthy D. T., Deletic A. and Rauch W. (2013). Modelling cities and water infrastructure dynamics. *Proceedings of the Institution of Civil Engineers-Engineering Sustainability*, **166**(5), 301–308.

Wang X. H., Wang X., Huppes G., Heijungs R. and Ren N. Q. (2015). Environmental implications of increasingly stringent sewage discharge standards in municipal wastewater treatment plants: case study of a cool area of China. *Journal of Cleaner Production*, **94**, 278–283.

Wang Y., Liu X., Huang M., Zuo J. and Rameezdeen R. (2020). Received vs. given: willingness to pay for sponge city program from a perceived value perspective. *Journal of Cleaner Production*, **256**, 120479.

Wilderer P. A. and Schreff D. (2000). Decentralized and centralized wastewater management: A challenge for technology developers. *Water Science and Technology*, **41**(1), 1–8.

Xia J., Zhang Y., Xiong L., He S., Wang L. and Yu Z. (2017). Opportunities and challenges of the Sponge City construction related to urban water issues in China. *Science China-Earth Sciences*, **60**(4), 652–658.

Xie M., Nghiem L. D., Price W. E. and Elimelech M. (2013). A forward osmosis-membrane distillation hybrid process for direct sewer mining: system performance and limitations. *Environmental Science and Technology*, **47**(23), 13,486–13,493.

Zhang K. and Chui T. F. M. (2019). Linking hydrological and bioecological benefits of green infrastructures across spatial scales – A literature review. *Science of the Total Environment*, **646**, 1219–1231.

Zhang Z., Szota C., Fletcher T. D., Williams N. S. G., Werdin J. and Farrell C. (2018). Influence of plant composition and water use strategies on green roof stormwater retention. *Science of the Total Environment*, **625**, 775–781.

Zhang L., Sun X. and Xue H. (2019). Identifying critical risks in Sponge City PPP projects using DEMATEL method: A case study of China. *Journal of Cleaner Production*, **226**, 949–958.

Chapter 4

US version of water-wise cities: Low impact development

Peter T. Weiss[1], *John S. Gulliver*[2] *and Ali Ebrahimian*[3]
[1]*Valparaiso University, Valparaiso, Indiana, USA*
[2]*University of Minnesota, Minneapolis, Minnesota, USA*
[3]*Florida International University, Miami, Florida, USA*

4.1 INTRODUCTION TO REGULATORY HISTORY

The population of the United States has continually increased and has become vastly more urbanized. The US population rose from 5.3 million people in 1800 to 76 million in 1900, to over 248 million in 2000, and in 2019 it was approximately 329 million people. Over this time period the population has also shifted from primarily rural to primarily urban. In 1800 approximately 5% of Americans lived in urban areas (defined as populations over 2500 for that time) but, due in large part to industrialization, that value increased to approximately 35% in 1900 and now, in 2019, approximately 80% of Americans live in urban areas, which is currently defined as having a population of 50,000 or more.

With the increase in population and urbanization have come added stresses on the environment. These stresses include large amounts of wastewater generated from relatively small land areas, increased stormwater runoff volumes due to an increase in impervious land surfaces, and a degradation in the water quality of stormwater runoff. The water quality of stormwater runoff is degraded due to human activities such as widespread automobile use, erection of buildings, and increased fertilizer use. Automobiles generate pollution through tire wear, rusting

Editors: Xiaochang C. Wang and Guangtao Fu
doi: 10.2166/9781789060768_0077

automotive parts, oils, gasoline and associated by-products. Buildings mostly contaminate runoff through roofing and siding materials that leach metals and other pollutants. Fertilizers add nutrients such as phosphorus and nitrogen to stormwater runoff. These nutrients can cause algae blooms that lead to the eutrophication of surface waters. As a result of these increasing human activities and the increase in population and urbanization, the trend is that the surface waters of the United States have become more polluted.

As the negative environmental impact has increased, people have become more aware of the problem and laws have been enacted at the federal, state, and local levels to mitigate the negative environmental impact. The first attempt at federal legislation addressed wastewater treatment plant discharges through the Federal Water Pollution Control Act of 1948, which, when amended in 1972, became known as the Clean Water Act (CWA). The CWA established the framework for regulating pollutant discharges into waters of the United States and:

- gave the US Environmental Protection Agency (US EPA) authority to enact pollution control programs;
- maintained existing policy to establish water quality standards in US surface waters;
- made it illegal to discharge pollutants into navigable waters without a permit;
- funded the construction of wastewater treatment plants; and
- recognized that planning is necessary to address problems caused by non-point source pollution, such as stormwater runoff.

The CWA also established the National Pollution Discharge Elimination System (NPDES) permit program, which regulates point source discharges into US waters. In most cases the US EPA authorizes state governments to perform permitting and enforcement aspects of the NPDES program. An NPDES permit is essentially a license for a facility such as a wastewater treatment plant or an industrial process to discharge a certain amount of pollutant into a water body.

The first sentence of the CWA reads "The objective of this Act is to restore and maintain the chemical, physical, and biological activity of the Nation's waters." The establishment of the NPDES permit system helped to reduce pollutant loading to surface waters of the United States by initially regulating discharges from wastewater treatment plants, industrial processes, and other point sources throughout the country. As defined in the CWA and shaped by years of litigation, a point source includes a pipe, channel, ditch, conduit, or container. It includes boats that may discharge pollutants and it also includes concentrated animal feed lots where animals such as pigs or cows are kept and fed. By law, however, agricultural stormwater and return flows from agricultural processes other than feed lots (e.g. crops) are not considered point sources. The NPDES permit program has helped improve water quality in many US surface waters. For example, as reported by the US EPA Region 5 (1984), in the 10 years that followed the passage of the CWA of 1972, water quality in 75,000 km of US streams improved and water quality in 474,000 km remained the same despite an overall increase in the number of sources of pollution. In 18,000 km of streams, however, the water quality was degraded. With regards to publicly owned lakes and reservoirs over the same time span (i.e. 1972–1982), 390,000 acres experienced improved water quality, four million hectares had no change in water quality, and 0.68 million hectares were degraded. These values were estimated from reports that were provided voluntarily from states to the US EPA. Overall, the reports covered the assessment of 42% of all streams and 50% of all the publicly owned lakes and reservoirs in the United States.

Although water quality improved in many streams and lakes in the 10 years following the CWA of 1972, most of the water bodies assessed did not change at all, and for lakes and reservoirs, the area that degraded over that time span was over four times the area that improved. Thus, although some water bodies did improve in quality after the CWA was enacted, many more either did not improve at all or were degraded. A likely reason for this outcome was determined to be urban stormwater runoff, which was found to be a leading source of surface water contamination (US EPA, 2005). Thus, in response, the 1987 amendments to the CWA were passed, which focused on urban stormwater runoff.

Those amendments required the US EPA to address stormwater runoff in two phases. Phase I of the program began in 1990 and regulated municipal separate storm sewer systems (MS4s) with populations of 100,000 or more, ten different categories of industrial sites, and construction sites that disturbed five or more acres (2 hectares or greater). By law, an MS4 is a conveyance or a system of conveyances that:

- is owned by a public entity such as a state, city, or town that discharges to waters of the United States;
- collects or conveys stormwater but not wastewater (i.e. not combined sewers), and;
- is not part of a wastewater treatment plant.

After Phase I regulations were adopted, many surface water bodies in the United States still did not meet water quality standards. For example, in 2000, approximately 40% of assessed water bodies remained polluted above water quality standards, due in large part to stormwater runoff (US EPA, 2005).

Phase II began in 1999 and regulated smaller MS4s (which met certain criteria) and construction sites as small as 1 acre (0.4 hectares). The MS4s that are under jurisdiction of the Phase II requirements are those designated as a regulated small MS4. The three ways in which a small MS4 can be designated as regulated are:

(1) automatic designation if it falls within an urbanized area as defined by the US Census Bureau;
(2) if the NPDES permitting authority determines that its discharge has caused or has the potential to cause adverse impact on water quality; and

(3) if it contributes substantially to the pollutant loads of a physically connected regulated MS4.

To meet the requirements of the CWA a regulated small MS4 must develop, implement, and enforce a stormwater management program designed to reduce pollutant loading from their MS4 to the "maximum extent practicable." The stormwater management program must include the following six minimum control measures:

(1) public education and outreach;
(2) public participation;
(3) illicit discharge detection and elimination;
(4) construction site runoff control;
(5) post-construction runoff control; and,
(6) pollution prevention/good housekeeping.

A regulated small MS4 must identify its selection of stormwater management practices for each of the above six minimum control measures along with measurable goals for each. When reporting to the NPDES authority, each of the minimum control measures and measurable goals must be evaluated and assessed.

Under the CWA, every two years states must develop lists of rivers, lakes, coastal waters, estuaries, and other surface water bodies that do not meet water quality standards. Any water body that is put on the list is deemed "impaired." For each impaired water body and all corresponding pollutants, a total maximum daily load (TMDL) must be developed. The TMDL specifies the maximum amount of a pollutant a water body can receive, usually in pounds per day, that will allow the water body to meet the corresponding water quality standards. In other words, the TMDL specifies target maximum pollutant loading rates that the water body can receive that will result in a reduction of pollutant concentrations within the water body so that the corresponding water quality standards will be met. If or when this is achieved, the water body will no longer be considered impaired and will be removed from the impaired list. The TMDL further breaks down the total maximum daily pollutant load to specify the amount of the daily pollutant load that can be contributed from each loading source (e.g. wastewater treatment plants, internal sediment loading, stormwater runoff, etc.). If an impaired water body has an approved TMDL and an MS4 contributes the corresponding pollutant to that water body, steps must be taken by the MS4 to ensure that the TMDL target loading rate for stormwater runoff will be achieved.

For comparison, median stormwater runoff concentration values for typical pollutants (Maestre & Pitt, 2005) are compared to US EPA drinking water and/or aquatic life criteria in Table 4.1. In this table, the maximum contaminant level (MCL) is the legal threshold limit of the amount of a substance that is allowed in public water systems under the Safe Drinking Water Act (SDWA). Copper and lead are regulated by a US EPA rule that requires drinking water providers to

Table 4.1 Comparison of median stormwater runoff concentrations to US EPA regulatory limits.

Pollutant	Median values* (µg/L)	US EPA drinking water MCL (mg/L)	US EPA chronic aquatic life criteria for freshwater (µg/L)
Aluminum			
Cadmium	1.0	0.005	0.72
Chloride			230,000
Chromium	7.0	0.1	
Chromium(III)			74
Chromium(VI)			11
Copper	16	1.3**	
Cyanide		0.2	5.2
Lead	17	0.015**	3.2
Nickel	8		52
Nitrate		10	
Nitrite		1	
Oil	7.5 (mg/L)		
PCBs			
Phosphorus (total)	270		
Silver			
Zinc	116		120
		Secondary MCL (mg/L)	
Chloride		250	
Zinc		5	
Iron		0.3	
pH		6.5–8.5	6.5–9.0
TSS(mg/L)	59		
Fecal coliform (mpn/100 mL)	5091		
Fecal streptococcus (mpn/100 mL)	17000		
Total coliform (mpn/100 mL)	12000		
Total *E. coli* (mpn/100 mL)	1750		
Nitrate + Nitrite (mg/L)	0.6		

*Data source: Maestre and Pitt (2005).

**Lead and copper are regulated by a treatment technique that requires systems to control the corrosiveness of their water. If more than 10% of tap water samples exceed the action level, water systems must take additional steps. Listed values are the action levels.

control the corrosiveness and subsequent release of copper and lead in their water system. If more than 10% of tap water samples, sampled at the customers tap, exceed the action level, water providers must take additional steps. Listed values for copper and lead are those action levels. Also, the aquatic life criteria values are the highest concentration in water that are not expected to pose a significant risk to the majority of species in a given environment. Secondary MCLs (SMCLs) set non-mandatory water quality standards but the US EPA does not enforce these values. Rather, they are established as guidelines to assist public water systems in managing their drinking water for aesthetic considerations, such as taste, color, and odor. These contaminants are not considered to present a risk to human health at the SMCL.

In 2019, according to the US EPA (2019), of the water bodies assessed, approximately 53% of rivers and streams, 71% of lakes and reservoirs, 80% of bays and estuaries, 54% of wetlands, and 72% of coastal shoreline were impaired due to at least one pollutant. Other water bodies such as the Great Lakes, Great Lakes shoreline, and coastal shoreline show similar or higher values. Overall, 69% of all United States water bodies that have been assessed were impaired. The leading causes of impairment vary by the type of water body but typically include pathogens, sediment/turbidity, nutrients, metals, temperature, polychlorinated biphenyls (PCBs), pH, salinity or total dissolved solids (TDS), and pesticides.

For each of the impaired water bodies and for each pollutant causing an impairment, a TMDL plan must be developed. Stormwater target loading rates in TMDLs often seek a reduction in pollutant loads upwards of 60%. If pollutant loading reductions to this extent are to be achieved, multiple methods or strategies will likely need to be employed. In response to this need, a wide range of management strategies have been developed that have shifted the focus towards more sustainable development. The next section first discusses this shift in focus and includes topics such as public education, pollution prevention, and source control as stormwater management strategies. These non-structural stormwater management methods seek to reduce the amount of pollution in runoff prior to a runoff event and can be much more cost-effective than treating stormwater after it has already been polluted. Following this, the next section continues with a discussion on stormwater runoff volume reduction via infiltration and evapotranspiration and the potential of soil and groundwater contamination due to the increased infiltration of polluted stormwater. Specific structural practices to manage and treat stormwater runoff are discussed later in this chapter.

4.2 A SHIFT IN STORMWATER MANAGEMENT IN THE UNITED STATES

The CWA of 1972 and the 1987 amendments to this act have forced a change in thinking regarding the management of stormwater runoff in the United States.

Historically, stormwater management was achieved by removing runoff from the land surface as quickly as possible, reducing the peak flow rate (if required by local regulations), transporting runoff through concrete lined channels and/or concrete pipes, and discharging it to receiving surface waters. Little thought was given to the quality of the runoff or the impact the runoff would have on the receiving water body. Similarly, not much thought was given to reducing runoff volumes or treating the runoff to improve its water quality. Phase I and II of the 1987 amendments to the CWA motivated stormwater managers throughout the United States to focus on water quality in addition to water quantity. Initially, when trying to improve stormwater runoff quality, many sought to treat runoff at the downstream end of a watershed but, through years of experience, that focus has shifted to targeting runoff in the upstream reaches of the watershed. Now stormwater management methods encourage infiltration of stormwater runoff and/or the improvement of runoff quality as far upstream as possible, utilizing small, decentralized practices to have an urbanized watershed mimic the undeveloped watershed as much as possible. These efforts have resulted in stormwater management practices that, as a whole, are called low impact development (LID), or green stormwater infrastructure (GSI). Such practices will be referred to as LID in this chapter.

According to the US EPA (2019), LID stormwater management practices "refers to systems and practices that use or mimic natural processes that result in the infiltration, evapotranspiration or use of stormwater in order to protect water quality and associated aquatic habitat." Typically, a large component of a LID practice is the infiltration of stormwater instead of conveyance to receiving waters through sewers. LID gained popularity because it promotes more sustainable water resources management while recognizing the needs of economic growth within local communities (Coffman, 2002). Additionally, LID may be beneficial to air quality and the quality of life (Coffman, 2002). Several modeling experiments have shown that LID – when properly implemented – is capable of nearly restoring the predevelopment hydrologic regime for storms ranging from about 2–8 cm of rainfall depth, depending on the infiltration capacity of the soil (Brander *et al.*, 2004; Holman-Dodds *et al.*, 2003).

While LID practices are a key component of stormwater management, thorough and knowledgeable stormwater management also includes practices such as pollutant source reduction, public education, and other corresponding issues such as awareness of the potential for soil and groundwater contamination due to increased infiltration of polluted stormwater.

4.2.1 Pollution prevention, source control, and public education

A cost-effective means of reducing stormwater pollutant loading to a surface water body is to prevent the pollutant from being in the runoff in the first place. Strategies that use this approach are typically divided into one of two categories. One category is pollution prevention, which involves preventing the pollutant from being generated or from being present on the land surface in the first place. Examples include illicit discharge detection and elimination, pet ordinances that require pet owners to pick up and dispose their pet's feces, and ordinances that require grass clippings to be collected in bags attached to lawn mowers. The second category is source control, which involves removing or retaining the pollutant from the land surface at its source before a runoff event occurs, so that the pollutant cannot be transported by the runoff. Examples of source control include street sweeping to remove sediment and other gross solids from road surfaces prior to rainfall or snowmelt events and growing vegetation on disturbed soils at construction sites to prevent erosion. Methods in both categories typically benefit from public education that increases awareness of stormwater pollutants, their sources, and their effect on surface water bodies. Thus, public education, outreach, and involvement is a requirement of all MS4 NPDES permits. Also, although not a CWA requirement, any stormwater management plan should strongly consider incorporating these two methods due to the cost-effectiveness of pollution prevention and source control.

4.2.2 Volume reduction

A developed watershed typically includes such items as buildings, roads, parking lots, sidewalks, and possibly other structures that alter the land surface from its native state. These changes to the land surface have historically been made with methods that increase the imperviousness of the watershed and, as a result, also increase runoff volumes. Not only do runoff volumes increase, but the water quality of the runoff is also degraded, and the water typically contains many pollutants. As previously mentioned, stormwater runoff is a major contributor to impaired surface water bodies across the United States. Thus, if CWA and TMDL requirements are to be met and developed watersheds are to more closely mimic the undisturbed watershed, runoff volumes must be reduced to near predevelopment levels. This essential part of the shifting of stormwater management focus in the United States has been incorporated into many LID practices, usually by two methods: infiltration and evapotranspiration. Infiltration is the process of the runoff entering the soil structure where it can move as shallow groundwater flow or move deeper into the soil towards the water table. Evapotranspiration is the process of water transferring from the land to the atmosphere via evaporation and plant transpiration.

LID practices that achieve volume reduction include green roofs, bioinfiltration systems (i.e. rain gardens), infiltration chambers/trenches/basins, and constructed wetlands. Green roofs are planted atop buildings and slow runoff flows while enabling evapotranspiration (Teemusk & Mander, 2007). Rain gardens are shallow vegetated depressions into which stormwater is directed for infiltration and groundwater recharge (US EPA, 2000), while also allowing evapotranspiration. Infiltration chambers are sub-surface (i.e. underground) chambers that store runoff and allow it to infiltrate into the existing soil. Infiltration trenches are trenches that are typically backfilled with large gravel. The void spaces in the gravel provide storage space for runoff that infiltrates into the existing soil. Infiltration basins are similar to detention ponds but, due to porous soil, are able to infiltrate larger volumes of runoff. They are typically covered with vegetation which will also dry out the soil through evapotranspiration. Although initially constructed to treat municipal wastewater, constructed wetlands are now used to treat stormwater (Shutes *et al.*, 1997; Walker & Hurl, 2002) and are also considered an LID practice because they can reduce runoff volumes by evapotranspiration and, in some cases, through infiltration.

4.2.3 Pollution retention by soil and potential for soil and groundwater contamination

Pollutants carried by infiltrated stormwater, whether particulate or dissolved, can be retained in the soil structure via physical straining, cation exchange, adsorption to the surface of soil particles, microbial activity, precipitation followed by physical straining, and the formation of surface complexes. The ability of a soil to retain pollutants varies with the soil properties that affect these removal processes such as particle size distribution, mineral composition, cation exchange capacity, etc. Also, pollutant transport through the soil is affected by the soils organic content, microorganism activity in the vadose zone, porosity, infiltration capacity, moisture content, and other factors (Clark & Pitt, 2007, 2011). Thus, the transport and retention of pollutants through and in soil media is highly site specific.

Although infiltration can reduce pollutant loading to surface water bodies, the pollutants will remain in the environment and may move through the soil media. Thus, their destination should be greatly considered with any LID practice that infiltrates stormwater. Some pollutants, such as dissolved metals, may be adsorbed to the surface of soil particles and thereby removed from the transport process. They remain in the soil, however, and may become mobile again in the future, especially when the runoff is high in chloride content from road salt. If the soil continues to adsorb metals or any other pollutant, the soil pollutant concentration may, at some point, become contaminated above accepted levels. Also, soil has a finite capacity to adsorb pollutants and that capacity may become exhausted. In such cases the pollutant(s) would not adsorb to the soil solids but would continue to move with the water and/or diffuse through the soil structure. Other contaminants, such as nutrients or PAHs, may be utilized by microorganisms in the soil if conditions are suitable to support the appropriate microbial population. If not, the pollutants may move with the water through the soil to an unknown destination. Thus, it is possible that any infiltrated pollutant could reach a groundwater source or contaminate the soil to unacceptable levels.

This means that, while infiltrated pollutants are usually assumed to be removed from the conveyance system and are not able to contaminate surface water bodies, it is possible that they will contaminant soil and/or groundwater sources. Such potential contamination and corresponding ramifications must be considered when implementing any LID practice that infiltrates stormwater. This section discusses some common stormwater pollutants and their fate in infiltration systems along with their potential for soil and groundwater contamination.

4.2.3.1 Nutrients

The most common and influential forms of nutrient pollution to stormwater are phosphorus and nitrogen, the latter of which may be in many forms. Upon infiltrating into a soil, a majority of the particulate phosphorus will likely be physically strained by the soil particles (i.e. media). Once in the soil media, orthophosphate (PO_4^{-3}) can be removed from infiltrated stormwater via precipitation or chemical adsorption onto soil particle surfaces through reactions with iron, calcium, or aluminum. It may also be utilized in the biological processes of bacteria or vegetation. Also, in some cases soils can be a source of phosphorus and it is not uncommon for an LID practice such as a rain garden or detention pond to release rather than retain phosphorus (Dietz & Clausen, 2005, 2006; Wu *et al.*, 1996). This is especially true for LID practices containing organic material such as plants or mulch. As the organic material breaks down over time, phosphorus may be released and the soil or LID practice may act as a source of phosphorus.

Nitrogen in stormwater may be present in many forms with ammonium (NH_4^+) being most toxic to aquatic life. Ammonium can be converted through a microbially-mediated nitrification reaction to nitrite (NO_2^-) and nitrate (NO_3^-). Nitrate is the most common nonpoint-source groundwater contaminant in the world (Gurdak & Qi, 2012). Nitrate, due to its high solubility, can leach from decaying plant material and other organics in the soil to infiltrating stormwater and eventually contaminate groundwater aquifers. Naturally occurring ammonium in soils can be oxidized to nitrite and nitrate and cause groundwater contamination. Areas in the United States with nitrate groundwater contamination are often those with large farm animal populations that have dairy and poultry industries and irrigated agricultural areas. In urban areas, the major source of groundwater nitrogen contamination is from road runoff, which is contaminated by nitrates in vehicle exhausts and the soil brought onto the roadway by vehicle traffic.

According to Pitt *et al.* (1996), nitrate, due to its typically low concentrations in urban runoff, has low to moderate groundwater contamination potential for both surface infiltration practices (e.g. bioinfiltration practices, infiltration basins, filter strips, swales, pervious pavement, etc.) and subsurface infiltration/injection systems (e.g. infiltration chambers, infiltration trenches, etc.). If nitrate concentrations in runoff are high, however, the groundwater contamination potential is also high.

4.2.3.2 Metals

Due to frequency of occurrence and toxicity, cadmium, copper, lead, and zinc are the primary metals of concern in urban stormwater runoff. Metals are present in stormwater in dissolved phases, but a fraction of most metals are bound to suspended solids (Davis & McCuen, 2005; Marsalek *et al.*, 2001). Thus, metals may be removed by adsorption to soil media. Of the metals typically found in stormwater, lead has the largest tendency to adsorb to solids. The ranking of adsorption potential for some common metals to soil particles is, with lead having the highest potential, as shown below (Pitt *et al.*, 1995):

$$\text{Lead} > \text{Copper} > \text{Nickel} > \text{Cobalt} > \text{Zinc} > \text{Cadmium} \tag{4.1}$$

Other metal removal mechanisms include precipitation, occlusion with other precipitates, diffusion into solid particles, and biological uptake. Overall, dissolved metals are typically removed through adsorption to soil particles in the vadose zone, while metals associated with the particulate phase are usually removed via physical straining at or near the soil surface (Pitt *et al.*, 1999).

Since metals are often bound to solid particles, removal of suspended solids can be an effective method of reducing total metal concentrations in stormwater. Metals do not generally degrade to another element in the environment, however, and stormwater loading into LID practices will result in metal accumulation. Using typical stormwater pollutant loading and soil capacity estimates for bioretention practices, Davis *et al.* (2003) estimated that, after 20 years, concentrations of cadmium, lead, and zinc would reach or exceed levels permitted by the US EPA biosolids land application regulations. Once adsorbed to the surface of a soil particle, metals may not necessarily remain stationary. Furthermore, depending on soil conditions such as pH and Eh, metals may be released from the solid surface.

Some metals (e.g. copper, iron, manganese, molybdenum, nickel, zinc) are also micronutrients needed by plants and may be accumulated into plant biomass as plants grow. Some plants uptake metals at a much higher rate than other plants (known as hyperaccumulators) and/or have greater tolerances to high metal concentrations. Ideally, if used for removal of a metal from stormwater or soils, a plant would have a high uptake rate and a tolerance to high metal concentrations within the plant material. Sun and Davis (2007) ranked the general tendency of metals to accumulate in plants, with zinc having the highest tendency, as:

$$\text{Zinc} > \text{Copper} > \text{Lead} > \text{Cadmium} \tag{4.2}$$

Overall, metals are typically retained by soils within the top 30–50 cm, with some suggesting that at least 40 cm of unsaturated soil be present between the bottom of an infiltration practice and a groundwater source (Dierkes & Geiger,

1999). Metals have been detected in groundwater under infiltration practices, but the concentrations have been typically below water quality standards and, thus, have not been deemed to be a threat. As mentioned previously, however, depending on conditions within the water and soil (e.g. low dissolved oxygen), the previously retained metals may become mobile at some time in the future. For example, increased metal concentrations have been detected in the groundwater beneath infiltration practices when the groundwater is acidic (Pitt *et al.*, 1996). Also, O'Connor *et al.* (2012) showed that maintenance activities that disturbed sediment of a constructed wetland caused a temporary (<9 months) increase in the mobile fraction of some metals and a corresponding increase in metal concentrations in the tissues of macroinvertebrates.

According to Pitt *et al.* (1996), nickel and zinc would have high groundwater contamination potential in infiltration/injection systems and chromium and lead would have moderate potential. They assert that if sedimentation pre-treatment were used, all metals would probably have low groundwater contamination potential.

4.2.3.3 Suspended solids

The primary removal mechanisms of suspended solids are physical filtration and sedimentation. Infiltration systems provide filtration of runoff, but the percent removal of solids depends on, among other variables, the particle size distribution of the suspended solids and the size of the pore opening between soil particles. Davis (2007) found that only 43 and 47% of the suspended solids were removed (on average) from two bioretention cells monitored for 12 storm events. In a field study, Hunt *et al.* (2006) observed a rain garden acting as a source of suspended solids, which may have been due to the uncertainties involved in monitoring research or erosion of soil surfaces within the practice. Suspended solids alone, however, present very little concern for soil or groundwater contamination.

4.2.3.4 Organic compounds

Organic compounds can be naturally occurring (e.g. animal waste, vegetation, soil organisms) or anthropogenic in origin (e.g. petroleum hydrocarbons, automobile tire particles). Removal of organic compounds may occur through volatilization, sorption, and degradation (Pitt *et al.*, 1999). All organics are listed as having low or moderate contamination potential for subsurface infiltration with sedimentation as a pre-treatment mechanism. With respect to potential groundwater contamination, it appears that most hydrocarbons are trapped in the first few centimeters of soil in infiltration basins (Dierkes & Geiger, 1999). The hydrocarbons may then be utilized by microorganisms in the soil. This has led some to conclude that hydrocarbons pose little risk to groundwater contamination. LeFevre *et al.* (2012a, 2012b) found that bioretention with microbial activity can be suitable for hydrocarbon removal and that vegetation can increase removal rates and stimulate biodegradation.

Pitt *et al.* (1996) reported that 1,3-dicholobenzene, pyrene, and fluoranthene may have high groundwater contamination potential in subsurface infiltration/injection systems without pre-treatment but would probably have lower contamination potential in surface infiltration systems. A series of other organics (benzo (a) anthracene, bis (2-ethylhexyl) phthalate, butyl benzyl phthalate, pentachlorophenol, phenanthrene) are listed as having moderate groundwater contamination potential for subsurface injection when no pre-treatment system is used. Anthracene, fluorine, and naphthalene are listed as having low potential for groundwater contamination for subsurface injection with minimal pre-treatment.

In studies that have investigated the potential contamination of groundwater from infiltration of stormwater it was not uncommon for groundwater to contain organic compounds, presumably from infiltrated stormwater (Ku & Simmons, 1986; Pitt *et al.*, 1999; Plaza *et al.*, 2007; Wilde,, 1994). In some cases, pollutant concentrations exceeded drinking water standards. Thus, considering the use of an infiltration practice should not occur without also considering the potential for groundwater contamination.

4.2.3.5 Pathogens

For residential and light commercial developments, pathogens (i.e. bacteria and viruses) in stormwater are a primary pollutant of concern. They may be present in high concentrations and not retained well in the soil (Pitt, 1999). Bacteria may be removed by straining at the soil surface and sorption to solid particles. Not all pathogens are removed, however, as evidenced by the fact that the highest bacteria and virus concentrations in groundwater are found to occur when the water table is near the land surface (Clark *et al.*, 2006; Pitt *et al.*, 1999).

Documented virus contamination of groundwater due to infiltration practices has occurred at sites on Long Island, New York, USA where the water table was less than 11 m below the infiltration practice. If removed from the water by soil, the ability of bacteria to survive is a function of factors such as temperature, pH, presence of metals, etc. Bacteria survive longer in acidic soils and in soils with large amounts of organic matter. Bacteria survival may typically be between two and three months but survival for up to five years has been documented (Pitt *et al.*, 1999).

Groundwater contamination potential by bacteria and viruses depends on the soil chemical properties, adsorption capability, the ability of the soil to physically strain the pathogens, and pathogen survival characteristics. Bacteria and viruses can move through soil media and may be transported to aquifers by infiltrating stormwater. The transport distance of bacteria seems to be partially a function of bacteria density and water velocity through the soil (Camesano & Logan, 1998; Unice & Logan, 2000). Pitt *et al.* (1996) state that enteroviruses have high groundwater contamination potential for all surface and subsurface infiltration/injection systems and that many other pathogens have high groundwater contamination potential for subsurface infiltration/injection systems.

4.2.3.6 Chloride

Because chloride is soluble, easily transported in surface and sub-surface flow, non-filterable, and does not readily adsorb to solids, it has a high potential for groundwater contamination (Pitt *et al.*, 2002). Rather than being reduced, chloride concentrations typically increase as water moves through soil due to leaching of salts into the water (Pitt *et al.*, 1999).

Novotny *et al.* (2008) found that in the Twin Cities Metropolitan Area of Minnesota, USA, chloride from winter road salt application was washed to nearby lakes where it significantly increased chloride concentrations at the bottom of the lakes. Due to the increased density of the water with a high chloride concentration, fingers of saltwater moved further into the sediment through advection and dispersion. High chloride concentrations can cause the release of metals that are fixed to soil particles. Once released, the metals are free to diffuse upwards towards the interface between the sediment and lake water or to move with the saltwater farther into the sediment below the interface.

4.2.4 Summary of groundwater contamination due to stormwater infiltration

Although every infiltration practice must be considered on an individual basis with pollutant concentrations, runoff volumes, soil properties, and other factors taken into account, research indicates that the soil has the ability to capture many pollutants. Evidence suggests that hydrocarbons will be degraded by bacteria surrounding plant roots and metals will accumulate on the media, which will eventually need to be treated or excavated. In most cases, due to the large capacity for metals that most soils have, such action may not be needed for decades or more. Pathogens are also typically filtered by soil media, but this may not be true for viruses. Nitrate, although not retained by soil, is typically at concentrations in stormwater that raise little overall concern for nitrate contamination. Chloride is not retained in significant amounts by soil and is a long-term pollutant of concern for areas that use road salt during the winter to remove snow and ice from road surfaces. The potential for groundwater contamination is usually greater for sub-surface infiltration practices (e.g. infiltration chambers, infiltration trenches, etc.) as compared to surface infiltration practices (bioinfiltration practices, infiltration basins, etc.) due to added treatment mechanisms in the soil surface layer (microbial activity, organic material, etc.)

One must keep in mind that, for most pollutants discussed herein, cases of groundwater contamination due to stormwater infiltration have been documented. If care and consideration are given to the selection and placement of an infiltration practice, groundwater contamination can be avoided. In some cases, modeling of groundwater and pollutant transport may be warranted to verify that the potential risk of groundwater contamination is low.

4.3 LID APPLICATIONS

There are different reasons for the application of LID practices, and the predominant practices are selected due to these reasons. This section discusses the rationale behind LID implementations and the objective of the implementations.

4.3.1 Combined sewer overflows

Combined sewer overflows (CSOs) occur when a community has a sanitary sewer that doubles as a stormwater sewer. Overflows installed in the sewer system bypass a fraction of the combined flow during large rainstorms that would otherwise overload the sewer system and/or the sewage treatment plant. These overflows typically discharge into a river as CSOs. A major concern with CSOs is the human adapted bacteria that they contain, and CSOs are to be avoided when possible. In fact, the US Environmental Protection Agency can fine communities for each CSO event, although generally fines are not levied as long as the community is taking action to implement best practices that will reduce the frequency of CSOs.

From 1960 through 1995, the Cities of Minneapolis and St. Paul, Minnesota installed separated storm sewer systems in order to reduce CSO events. Since 1995, work has been targeted towards disconnecting stormwater drains to sanitary sewer pipes. While CSO events are still possible, in the City of Minneapolis, for example, there were zero overflow events in nine of the 10 years between 2008 and 2017. In 2006 the City of Chicago completed Phase I of the Tunnel and Reservoir Plan (TARP) system that consisted of 179 km of 2.4–10 m diameter tunnels in rock 46–107 m below the ground surface. These tunnels have the capacity to temporarily store 8.7 billion liters of combined sewer discharge and release it to nearby wastewater treatment plants after the storm has passed (Scalise & Fitzpatrick, 2012). Phase II of the TARP project, which is expected to be completed in 2029, consists of three large reservoirs that will increase the capacity of the system by 66 billion liters. These examples in Minnesota and Chicago were expensive, relative to methods that incorporate LID practices to reduce stormwater runoff. In the cases of Minnesota and Chicago, the primary goal was to reduce the quantity of runoff in order to reduce the frequency of CSO events. LID practices can also help achieve these goals. For example, Kansas City, Missouri, Philadelphia, Pennsylvania and New York, New York have recently adopted LID practices as a primary solution to their CSO concerns as opposed to large construction projects like the kind undertaken in Minnesota and Chicago.

4.3.2 Eutrophication in fresh surface water bodies

Eutrophication is a substantial problem in freshwater bodies, primarily lakes, reservoirs and slow moving streams where an excessive amount of plants, algae, and bacteria can reduce water quality, recreational opportunity, and cause the release of toxic substances. Eutrophication is caused by an over-abundance of nutrient supply to water body. Filamentatious algae or invasive plants can choke off a water body, duckweed can result in an unattractive water body, and harmful "algae" blooms of cyanobacteria can develop.

The control of nutrients is most easily achieved through a reduction of phosphorus to the water body, because phosphorus tends to be attached to or integrated into solids that can be settled and filtered. Nitrogen is not easily adsorbed to solids and, therefore, is not easily removed, and the other plant nutrients are typically available in abundance. Phosphorus is the primary limiting nutrient and can be settled, filtered, or adsorbed, and thus most LID applications associated with freshwater bodies will seek to control phosphorus in runoff.

4.3.3 Hypoxia in coastal waters

Phosphorus is not a limiting nutrient in the ocean because the ocean-based environment does not have sufficient iron to sequester phosphate (Blomqvist *et al.*, 2004). However, the nitrogen concentrations in these systems are low, so that nitrogen becomes a limiting nutrient. The discharge by rivers and streams into the ocean environment often brings sufficient nitrates (NO_3^{2-}) to create algal and bacterial blooms that deplete the available dissolved oxygen in portions of the ocean, creating what is known as a "dead zone." The best known dead zone in the United States is caused by the discharge of the Mississippi River into the Gulf of Mexico, but there are smaller dead zones at most locations where rivers that transport urban and agricultural runoff merge with the ocean. In these cases, LID practices are designed to retain phosphorus (for brackish waters in estuaries) and convert nitrate to nitrogen gas (denitrification) through a microbially-mediated reaction. These denitrification reactions require a substantial holding time in an anaerobic environment and a carbon source, adding additional complexity to LID practices. LID practices designed for denitrification have been shown to be successful and are discussed later in this chapter.

4.3.4 Climate change adaptation

The downscaled predictions of future storms in most regions of the country predict that extreme storm events will occur more frequently. To adapt to this, gray infrastructure such as underground storage and larger stormwater sewers are needed. However, LID practices have a role in climate change adaptation because they are often a more cost-effective method of reducing stormwater runoff (Moore *et al.*, 2016). Thus, in response to climate change, the role of LID is again infiltration. In general, more LID practices installed in an urban watershed will lower the cost of increasing gray infrastructure to adapt to climate change.

4.3.5 Selection of an LID practice

Primary objectives of LID include runoff volume reduction and the reduction in nutrient loading, typically phosphorus or nitrogen, to receiving water bodies. Because infiltration reduces runoff volume and any nutrient load corresponding to that volume, it is often the primary mechanism of LID practices. Nutrient reduction can occur by other mechanisms, however, such as settling and filtration for phosphorus and microbial denitrification for nitrogen. Thus, the selection of an LID practice for stormwater treatment must consider the circumstances that has caused the need for stormwater management and the treatment mechanisms of each potential LID practice. Common and emerging LID practices are discussed in more detail in the following section.

4.4 TECHNOLOGICAL ASPECTS OF LOW IMPACT DEVELOPMENT PRACTICES

Since Phase I and Phase II of the 1987 CWA amendments have been enforced, planners, scientists and engineers have been focusing on water quality issues, This has resulted in the development or emergence of new stormwater management options, the realization that some long-standing stormwater management methods developed solely to address peak flow rates do, in fact, provide additional LID benefits, a host of business endeavors to develop and market commercially available stormwater treatment products, and research to develop advanced stormwater treatment technologies. This section discusses LID practices and technologies in the United States that span this entire range from common stormwater practices such as ponds to emerging technologies that have been developed since the 1987 CWA amendments, and finally to potential future technologies such as achieving controlled and deliberate biological nitrogen removal in a stormwater treatment practice.

4.4.1 Common practices

This section discusses common LID practices that are used throughout large portions of the United States. An LID practice, according to the previously discussed US EPA definition, is one that uses or mimics natural processes that result in infiltration, evapotranspiration, or the use of stormwater to protect the aquatic environment. Some common stormwater management practices such as wet ponds, dry ponds, wetlands, and sand filters do not meet this definition and, although they are important tools for stormwater management, are not discussed herein.

4.4.1.1 Infiltration basins, trenches, and chambers

The primary treatment process of many LID practices is infiltration with the goal of reducing runoff volumes. Three such practices are infiltration basins, infiltration trenches, and infiltration chambers. All are similar in that they store runoff with no designed outlet structure, other than an emergency overflow bypass, and rely primarily only on infiltration as their draining mechanism.

Infiltration basins (Figure 4.1) are similar to dry ponds in appearance except they have no low-level outlet structure. They have a media (inherent or imported) with a typical infiltration rate greater than 2 cm/hr and are designed to allow the standing water to infiltrate into the ground. Infiltration trenches (Figure 4.2) are excavated trenches backfilled with gravel, where the void spaces (typically ~40%) within the gravel are used to store runoff. Infiltration trenches appear to be appropriate solutions in dense urban areas where the construction of other LID practices is difficult due to the limited available space and the existence of subsurface utilities. As an example, infiltration trenches with embedded tree pits underneath sidewalks, also known as tree trenches (Figure 4.3(a) and (b)), are popular LID practices in Philadelphia, Pennsylvania and other locations for stormwater runoff volume removal and urban forest expansion (Caplan et al., 2019; Tu & Traver, 2019). There are two inlets at each system: a green inlet, that is the primary inlet for directing stormwater runoff to the tree trench, and a city inlet, that is a regular street inlet connected to the city sewer system. Runoff enters the green inlet, which includes a pre-treatment filter bag and a sump, and is distributed into the

Figure 4.1 An example of an infiltration basin which can fill to the invert of the outlet structure (source: www.rivanna-stormwater).

Figure 4.2 An example of an infiltration trench that receives water from roof runoff. Infiltration trenches are most often placed under roadways (source: chesapeakestormwater.net).

trench through a perforated pipe. Runoff is designed to be stored in the stone trench, provide water for the trees, and exfiltrate into the surrounding in-situ soil through the bottom and the unlined side walls of the trench. If the system exceeds its capacity, runoff will bypass the green inlet and flow into the city inlet. Depending on the infiltration condition of the surrounding in-situ soil, an underdrain may be installed to collect the infiltrated water and discharge it to the city sewer system. The underdrain may be capped while the system's capacity permits and when needed, a hole in the cap can be drilled to allow a slow release to the city sewer system. Infiltration chambers (Figure 4.4) are underground chambers or vaults, many produced commercially for this purpose, with a large void fraction (typically 80–100%), and a means to infiltrate water, such as an open bottom or perforated walls. Infiltration basins, trenches and chambers all typically have emergency bypass outlets to convey flow in case storm discharges exceed the design discharge, but none have standard outlet structures to pass runoff volume at or below design levels. Rather, all rely primarily on infiltration to pass the runoff from the design storm or smaller events. As with any LID practice, infiltration practices require regular maintenance. Due to the susceptibility of infiltration trenches to clogging and the difficulty accessing clogged portions of soil that are covered with gravel, their maintenance costs are typically higher than other LID practices.

4.4.1.2 Permeable pavements

Permeable pavements (Figure 4.5) are systems that allow water to infiltrate through a pavement layer as part of a stormwater management strategy. There are two

Figure 4.3 (a) Schematic configuration of an infiltration trench with tree pits under sidewalks. Image credit: Philadelphia Water Department (phila.gov/water). (b) An infiltration trench with tree pits in Philadelphia, Pennsylvania. Image credit: Google Maps (google.com/maps).

Figure 4.4 Underground infiltration chambers typically placed under a parking lot (www.estormwater.com).

categories of permeable pavements; full-depth and open graded (or permeable) friction course (OGFC). Full-depth permeable pavements allow runoff to infiltrate vertically through the entire pavement system and into the existing subbase below. OGFC pavements consist of a porous asphalt layer approximately 2.5–5 cm thick placed on top of a nonpermeable, conventional pavement. In the OGFC system, runoff infiltrates vertically into the OGFC layer and horizontally over the conventional pavement layer until it exits the pavement at the edge. The infiltration provided by full-depth permeable pavements reduces runoff volumes and both full-depth and OGFC permeable pavements can improve water quality. Permeable pavements have additional benefits beyond volume reduction and

Figure 4.5 Permeable pavement (in the distance) is frequently placed on residential streets and parking lots (source: www.vaasphalt.org).

pollutant retention. For example, studies have shown that full-depth porous asphalt can receive less winter salt application than conventional asphalt pavement while achieving the same level of driving safety (Roseen *et al.*, 2014). Salting is not recommended for pervious concrete pavements, however, because it can cause damage the pavement. For full-depth porous asphalt pavements receiving less salt during the winter, the frequency of salt applications may need to increase because as water infiltrates vertically down through the pavement system, it takes a portion of the applied salt load with it. The fact that water infiltrates and does not remain on the pavement surface, however, can also result in bare pavement, which is in contrast to conventional asphalt where water may remain on the pavement surface and refreeze.

Permeable pavements also offer a quieter ride than conventional pavements because sound waves penetrate the pavement rather than being deflected back up towards the vehicle. Permeable pavements are safer because less water on the road results in a lower chance of hydroplaning and better visibility for the driver. Preliminary results from a study in Tennessee in the United States have shown a 32% drop in wet weather-related accidents on an OGFC pavement compared to a conventional asphalt control section (TN DOT, 2019).

Permeable pavements have also been found to be viable options for cold weather climates with harsh winters and heavy snowfalls. If the pavement is constructed properly, water will pass through the pavement layer and into the reservoir system underneath. In this case, freezing will not damage the permeable pavement layer because no water remains in it. Even if some water did remain, it would have more than enough room to expand when frozen due to the large void ratio of the pavement. Research suggests that the air in the pavement layer and porous stone layers of the reservoir system can insulate the existing subbase, which keeps it warm and allows for infiltration to occur throughout the winter.

4.4.1.3 Bioretention

Any stormwater management practice designed to store runoff that also incorporates plant life (i.e. vegetation) in a growing medium within the practice can be classified as a bioretention practice (Figure 4.6). Bioretention practices can be further classified into bioinfiltration practices and biofiltration practices. Bioinfiltration practices store runoff above ground and allow it to be removed from the stormwater conveyance system by infiltration into the soil and/or evapotranspiration. Of all the bioretention systems, these typically infiltrate the largest volume of runoff. Biofiltration practices temporarily store runoff above ground but have an underdrain collection system, typically of perforated plastic pipe, below the ground surface that collects runoff that has been filtered by the soil media above the underdrain. After collection, the runoff is conveyed downstream in the storm sewer system. Thus, the difference between the two

Figure 4.6 Bioretention facility draining parking lot (photo by authors).

practices is that bioinfiltration systems are designed to infiltrate runoff into the existing soil whereas biofiltration systems essentially act as a filter with a drain tile collection system that collects runoff and conveys it downstream. In the latter case, a fraction of the water may infiltrate into the existing soil but the primary mechanism is filtration.

Some biofiltration systems have underdrains buried under engineered media and just above the existing subgrade, while others have underdrains elevated above the bottom of the engineered media or with an upturned elbow at the downstream end. The latter two systems create an underground storage space for runoff because it will fill the voids in the engineered media before being conveyed by the drain tile system. Biofiltration systems with underdrains just above the existing soil will filter most or all of the runoff entering the practice with a small portion possibly infiltrating into the existing soil if there is no impermeable barrier installed. Systems with elevated underdrains or underdrains with upturned elbows will filter the runoff but a portion will be stored in the media below the outlet. This portion will eventually infiltrate into the existing soil and/or be displaced by the next runoff event. Thus, if no impermeable barrier is used, these systems typically infiltrate more runoff volume than biofiltration systems with lower underdrains but not as much as bioinfiltration systems. Some biofiltration systems are lined with an impermeable liner below the collection system, which eliminates infiltration and all runoff is either filtered or lost through evapotranspiration.

In some instances, bioretention systems are also classified according to their size or maximum designed water depth. Practices with a maximum design depth of no deeper than 40–60 cm are typically called rain gardens and much deeper (and usually larger) practices are called bioretention (i.e. bioinfiltration or biofiltration) systems or basins. Rain gardens (Figure 4.7) are essentially small depressions

Figure 4.7 Rain garden with curb cut on a residential street (photo by authors).

planted with select vegetation that can withstand frequent periods of inundation, have deep root systems to keep porosity of the soil high, and for aesthetic reasons. Like all bioretention practices, rain gardens may also have an underdrain system and/or an impermeable liner that prevents infiltration.

Bioretention facilities treat stormwater runoff via the processes of solid settling and infiltration and can achieve retention rates of up to 80% for suspended solids, 44% for total phosphorus, and 35% for total metals (MPCA, 2019). Pollutants corresponding to the infiltrated runoff volume, however, may sometimes be assigned full removal because it is assumed that they are captured by the soil. A large part of the metal removal often corresponds to the particulate fraction of metals that are retained with the settling of solids. In bioretention practices, due to the breakdown of dead and decaying organic matter (i.e. vegetation), long-term retention of dissolved phosphorus and nitrogen is not likely. In fact, bioretention facilities can often be a source of these pollutants depending on the time of year (i.e. season), water temperature, and dissolved oxygen concentration in the water, etc. (Dietz & Clausen, 2005; Hatt *et al.*, 2008; Li & Davis, 2009; Paus *et al.*, 2014).

Bioretention facilities and, in particular, rain gardens, have become common in the United States, likely due to their relatively basic maintenance requirements (e.g. weeding and removing trash), their ability to reduce runoff volume and improve water quality and, perhaps most important, the fact that many people consider them to be aesthetically pleasing. Other advantages of rain gardens are that they can be placed anywhere in a watershed, including scattered throughout the upstream reaches, and they can be placed on just about any property including residential and business. Infiltrating stormwater runoff in the upstream reaches of a watershed is essential when trying to have a developed watershed mimic an undeveloped watershed. In fact, some communities have encouraged homeowners to plant rain gardens by providing the vegetation and expertise needed to install the rain garden if the homeowner agrees to maintain the practice. For all the above reasons, rain gardens have become a key feature in stormwater management in the United States.

4.4.1.4 Swales and roadside ditches

A swale is a channel or ditch designed to carry stormwater runoff and the terms are often used interchangeably. They often run adjacent and parallel to roads and accept runoff from the road surface, conveying it downstream in the stormwater conveyance system. In some instances, vegetation is planted or allowed to grow without being cut within the swale. In these situations, the swale is called a vegetative swale. The vegetation is beneficial because it can filter solids, slows the water down, and keeps the soil pores from clogging, which promotes additional settling and infiltration (Barrett *et al.*, 1998b).

Grassed swales (Figure 4.8) also have the capability to reduce runoff volume and improve water quality. Volume reduction occurs primarily through infiltration into the soil, most of which occurs as the water flows over the slide slope perpendicular to the roadway and into the swale (Ahmed *et al.*, 2015; Garcia-Serrana *et al.*, 2017). Pollutant removal can occur by sedimentation of solid particles onto the soil surface, filtration of solid particles by vegetation, or infiltration of dissolved pollutants (with stormwater) into the soil (Abida & Sabourin, 2006). When solid particles settle to the soil surface or are captured by filtration on vegetation, the runoff solids concentration is reduced, and overall water quality is improved as long as solids do not become resuspended. Such resuspension has been determined to be

Figure 4.8 Grassed swale at the edge of a residential street without curb and gutter (source: bluegreenbldg.org).

negligible (Barrett *et al.*, 1998a, 1998b) due to such factors as the vegetation constraining detachment of settled solids (Cutierrez & Hernandez, 1996) and plant roots stabilizing soils (Mazer *et al.*, 2001). If a swale is not maintained and erosion develops, however, resuspension of solids and erosion of new bank material will occur.

Swales can also remove a significant fraction of metals because a portion of the metal load is adsorbed to the solids that are removed through filtration and sedimentation. Also, any infiltration that occurs takes with it dissolved metals (Barrett *et al.*, 1998b). Nutrient (i.e. phosphorus and nitrogen) removal, however, is limited to the water infiltrated and sometimes is negative due to decaying plant matter, which can release nutrients that were previously bound in plant matter. Due the fact that a fraction of the phosphorus load will adsorb to solids, but nitrogen does not, total phosphorus removal in swales is typically greater than nitrogen removal.

The performance of a swale depends on many factors, including the health and abundance of the vegetation within the swale. Mazer *et al.* (2001) reported several causes of poor vegetative cover in swales including standing water in the swale for prolonged periods of time, high flow velocities, large fluctuations in surface water depth and soil moisture, excessive shade, and improper installation. Improper installation could be the result of poor design or poor construction practices. For the eight bioswales in western Washington State, USA, investigated by Mazer *et al.* (2001), heavy shade was more important than other environmental factors that limited vegetation. The second most important factor was inundation with water. If water was present for more than 35% of the summer the bioswale had significantly less vegetation.

In summary, if healthy and dense vegetation is maintained in a swale, the swale can reduce runoff volumes and improve the water quality of the runoff. Water quality is improved through filtration and settling of solids (which includes adsorbed contaminants such as metals and phosphorus) and infiltration of polluted water that is removed from the conveyance system.

4.4.1.5 Green roofs

Green roofs are vegetative coverings placed on top of buildings (Figure 4.9). The coverings can be as thin as 2–5 cm or as thick as 15–30 cm. Extensive green roof systems are thin, simple, require little maintenance, and usually support vegetation that survives well in extreme climates. Thicker systems are called intensive green roofs and can serve as gardens or parks and can contain a variety of vegetation including shrubs and trees. Intensive systems require more maintenance, are more expensive to install, and usually require additional structural capacity of the building due to the heavy loads of soil and vegetation. Green roofs can reduce stormwater runoff volumes by storing runoff in soil where it can be transferred back to the atmosphere through evapotranspiration.

Figure 4.9 Example of a green roof (source: greenroofs.org).

Green roofs have benefits in addition to their effect on stormwater. They help insulate buildings in the winter and, through shading and evaporative cooling, cool buildings in the summer, thus saving money. They can provide space for gardening in urban settings, habitat for insects and birds, or a park-like space for users of the building. They provide insulation from sound and can make buildings that are subject to noise, such as those near airports, quieter for its occupants. Green roofs can improve air quality by removing carbon monoxide, carbon dioxide, and compounds that form smog. Finally, because green roofs protect the roof structure from harsh weather conditions, green roofs can extend the life of the roof by a factor of two or more (Green Roofs for Healthy Cities, 2020).

One of the largest green roof systems in the world was installed at the Ford Motor Company's River Rouge Truck Plant in Dearborn, Michigan, in 2000 (Aquilina, 2019). Over 4.2 hectares of roof were covered with sedum plants growing in a four-layer vegetative mat. The mat system was selected because, with only 3.2 cm of growing medium, it was thinner and lighter than more conventional systems, it requires minimal maintenance, and the plants were grown in the mats prior to installation so the system was installed with full mature plants. Designed for the 10-year, 24-hour rainfall event (~9 cm), the system treats over 16,000 m^3 of rainwater on an annual basis and reduces TSS concentrations by 85%. It takes 48 hours for rain to pass through the roof system, which reduces peak runoff rates from the site.

The roof serves as habitat for birds, butterflies, bees, and other insects and keeps the building 5°C warmer in the winter and 5°C cooler in the summer. The insulation effect has reduced the buildings energy consumption by 5%. The system saves the Ford Motor Company money associated with stormwater treatment and it is expected to at least double the life of the roof. This project has shown that green roofs can be a cost-effective and environmentally friendly option and has helped promote the use of green roofs throughout the United States.

4.4.1.6 Rainwater harvesting

Rainwater harvesting involves the capture of rainwater from impervious surfaces such as roofs, roads, sidewalks, commercial areas, etc., and the temporary storage of the rainwater in cisterns or other containers for later use. Potential uses include irrigation, toilet flushing, fire suppression, cooling, or the washing of buildings, vehicles, or streets, among other uses. Rainwater harvesting reduces stormwater runoff volumes and the demand for typical well or municipal water supply. Storage containers can be as small as a 190 L barrel used at a residential property or they can be large scale systems that capture and store thousands of liters such as those implemented at commercial and industrial sites.

Steffen *et al.* (2013) found that, depending on the region's climate, a single 190 L barrel installed at a residence can supply approximately 50% of a family's indoor non-potable water use needs. In arid regions with limited rainfall, however, less than 30% of indoor non-potable water needs could be supplied with rainwater harvesting. In the areas with limited rainfall, such as the southwest United States and California, rainwater harvesting can reduce stormwater runoff by up to 20% but the reduction was less in other climates.

The performance of a rainwater harvesting system can be improved by a Continuous Monitoring Adaptive Control (CMAC) approach. CMAC uses sensors to collect information, such as water level in rainwater cisterns and local weather forecasts, to automatically control equipment (such as valves to drain cisterns) based on previously programmed user criteria, for example, when full, rainwater cisterns must bypass additional rainwater. Rather than having a cistern remain full throughout a future rain event and thereby bypass the entire event, a CMAC system can be used to slowly release stored rainwater in advance of the next rainfall event so that runoff from the upcoming event can be stored. This process can help optimize both the use of stored rainwater (by allowing for its use only when needed or when additional runoff is expected) and the collection of new rainwater (by minimizing or eliminating bypass events) in a harvesting system.

Roman *et al.* (2017) showed, through modeling, that a CMAC system integrated with a rainwater harvesting system in a highly urban area in New York City, New York could capture 76.6% of the annual roof runoff compared to a 14.8% capture rate for a conventional moisture-based system and 41.3% for a conventional timer-based system. A moisture-based system releases stored rainwater when soil moisture drops below a predetermined level and a timer-based system releases a set amount of rainwater at a predetermined interval (e.g. daily). Because of optimized irrigation schedules, the CMAC system would also use 81.4% less harvested rainwater for irrigation than a conventional moisture system and 18.0% less than a timer based system.

4.4.1.7 Maintenance and pre-treatment

Maintenance is any action or activity taken on a stormwater treatment practice to preserve or restore its function, its efficiency, or to extend its useable life. All stormwater treatment practices need maintenance if they are to function as intended over their design life. For example, over time, sedimentation practices fill with sediment and the remaining sediment storage capacity of the practice is reduced. At some point, sediment will need to be removed from the practice in order to restore capacity. As another example, infiltration and filtration practices can clog with particles over time and the corresponding hydraulic capacity is reduced. At some point, the hydraulic capacity will need to be restored. Bioretention practices also require maintenance to ensure the health and abundance of the desired vegetation and the elimination of weeds and invasive plant species. Additionally, almost all stormwater treatment practices may collect trash and debris or periodically have other issues that require attention such as soil erosion, slope stability, or the integrity of pipes or other structural components. Thus, the construction of any stormwater treatment practice should be accompanied by a continual maintenance plan and corresponding budget to accomplish the required maintenance (Erickson *et al.*, 2013).

In addition to maintenance, every structural stormwater treatment practice should have a pre-treatment system. Pre-treatment, as the name implies, is the treatment of stormwater runoff prior to the runoff entering the main treatment practice. For example, an infiltration basin may have a sedimentation basin or chamber just upstream of its inlet. The basin or chamber, as with any pre-treatment device, is designed to remove the large solids, trash, and debris that would otherwise clog the main practice and/or lead to its early disfunction. Pre-treatment units remove the large, easy to remove solids and preserve the function and capacity of the main practice for the removal of the smaller solids. This prolongs the life of the main practice and reduces its maintenance needs and/or the corresponding frequency of those needs. Although pre-treatment devices also require maintenance, their corresponding maintenance activities are usually less complicated, less time consuming, and less expensive. Thus, pre-treatment can provide significant cost-savings over the life of a stormwater treatment practice. In a study that demonstrated the importance of pre-treatment, Galli (1992) investigated 12 infiltration basins, all of which had no pre-treatment systems. All 12 had failed within their first two years of service.

4.4.1.7.1 Maintenance

Kang *et al.* (2008) breaks maintenance activities into three categories: routine, non-routine, and major as shown in Figure 4.10. Routine maintenance occurs frequently and, on an event basis, is typically less time and/or labor intensive than the other categories. Examples of routine maintenance are mowing grass, raking the surface of a sand filter, litter and trash removal, and weeding. Non-routine maintenance is less frequent and is more time and labor intensive, per event, than routine maintenance. Examples of non-routine maintenance include structural repairs, sediment removal, and partial rehabilitation of the practice. Major maintenance actions are rare and very time and labor intensive

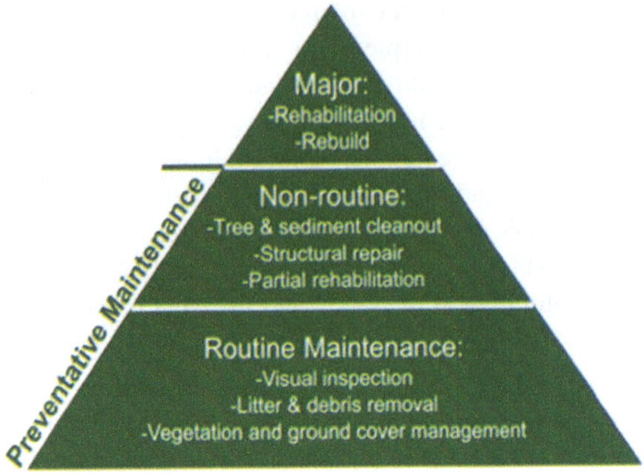

Figure 4.10 Stormwater treatment practice maintenance pyramid (source: Erickson *et al.*, 2013).

and include such actions as replacing an entire media bed in a filtration practice, full rehabilitation of the practice, or total reconstruction.

The frequency of any maintenance activity, regardless of its level on the maintenance pyramid, depends on a vast array of variables including type and characteristics of the particular practice, watershed size, land use and soil characteristics in the watershed, type and number of trees nearby, amount of construction in the watershed, rainfall patterns and amounts, and a host of other variables. Thus, the required frequency of maintenance activities should be determined for each individual practice but, typically, routine maintenance occurs at least once per year to once per season (i.e. 3–4 times per year).

In order to determine if a stormwater treatment practice needs maintenance it must be assessed, which is the evaluation of the practice to determine if it is functioning as desired. Thus, all maintenance plans and budgets should include the regular assessment and corresponding resources for the assessment of stormwater treatment practices. Like maintenance, there are various levels of assessment that span a wide range of difficulty and cost. Erickson *et al.* (2013) list four levels of assessment, in increasing level of time commitment and cost, as 1) visual inspection, 2) capacity testing, 3) synthetic runoff testing, and 4) monitoring.

Visual inspection is simply a visual evaluation of a stormwater treatment practice. It typically should include photographs and/or video of the practice and detailed field notes related to site conditions at the time of inspection and weather conditions at the same time and on the previous days. Visual inspections can usually be completed in less than half an hour and can sometimes be used to identify obvious problems with a practice. Visual inspection, however, is limited in the information that can be provided. For example, if it has not rained in the previous 2 days, visual inspection of a rain garden may reveal that it contains standing water and dead or dying vegetation. This would lead one to conclude that the rain garden is not infiltrating water as designed and needs maintenance. If upon visual inspection a rain garden does not have standing water and the vegetation looks healthy and normal with no invasive species, it does not guarantee that the rain garden is functioning properly. Stormwater entering the rain garden could be short-circuiting through the rain garden or its media or, if it has been several days since it rained, it may not be infiltrating water quickly enough. In the latter case, additional assessment at a higher level would be required to adequately assess the rain garden.

The second level of assessment, capacity testing, is performed by making a series of point measurements throughout the treatment practice in order to gain information related to the issue being assessed. For example, in order to assess the overall infiltration capacity of a treatment practice, an infiltrometer can be used to measure the infiltration capacity at a series of spatially distributed points throughout the practice. Capacity testing can be used to determine if maintenance is required but, like visual inspection, this level of assessment has limitations. For example, when determining the infiltration capacity at a series of locations in a treatment practice, areas of low capacity may be missed if those areas are not specifically tested. It may also miss areas where infiltration occurs too quickly through cracks or fissures in the media. Finally, although capacity testing can be used to estimate the overall effectiveness of a practice, there are associated errors and uncertainties in those estimates because a series of measurements over a fraction of the practice are used to represent the entire practice. Furthermore, capacity testing cannot be used to measure the pollutant removal effectiveness of a practice because flows and pollutant concentrations into and out of the practice are not measured.

Advantages of capacity testing are its lower time requirement, cost, and difficulty compared to higher levels of assessment. It can also identify specific areas within a practice that need maintenance and areas that do not. Thus, maintenance activities can be focused only on those areas needing attention, thereby saving time and resources.

The next level of assessment, synthetic runoff testing, is accomplished by adding synthetic stormwater to a treatment practice to evaluate the hydraulic characteristics of the practice and/or the pollutant removal effectiveness. Typically, the water source is either a fire hydrant or a water truck, which requires prior approval from the local fire department and/or municipality. The size of the practice and corresponding required volume of water must also not exceed the availability of water. When performing synthetic runoff testing, the water may be unaltered (e.g. straight from a hydrant) if only assessing the hydraulic characteristics of the practice or, if pollutant removal effectiveness is to be assessed, pollutants must be added to the water in an amount or dosage rate that results in the desired influent pollutant concentration. Unaltered water is added to a bioretention facility from a fire hydrant in Figure 4.11. The topography and soil moisture content has been

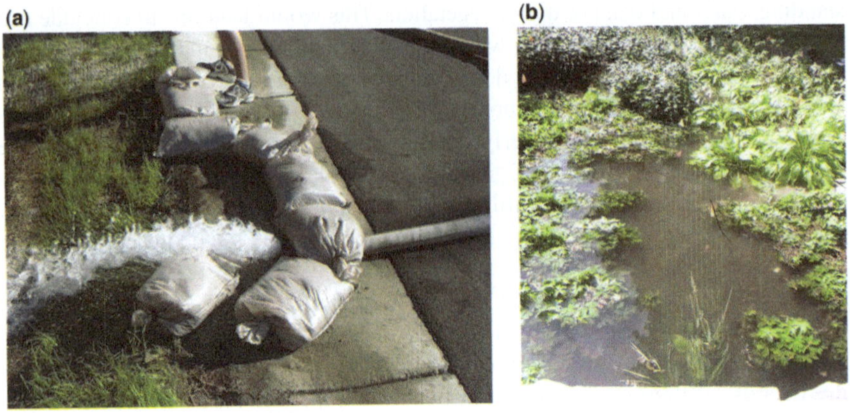

Figure 4.11 Synthetic runoff test using a fire hydrant at a rain garden. After water fills the rain garden (left), the recession of the water surface is measured over time (source: Erickson *et al.*, 2013).

measured prior to adding the water, and the speed of infiltration is recorded, resulting in the information required to compute the resultant saturated hydraulic conductivity for the bioretention facility. A predetermined mass of pollutant is typically added to a water truck (and mixed) to give the desired pollutant concentration in the truck or, if using a fire hydrant, pollutants are fed to the influent stream upstream of the practice at a predetermined pollutant mass flow rate to give the desired pollutant concentration in the water entering the practice. Figure 4.12 shows a synthetic runoff test of an underground pre-treatment practice that is using a fire hydrant as the water source. A sediment feeder, located downstream of the fire hydrant and before the treatment practice, is used

Figure 4.12 Synthetic runoff testing of sediment capture by an underground pre-treatment chamber (source: Erickson *et al.*, 2013).

to dose the influent stream with well characterized sediment. The mass discharge rate of the sediment feeder has been set to correspond with the volumetric flow rate from the fire hydrant to give the desired sediment concentration in the synthetic stormwater entering the practice.

Synthetic runoff testing can be used to assess the overall performance of a practice. Because it does not rely on point measurements to represent the entire practice, it is more accurate in this regard than capacity testing. Unlike capacity testing, however, synthetic runoff testing cannot identify specific areas within a practice that need maintenance. It can only assess how the overall practice is performing. Thus, it is possible that some areas of a practice that are underperforming will go undetected.

The fourth and last level of assessment, monitoring, is the most time consuming, expensive, and involves a large degree of uncertainty compared to synthetic runoff testing. Monitoring can be used to assess runoff volume reduction, peak flow reduction, and pollutant removal efficiency. It can also be used to assess the performance of the LID practice in the watershed. Monitoring is performed by measuring all influent and effluent volumetric flow rates entering or exiting a practice during natural rainfall or snowmelt events. If pollutant removal efficiency is to be assessed, pollutant concentrations of all volumetric flow rates must also be sampled and determined throughout the runoff period. Unlike synthetic runoff testing that can be controlled to maintain constant volumetric flow rates and constant pollutant concentrations for relatively short testing periods, natural runoff events have large variability in flow rate and pollutant concentration within events and separate events can have vastly different durations. As a result, flow and pollutant concentration must be measured throughout a runoff event. Also, prior to an event, the duration and volume of the runoff is unknown, yet automatic samplers are typically pre-programmed to collect samples at a user specified time or flow volume increment. This can lead to automatic samplers becoming full prior to the end of a runoff event, which causes some runoff to not be sampled, or it can lead to unused sample capacity, which results in a less accurate representation of pollutant concentrations and loads. The result of the variability is a larger amount of uncertainty in the results of a monitoring project compared to those obtained by synthetic runoff testing. Monitoring also often involves a larger time commitment due to unpredictable weather conditions which can result in long periods between rain events where the monitoring equipment must be maintained and ready to implement. Monitoring, however, may be more representative than synthetic runoff testing because it relies on natural runoff events with the full range of pollutants that occur in the watershed. Synthetic runoff testing is typically dosed with only the pollutant(s) of interest and this can impact pollutant removal effectiveness when processes such as adsorption and ion exchange are involved.

In summary, all stormwater treatment practices require maintenance to maintain effectiveness. Too much maintenance, however, is a waste of time, money, and other resources but too little maintenance will lead to the degradation and failure of the practice. Assessment can be used to determine when maintenance is necessary and can help to optimize the use of limited resources and budgets. Assessment should occur, at a minimum, at least once a year but many more occurrences may be warranted depending on watershed characteristics, the kind of treatment practice, and other variables. Different levels of assessment that vary in cost and time commitment are available. They range from simple visual inspection and documentation that may take 30 minutes to monitoring of natural runoff events that may last for months or years. Thus, the construction of any stormwater treatment practice must be accompanied by an assessment and maintenance plan with a corresponding budget to support the plan for as long as the practice is to remain in service. For more details about conducting any of the four levels of assessment and recommended maintenance activities for specific stormwater treatment practices, see Erickson *et al.* (2013).

4.4.1.7.2 Why pre-treatment

The purpose of pre-treatment is to remove larger sized solids (typically 80 μm or more in diameter including leaves, sticks, and other large objects) and store them in a location that can be more easily maintained than the primary treatment practice. This reduces the maintenance frequency required for the main treatment practice and associated costs because it receives less solid load. It also extends the life of the main treatment practice. Pre-treatment practices generally have smaller hydraulic residence times than primary treatment practices and, therefore, remove only large particles. Thus, in addition to its own design variables, the fraction of solids removed by a pre-treatment practice is dependent on the particle size distribution of solids in runoff. The higher fraction of larger particles contained in runoff, the higher the fraction that will be removed by the pre-treatment practice. Although the particle size distribution of solids in runoff depends on many variables including rainfall intensity, soil characteristics, percent impervious, land use, and others, pre-treatment practices are typically designed to remove approximately 50% of solids. They should also be designed to prevent resuspension of settled solids, which can be accomplished by providing enough depth so that incoming flows do not dislodge settled solids and/or by including energy dissipaters to calm influent (Howard *et al.*, 2011). Some large particles, such as sticks, plastic cups or bags, metal cans, or organics such as leaves, etc., will float. These should also be targeted for removal in pre-treatment practices and can usually be achieved through a submerged outlet. The submerge outlet causes water leaving the practice to come from below the surface, which causes the floatables to remain within. Of course, pre-treatment practices require their own maintenance but, if designed and constructed well, maintenance for pre-treatment practices is much easier and less expensive than on larger, primary practices.

4.4.1.7.3 Commercial products

The need for stormwater treatment has spurred the growth of an industry in the United States that provides commercial products designed to treat stormwater runoff. The products are available for purchase with the sizing/design usually performed by the manufacturer rather than the user of the product. These products achieve treatment through settling, filtration, infiltration, or vegetative processes, along with typically providing a mechanism for removal of floatables. Most are underground units that require little to no additional land space on the site. For this reason they are often used in urban and suburban areas where little land space is available or land costs preclude the purchase of land solely for stormwater treatment.

Many commercial units that use settling are classified as hydrodynamic separators, which are self-contained units characterized by having a design or an internal structure that induces a swirling pattern of the water within a chamber inside the unit. These devices are typically flow through devices with low hydraulic residence times (on the order of minutes) that also have a mechanism to separate floatables including oils and greases. Due to their low hydraulic residence time they are typically not effective in removing small particles (i.e. $<$80 μm) and, therefore, are most suited to serve as pre-treatment units for other stormwater treatment practices.

Commercial products available for stormwater treatment that improve water quality through filtration are also self-contained units. In this case, the units have an internal filtration mechanism that can work via physical straining or, depending on the filter media used, may also include adsorption and/or ion exchange. The filter media used is typically selected to target a particle size and/or specific pollutants for removal. Filtration units may also incorporate a settling mechanism and a means to remove floatables, including oils and greases.

Commercially available stormwater treatment products that promote infiltration are typically underground chambers with high void ratios (90–100%) that temporarily store runoff under the ground surface and allow it to infiltrate into the existing soil. Settling of solids may also occur in the chambers so, as with all commercially available stormwater treatment products, the chambers must be accessible for maintenance actions, such as sediment removal.

Systems that contain vegetation include tree box units, which typically receive runoff from parking lots, sidewalks, and other highly impervious surfaces. The units are typically designed to receive the first flush and treat it before it is conveyed to the storm sewer system or existing soil. The units consist of a prefabricated concrete box that is installed in the ground, filled with soil, and planted with trees and/or shrubs. Tree box units may have their own pre-treatment mechanism in the form of a sump chamber that receives the stormwater and allows large particles to settle. Some commercially available tree boxes are concrete boxes with solid walls and an underdrain to collect stormwater and convey it downstream. Others may also have open walls or at least openings in the walls that allow infiltrated water to pass into the native soil.

Stormwater treatment has developed into its own industry in the United States and it is evolving as the requirements and expectations of stormwater management change and new technologies are investigated and develop. The previous discussion reflects common commercially available options for stormwater treatment but, because the field is always changing and developing, new units and types of units will likely be available in the future.

4.4.2 Emerging LID practices

Since the passage of the CWA amendments in 1987, new strategies and technologies have been developed to manage and treat stormwater runoff. These efforts have focused on reducing runoff volumes and improving its water quality. For example, rain gardens were essentially unheard of in the 1980s but now are a common stormwater management practice throughout the United States. As municipalities continue to strive to meet TMDL requirements and as overall public awareness of the negative environmental impact of stormwater runoff increases, these efforts will undoubtedly continue and new strategies and technologies will be researched, tested, and developed with the intent that they will one day become one of the strategies available to protect water resources in the United States. This section discusses such emerging technologies. Although they are not yet common practice, they have either shown potential as a stormwater treatment practice while in the development stage or have already been applied successfully in practice in some areas of the United States.

4.4.2.1 Enhanced media

As previously discussed, LID practices often do little to remove nutrients, especially the dissolved fraction. In fact, phosphorus and nitrogen are often released from decaying organic matter (e.g. vegetation or compost), making bioretention practices, for example, a nutrient source. Soil particles alone often have little to no capacity to retain nutrients, thus any capacity that may exist is quickly exhausted. In order to improve the performance of LID practices with regards to nutrients, efforts have focused on finding enhancing agents that can be added to the media and that will adsorb phosphorus and/or nitrogen. The intent is to make the LID practice a sink for nutrients rather than a source. For phosphorus, metal oxides have been found to have such potential. For nitrogen, which is more difficult to remove, some activated carbons have been found to have this ability. These potential enhancing agents are discussed below.

4.4.2.1.1 Iron

Elemental iron, which oxidizes to form iron oxides, mixed with engineered media can be used to retain phosphate in stormwater runoff. The positive charge on the iron oxides attracts the negatively charged phosphate ion and can remove phosphate ions from solution. In bioretention media, iron oxides have also shown the ability to stabilize soils and reduce leaching of dissolved organic carbon (Carter *et al.*, 2009; Schneider *et al.*, 2010), which may minimize the mobilization of phosphorus and some metals (i.e. copper, lead, and zinc) (Ingvertsen *et al.*, 2012). Although this technology has been used more with sand filters than with bioretention practices, especially in the State of Minnesota, it has potential for widespread use in LID practices containing engineering media.

Erickson *et al.* (2007) showed that chopped granular steel wool (i.e. elemental iron) when mixed into a sand filter media meeting ASTM C33 specifications has the ability to remove dissolved phosphorus from water passing through the media. Iron content ranging from zero up to 5% (by weight) was tested and, unsurprisingly, the more iron included in the media, the greater the amount of dissolved phosphorus that was removed. Later, an iron enhanced sand filter was constructed in Maplewood, Minnesota, a Minneapolis suburb, using approximately 5% scrap iron shavings by weight (Erickson *et al.*, 2012). Scrap iron shavings were used because they were less expensive than chopped granular steel wool. Although only tested for use in sand filters, this technology has potential for use in infiltration basins, rain gardens, and other LID practices.

When selecting an iron source, care must be taken to ensure that the metal contains no impurities, such as toxic metals, that could contaminant the runoff it is intended to treat. Other design requirements include ensuring that the media will dry after filtering runoff events. This enables the iron to further develop ferric oxides, which creates additional positively charged adsorption sites with the ability to retain negatively charged phosphate ions. Since dissolved phosphorus in stormwater is predominantly in the form of phosphate, the rusted iron shavings can significantly reduce dissolved phosphorus concentrations in the filtered runoff. Allowing the media to dry between events also prevents anaerobic conditions from developing, which would allow anaerobic bacteria to thrive and clog the media.

Although the ferric oxides create new adsorption sites, iron has a finite capacity which will eventually be exhausted. The expected useful life of iron a media mix depends on the amount of iron in the media, stormwater dissolved phosphorus concentrations, the depth of media, the depth of water treated, and perhaps the presence and concentrations of other anions in the runoff, which may outcompete phosphate ions for adsorption sites on the iron. Also, iron content of greater than 7% by weight is not recommended because, based on experience, if too much iron is present in the media, the iron tends to clump together and form large conglomerate masses which can reduce the filters' performance.

Currently there is no means to regenerate or restore the phosphate capacity of the iron once it has been used. As a result, once the phosphorus capacity of the iron is reduced and phosphorus removal falls below desired levels, the spent iron/media would have to be removed and replaced with new media. The used media could then be used or disposed in a manner that meets all governing regulatory requirements.

4.4.2.1.2 Aluminum oxide

Like iron, aluminum oxide can remove phosphate ions from stormwater runoff due to its positive charge and has the potential for use in the engineered media of LID practices. Hinman & Wulkan (2012) stated that aluminum is more suitable for bioretention systems than iron but Wu and Sansalone (2013a) found that aluminum oxides (and iron-coated perlite) leached aluminum (and iron) into water as it passed through the media. Thus, as with any enhancing media, care must be taken to ensure that other pollutants are not being released into the treated water.

Others have coated media particles (such as grains of sand, clay, pumice, and concrete-based particles) with aluminum oxide to improve their phosphate retention capabilities (Liu *et al.*, 2009; Wu & Sansalone, 2013b). Coating particles in this way has many advantages. It increases the surface area per mass of aluminum oxide by using sand or other particles, which are relatively cheap, and the aluminum oxide coated particles can be made the same size as other particles in the media mix, which promotes homogeneity, and even distribution of aluminum oxide, when mixing the media (Johannsen *et al.*, 2016). As with iron oxides, aluminum oxides have also shown the ability to stabilize soils and reduce leaching of dissolved organic carbon (Carter *et al.*, 2009; Schneider *et al.*, 2010), which may minimize the mobilization of phosphorus and some metals as previously mentioned (Ingvertsen *et al.*, 2012).

With the realization that bioretention practices typically do not remove nutrients from stormwater runoff, at least over the long-term, the focus is shifting towards the incorporation of enhancing agents mixed into bioretention media. The enhancing agents would be selected and/or engineered to increase the nutrient retention of the media. Aluminum oxide and aluminum oxide-coated particles have shown this ability and have the potential to greatly improve the performance of bioretention practices in this regard.

4.4.2.1.3 Water treatment residuals

Water treatment residuals (WTRs) are by-products from drinking water treatment processes that contain iron or aluminum hydroxides. The metal oxides (e.g. alum) are added to the source water to help coagulate suspended solids, which facilitates settling of the solids. The settled solids, which still contain metal oxides, are then removed from the system as sludge and often must be disposed of in a landfill for a fee. Rather than pay to dispose of the solids, research has shown that they may be used to remove phosphate from stormwater runoff. When added to the soil of a bioretention practice or to the media of a sand filter, the metal oxides, which are positively charged, have the ability to remove greater than 90% of the phosphate at typical stormwater runoff concentrations (Lucas &

Greenway, 2011, O'Neill & Davis, 2012a, 2012b; Yan *et al.*, 2018). Results indicate that these high levels of phosphate removal are possible for up to 20 years or more. Poor *et al.* (2019) found that WTRs were more effective when included in a bioretention system as a layer placed below a compost layer and above a sand layer as compared to when all bioretention media components are thoroughly mixed. Typically, however, the WTRs are mixed with the media. In a technical guidance document for Puget Sound in the State of Washington, Hinman and Wulkan (2012) suggested adding 10% WTR by volume to bioretention media and adding 15% by volume shredded tree bark to maintain sufficient permeability. Otherwise, the fine texture of the WTR may reduce permeability of media to unacceptable levels. Although more research and practical experience is needed with this technology, WTRs have shown potential to be a key component of stormwater management of the future.

4.4.2.1.4 Activated carbon and biochar

Activated carbon is a carbon-based material that is produced by heating carbon-rich sources such as coal or parts of plants (e.g. coconut shells, peach pits, etc.) at high temperatures (~100–800°C, or more) under low oxygen conditions. A similar product is biochar, which is made from biomass and is produced under similar conditions as activated carbon. In fact, these two terms are sometimes used interchangeably. Activated carbon and biochar have large surface areas and can be used to remove pollutants from water through surface adsorption. The product is sometimes impregnated with chemicals to increase surface area and add surface charge, which allows it to adsorb more pollutant molecules. Additional removal processes include filtration, ion exchange, electrostatic attraction (activated carbon and biochar are negatively charged), and precipitation (Tsang *et al.*, 2019).

Much recent research has focused on the use of activated carbon or biochar for the removal of dissolved pollutants in stormwater that are difficult to remove with conventional practices. In most cases the activated carbon or biochar was added to the media of a bioretention practice or sand filter to increase the removal of specifically targeted pollutants. For example, Erickson *et al.* (2016) showed that commercially available activated carbon could be used to remove nitrate from synthetic stormwater. The cost of the activated carbon that would be needed to remove sufficient amounts of nitrate from a single typical runoff event, however, would have a relatively high cost, and there would be no remaining nitrate removal capacity for the next runoff event. Erickson *et al.* (2016) suggested that biological processes within the media could be used between storm events to recharge the activated carbon through denitrification. This process has yet to be investigated further, however.

Potential carbon sources for activated carbon or biochar are almost limitless and different materials and different production processes (e.g. oxygen levels, temperatures, impregnation materials) produce materials with differing characteristics and removal capabilities and capacities. One such carbon source is wastewater treatment plant (WWTP) sludge, which has been used to produce activated carbon that has the ability to remove phosphate and nitrate (Yue *et al.*, 2018) and hydrophobic organic compounds (Bjorkland & Li, 2017). The use of WWTP sludge for this purpose would have the added benefit of creating a use for a waste product. WWTPs must often pay for sludge disposal and having a use for their sludge would likely reduce their operational costs. Activated carbon and/or biochar have also been shown to have the ability to increase removal rates of organic compounds (Ashoori *et al.*, 2019; Ulrich *et al.*, 2017), metals (Ashoori *et al.*, 2019; Ko *et al.*, 2018), and *E. coli* (Lu & Chen, 2018).

As discussed above, research suggests that media amended with activated carbon and/or biochar may enable bioretention practices to remove pollutants at rates they could not otherwise achieve. With different carbon sources and different production processes creating materials with different properties, activated carbon and/or biochar can be developed to target specific pollutants or group of pollutants. The goal would be to develop materials, ideally from local sources, with high capacity and rapid kinetics that can be used to target specific pollutants, as needed. Although much research needs to be performed, activated carbon and biochar show potential for stormwater management applications.

4.4.2.2 Floating islands

Floating islands are constructed buoyant systems that contain vegetation and are designed to float on the surface of a water body such as a stormwater pond. Floating islands can occur naturally when vegetative mats of wetland plants break off from the shore of a water body and drift on the water surface. These systems typically achieve buoyancy through gases, such as methane and nitrogen, that become trapped in their root systems (Hogg & Wein, 1988). Buoyancy of constructed systems, however, is typically achieved with a buoyant frame, foam, or some other, low density material in which the vegetation is planted. The frame or planting medium floats on or near the water surface so that the plant stems and leaves grow above the water surface with their roots submerged.

The roots decrease turbulence and slow water velocity, which promotes settling of solids (Hoban, 2019). Roots also filter solids and support biofilm growth that can reduce pollutants through adhesion, nutrient uptake, and sequestration (Stewart & Downing, 2008). The islands themselves also reduce algae growth by blocking sunlight (Hubbard, 2010). Potential negative impacts include the development of anoxic conditions in the soil beneath the roots and birds inhabiting the islands and increasing pollutant loads through defecation (Hoban, 2019). Also, although the roots can act as a filtering mechanism, if the end of the roots do not extend to the bottom of the water body, a preferential flow path may simply develop under the roots. In this case the system will act like a conventional pond (Hoban, 2019). Conversely, if the roots touch the bottom they may attach there and essentially anchor the island to the bottom of the water body. Then, when water levels rise during a subsequent runoff event, the roots will hold the island down and prevent it from rising with the water surface. This can drown the island and kill the plants (Hoban, 2019).

Floating islands have been shown to be effective at reducing sediment, nutrients, and metals (Chen *et al.*, 2016; Tanner & Headley, 2011) in stormwater and wastewater. All investigations, however, have not observed such positive results. Winston *et al.* (2013), for example, investigated the impact of floating islands on two different stormwater ponds and found that in one pond the floating island made a significant impact on solids and total phosphorus reductions but in the other pond there was no significant reduction in solids, nitrogen, or phosphorus (in any form). An increase in aquatic weeds was also observed after the installation of the floating islands. The specific reasons for the differences was not known.

The use of floating islands for stormwater treatment is relatively new and, while they have shown potential for improving water quality, much about their operation, effectiveness, and optimization remains unknown. For example, different plants remove different pollutants at different rates (Pavlineri *et al.*, 2017) and plant tolerance to pollutants also varies by pollutant and plant. Ideally, plants would uptake pollutants rapidly and have a high tolerance for the pollutant. Also, the effectiveness of the root biofilm depends on various geochemical processes that are impacted by weather conditions (Shahid *et al.*, 2018). Overall, floating islands do not have the capability to remove most pollutants, except some hydrocarbons. They simply store the pollutants and encourage settling. The stored pollutants are often released from the plants upon microbiological degradation in the dissolved form. The design parameters such as percent of vegetative cover on the island, the growth medium, the method of buoyancy (Pavlineri *et al.*, 2017), and the fraction of the water body area to be covered by floating islands also have yet to be fully investigated or optimized.

Thus, although floating islands have potential and have been used successfully in limited applications for stormwater treatment, much work needs to be completed in order to advance the understanding of long-term impacts and optimization of these systems before their use as a stormwater management tool becomes widespread.

4.4.2.3 Rain gardens for nitrogen removal

As previously discussed, conventional bioretention cells (e.g. rain gardens) typically do little to reduce phosphorus and nitrogen concentrations in stormwater runoff due to the breakdown of plant matter, mulch, and other organics. While phosphorus concentrations can be reduced by media enhancing agents such as iron, aluminum oxide, and WTRs, nitrogen is not removed by these agents. Also, although activated carbon can remove nitrate from water (Erickson *et al.*, 2016), it is currently cost-prohibitive to use in practice. Much of the nitrogen in stormwater is dissolved (ammonia, nitrite and nitrate) and cannot be removed by physical means such as filtration or sedimentation. An alternative method for nitrate removal involves the use of microorganisms (i.e. bacteria) through the process of nitrification-denitrification. Nitrification occurs when aerobic bacteria convert ammonia to nitrite and nitrate. This process occurs readily in LID practices because the aerobic conditions are typically present, and the conversion by the bacteria is in the order of hours. Denitrification occurs in an anaerobic environment with bacteria that respond slower than aerobic bacteria. It requires sufficient water storage, organic carbon source and lack of exposure to the atmosphere to achieve denitrification.

Although anaerobic zones do not typically exist in conventional bioretention practices, an anaerobic zone can be created by using an elevated drain tile or a drain tile with an upturned elbow at the downstream end. Both configurations have an elevated outlet, which creates an internal water storage zone below the outlet elevation. The stored water, which is not exposed to the air and thus is not aerated, will become anaerobic and can support the bacterial population needed for denitrification. For example, Igielski *et al.* (2019) investigated rain gardens with an internal water storage zone and found nitrate was reduced from 3.0 mg N/L to less than 0.01 mg N/L with a hydraulic residence time of 2.6 days. Qiu *et al.* (2019) used an internal water storage zone in a bioretention practice that also contained WTRs as an enhancing agent for phosphorus removal. With a retention time of 2 hours, nitrate removal was 85%, up from the 21% achieved by a control practice with no anaerobic water zone. In a modeling exercise, Sun *et al.* (2017) determined that complete microbial nitrogen removal is possible with a hydraulic retention time of 12 hours as long as there is enough organic carbon present. By observing data in the National Stormwater Quality Database, Sun *et al.* (2017) also estimated that 71% of the runoff in the United States met the conditions necessary for nitrogen removal via partial nitrification followed by anaerobic oxidation.

In order to meet the requirements for organic carbon, it could be added to bioretention cells or even to the influent stormwater. Possible carbon sources are plentiful but the rate of degradation is important. Kim *et al.* (2003) tested several potential organic sources and determined that newspaper (out of alfalfa, leaf mulch compost, newspaper, sawdust, wheat straw, wood chips, and elemental sulfur) yielded the highest nitrate removal rate and that it could support biological denitrification under intermittent stormwater loads. Ding *et al.* (2019) showed that denitrification can occur in the internal water storage zone provided there is adequate detention time even when subject to freeze-thaw cycling from –10 to 10°C, indicating that this process could be effective in cold climates.

The optimization and long-term performance of bioretention practices with internal water storage zones for denitrification is relatively new and there are still many unknowns. For example, Willard *et al.* (2017) investigated the performance of such practices after seven years of service and compared results to their performance immediately after they were constructed. The practices were still effective at reducing flows, nutrients, and fecal indicator bacteria. Based on analysis of soil samples, however, most denitrifying bacteria were present in the top soil layers despite the presence of an internal water storage zone below. The authors suggested that there were small anoxic pockets in the top soil layers and that the lack of carbon in the water storage zone limited microbial growth in that zone.

Although much work is needed to refine the process and sustain long-term denitrification, this method is an emerging method of stormwater treatment and, more specifically, nitrogen removal.

4.4.3 Future perspectives

Over the past 30 years, stormwater management in the United States has evolved from simply focusing on removing stormwater from a site and conveying it downstream, often with a reduction in peak flow, to a focus on water quality improvement and runoff volume reduction that has spurred the growth of an entire industry. The Clean Water Act of 1972, its amendments, and state and local regulations and/or initiatives have driven these changes and will continue to drive the future of stormwater management in the USA. This section discusses the potential factors and technologies that appear poised to significantly impact stormwater management over the next few decades and beyond.

4.4.3.1 Climate change

The global climate has been changing for decades and it is expected to continue changing for the rest of the century and even longer (Walsh *et al.*, 2014). Since 1895 the average temperature in the United States has increased by between 0.7 and 1.1°C, with most of the increase occurring since 1970. Large rainfall events are increasing across the United States, with the largest increases in the Midwest and Northeast. The continued increase in the frequency and intensity of extreme rainfall events is projected nationwide (Walsh *et al.*, 2014) and historical measured increases in daily extreme precipitation have exceeded model projections (Allan & Soden, 2008; Lenderlink & van Meijgaard, 2008). Thus, model projections for the future may also be low.

An increase in the frequency and intensity of extreme rainfall events will tax stormwater systems. In fact, portions of existing stormwater management systems are already undersized for their accepted levels of risk (Moore *et al.*, 2016) and further changes will only exacerbate the problem. Managing and lowering these risks will be paramount for municipalities as they strive to protect communities from future extreme precipitation events. LID practices can help meet this challenge because they help reduce runoff volumes. Although the benefits of LID practices are relatively small at extreme precipitation intensities, in many instances they can help because they are more cost-effective than grey infrastructure (Moore *et al.*, 2016).

4.4.3.2 Combined sewer overflows

As previously discussed in this chapter, CSOs exist when a sewer system that is designed to carry both sanitary sewage and stormwater runoff carries a flow rate that is too large for the municipal wastewater treatment plant to process. In these instances, which occur after rainfall or snowmelt events, untreated sewage or sewage with limited treatment is intentionally discharged into the receiving water body. Such combined sewer systems exist in approximately 860 communities in the United States and serve a total population of about 40 million people (US EPA, 2016). In the United States, most communities that experience CSO events exist in the Great Lakes Region or the Northeast portion of the United States.

In New York Harbor, for example, more than 102 billion liters of raw sewage are discharged annually, which amounts to a weekly average of about two billion liters (Riverkeeper, 2019). In the United States portion of the Great Lakes watershed, there are 184 combined sewer systems. In 2014 these systems reported 1482 CSO events with a total discharge of 83 billion liters of sewage. The actual values are likely higher because some systems had no data available, meaning that if a CSO event occurred, it was not reported.

According to US EPA requirements, communities must address CSOs, but they are allowed phased implementation that fits within their financial means and they are provided flexibility to find the most cost-effective methods (US EPA, 1994). LID practices can help address CSOs and will certainly play a role doing so in the future. The City of New York's Green Infrastructure Plan, for example, seeks to address CSOs by capturing the first inch of rainfall from 10% of the City's impervious surfaces (NYC undated). This will reduce CSO discharge by an estimated 14 billion liters per year, which is 7.6 billion liters a year more than if standard grey infrastructure was used alone. As another example, it has been estimated that climate change may cause up to a 12–18% increase in the occurrence, volume, and duration of CSO events in Toledo, OH, which is in the Great Lakes watershed (Tavakol-Davani *et al.*, 2016). In the same study it was shown that, for this range of CSO increase, a rainwater harvesting plan with a capacity of 0.76 m^3 implemented on half of the buildings in the area would be able to mitigate the increase, control peak flows, and meet toilet flushing demands. LID, however, should not be viewed as a panacea to all CSO or stormwater issues. This notion was supported in a study involving LID applications in the United States (Shamsi, 2018). The study suggested that hybrid systems of LID practices combined with smart (e.g. CMAC) and/or grey infrastructure had the most value when incorporating financial costs along with environmental and social benefits in a full life-cycle cost analysis.

4.4.3.3 Dynamic design

Traditional design of volume-reduction LID practices is based on static storage volumes (i.e. storing water in the surface ponding area and the available void space in subsurface stone and soil media). This design approach does not reflect the variable performance of LID practices over the year and under variable weather conditions. In general, LID practices are a complex system of components and processes related to climate, environment, soil, water, plant, site conditions, drainage area, and human factors in which components have several connections and interrelationships with each other (Traver & Ebrahimian, 2017). Malfunction or property change can affect the overall performance or longevity of the system. Understanding the dynamics of hydrologic processes, e.g. infiltration and evapotranspiration (ET), that affect the performance of LID practices is fundamental to the design process. Temporal variation of infiltration rates over a year has been observed in different LID practices because of the water viscosity changes with temperature (higher infiltration rates in warmer months and lower infiltration rates in colder months) (Emerson & Traver, 2008) and other factors including, but not limited to, soil composition, level of soil compaction, vegetation (plant root) condition, biological activities in the soil, inflow sediment characteristics, and quality of infiltrating water (Ebrahimian *et al.*, 2020). ET is a viable runoff reduction mechanism in vegetated LID practices that is generally greater after a rainfall event than during extended dry periods, but it continues as long as water is available between storm events (Nocco *et al.*, 2016; Wadzuk *et al.*, 2015). However, ET is mostly unaccounted for in the design of LID practices due to being a continuous process that is difficult to quantify and fit in the static design approach. ET is highly dependent on available water in the soil media. Therefore, ET and infiltration should be considered simultaneously in the design of vegetated LID practices (Ebrahimian *et al.*, 2019).

The static design approach underestimates the performance of volume reduction LID practices because it does not consider exfiltration and ET in the design process. A substantial portion of water may exfiltrate from the media to the in-situ soil during storm events (Traver & Ebrahimian, 2017) and ET can participate to the storage capacity recovery of LID practices between storm events (Ebrahimian *et al.*, 2019). A dynamic approach is needed for LID design to take full advantage of their capabilities. To address this need, a dynamic approach can be used to consider the full hydrologic cycle within a larger time horizon than traditional designs by a continuous simulation to select LID design parameters (Traver & Ebrahimian, 2017). Continuous simulation over a large time horizon allows for the consideration of regional climate and the effects of infiltration and ET variability, initial soil moisture conditions, and back-to-back storms in LID design. The potential benefits of moving toward a dynamic design include promoting a wider coverage of smaller LID practices in urban areas (making it possible to build more facilities with the same budget), eliminating or reducing the need for subsurface components that may cause failure or require costly maintenance (e.g. rock beds and underdrains), and improving soil media design by considering the balance between infiltration and ET to support plant growth and not just runoff volume removal. Also, the continuous simulation under the dynamic approach would enable the consideration of potential climate change scenarios in LID design.

4.4.3.4 Advances in enhanced media

As previously discussed, the addition of elemental iron, aluminum oxide, or water treatment residuals has been used to increase the removal of dissolved phosphorus from stormwater runoff. These enhancing agents adsorb the phosphate ion through their surface charge and/or ion exchange processes. Nitrogen and nitrate, however, are not removed. With regards to removing nitrogen through similar abiotic processes, activated carbon has been shown to be capable but has not been shown to be cost-effective. An enhancing agent for nitrogen/nitrate removal that is cost-effective would be a major advancement in stormwater management and could help many TMDL plans achieve their goals. There are at least two ways that this could be realized. One is for a new, low cost/high nitrogen capacity enhancing agent to be developed. The other would be for the development of a method that would recharge enhancing media by removing nitrogen from the media that was previously adsorbed in order to free those adsorption sites for a future runoff event. As suggested in Erickson *et al.* (2016), this process could be biologically based by having microorganisms utilize the adsorbed nitrate and remove it from the activated carbon between runoff events. Of course, other processes are also possible. For example, if the nitrate could be removed from the in-place activated carbon through chemical or ion exchange processes, the nitrogen could be used as a fertilizer and the capacity of the activated carbon would be restored. This cycle, which could continue indefinitely, would create a useful product from what previously would have been wasted or landfilled. A similar process for removing phosphorus from iron, aluminum oxide, WTRs, or a future enhancing agent, would also be beneficial.

4.4.3.5 Source reduction

Perhaps the most economical means of reducing stormwater pollution is to prevent the water from being polluted in the first place. This approach is called source reduction and current methods include street sweeping and bans on phosphorus containing fertilizer. The latter method has been implemented in some parts of the United States after the realization that many urban and suburban landscapes already contain an excess amount of phosphorus. The effective management of stormwater pollutant loads in the future will likely need to include more source reduction if TMDL goals are to be met.

In addition to the current efforts just mentioned, future efforts could include reducing atmospheric deposition, which is a significant pollutant source for some pollutants in some watersheds (Davis *et al.*, 2001). Other possibilities include reducing pollutants from cars, which are major sources of metals, oils and organic chemicals, and suspended solids, and from buildings, which, depending on their siding and roofing material, can be a major contributor of metals (Davis *et al.*, 2001). Another key component of source reduction may include public education, which may help reduce pollutant loads from homes (grass clippings, fertilizer, herbicides, oil spills, etc.) and other sources.

If the challenges of stormwater pollution are going to be met in the future, a multi-pronged approach will need to be taken and source reduction can play an important part of any stormwater management plan.

REFERENCES

Abida H. and Sabourin J. F. (2006). Grass swale-perforated pipe systems for stormwater management. *Journal of Irrigation and Drainage Engineering*, **132**(1), 55–63.

Ahmed F., Gulliver J. S. and Nieber J. (2015). Estimating swale performance in volume reduction. Proceedings of the World Environmental and Water Resources Congress: Floods, Droughts and Ecosystems, Austin, TX, May 17–21. American Society of Civil Engineers, Washington, DC, USA, pp. 255–260.

Allan R. P. and Soden B. J. (2008). Atmospheric warming and the amplification of precipitation extremes. *Science*, **321**, 1481–1484, doi: 10.1126/science.1160787

Aquilina D. (2019). Ford Dearborn Truck Plant Green Roof at Ten Years. Available from: www.buildings.com/buzz/buildings-buzz/entryid/180/ford-dearborn-truck-plant-green-roof-at-ten-years, [Accessed 1 July 2019].

Ashoori N., Teixido M., Spahr S., LeFevre G. H., Sedlak D. L. and Luthy R. G. (2019). Evaluation of pilot-scale biochar-amended woodchip bioreactors to remove nitrate, metals and trace organic contaminants from urban stormwater runoff. *Water Research*, **154**, 1–11.

Barrett M. E., Walsh P. M., Malina J. F. and Charbeneau R. B. (1998a). Performance of vegetative controls for treating highway runoff. *Journal of Environmental Engineering*, **124**(11), 121–128.

Barrett M. E., Keblin M. V., Walsh P. M., Malina J. F. and Charbeneau R. B. (1998b). Evaluation of the Performance of Permanent Runoff Controls: Summary and Conclusions, Center for Transportation Research, University of Texas at Austin, Report #: 2954–3F.

Bjorkland K. and Li L. Y. (2017). Adsorption of organic stormwater pollutants onto activated carbon from sewage sludge. *Environmental Science and Pollution Research*, **24**(23), 19,167–19,180.

Blomqvist S., Gunnars A. and Elmgren R. (2004). Why the limiting nutrient differs between temperate coastal seas and freshwater lakes: A matter of salt. *Limnology and Oceanography*, **49**(6), 2236–2241.

Brander K. E., Owen K. E. and Potter K. W. (2004). Modeled impacts of development type on runoff volume and infiltration performance. *Journal of the American Water Resources Association*, **40**, 961–969.

Camesano T. A. and Logan B. E. (1998). Influence of fluid velocity and cell concentration on the transport of motile and non-motile bacteria in porous media. *Environmental Science and Technology*, **32**(34), 1699–1708.

Caplan J. S., Galanti R. C., Olshevski S. and Eisenman S. W. (2019). Water relations of street trees in green infrastructure tree trench systems. *Urban Forestry and Urban Greening*, **41**, 170–178.

Carter C. M., van der Sloot H. A. and Cooling D. (2009). pH-dependent extraction of soil and soil organic amendments to understand the factors controlling element mobility. *European Journal of Soil Science*, **60**, 622–637. doi: 10.1111/j.1365-2389.2009.01139.x

Chen Z., Cuervo D. P., Müller J. A., Wiessner A., Köser H., Vymazal J., Kästner M. and Kuschk P. (2016). Hydroponic root mats for wastewater treatment – A review. *Environmental Science and Pollution Research*, **23**(15), 911–15,928.

Clark S. E. and Pitt R. E. (2007). Influencing factors and a proposed evaluation methodology for predicting groundwater contamination potential from stormwater infiltration practices. *Water Environment Research*, **79**, 29–36.

Clark S. E. and Pitt R. E. (2011). Filtered metals control in stormwater using engineered media. Proceedings of the 2011 World Environmental and Water Resources Congress, Palm Springs, CA. American Society of Civil Engineers, Washington, DC, USA.

Clark S. E., Baker K. H., Mikua J. B., Burkhardt C. S. and Lalor M. M. (2006). Infiltration vs. Surface Water Discharge: Guidance for Stormwater Managers. Project Number 03–SW-4. Water Environment Research Foundation, Alexandria, Virginia.

Coffman L. (2002). Low impact development: smart technology for clean water. Proceedings of the Ninth International Conference on Urban Drainage. American Society of Civil Engineers, Portland, OR, September 8–13, 2002.

Cutierrez J. and Hernandez I. I. (1996). Runoff inter rill erosion as affected by grass cover in a semi-arid rangeland of Northern Mexico. *Journal of Arid Environments*, **34**(4), 435–460.

Davis A. (2007). Field performance of bioretention: water quality. *Environmental Engineering Science*, **24**, 1048–1064.

Davis A. P. and McCuen R. (2005). Stormwater Management for Smart Growth. Springer, New York, 368 pp.

Davis A. P., Shokouhian M. and Ni S. (2001). Loading estimates of lead, copper, cadmium, and zinc in urban runoff from specific sources. *Chemosphere*, **44**(5), 997–1009.

Davis A. P., Shokouhian M., Sharma H., Minami C. and Winogradoff D. (2003). Water quality improvement through bioretention: lead, copper, and zinc removal. *Water Environment Research*, **75**, 73–82.

Dierkes C. and Geiger W. (1999). Pollution retention capabilities of roadside soils. *Water Science and Technology*, **39**, 201–208.

Dietz M. and Clausen J. (2005). A field evaluation of rain garden flow and pollutant treatment. Water, Air and Soil Pollution, **167**, 123–138.

Dietz M. E. and Clausen J. C. (2006). Saturation to improve pollutant retention in a rain garden. *Environmental Science and Technology*, **40**, 1335–1340.

Ding B., Rezanezhad F., Gharedaghloo B., Van Cappellen P. and Passeport E. (2019). Bioretention cells under cold climate conditions: effects of freezing and thawing on water infiltration, soil structure and nutrient removal. *Science of the Total Environment*, **649**, 749–759.

Ebrahimian A., Wadzuk B. and Traver R. (2019). Evapotranspiration in green stormwater infrastructure systems. *Science of the Total Environment*, **688**, 797–810.

Ebrahimian A., Sample-Lord K., Wadzuk B. and Traver R. (2020). Temporal and spatial variation of infiltration in urban green infrastructure. *Hydrological Processes*, **34**(4), 1016–1034.

Emerson C. H. and Traver R. G. (2008). Multiyear and seasonal variation of infiltration from storm-water best management practices. *Journal of Irrigation and Drainage Engineering*, **134**(5), 598–605.

Erickson A. J., Gulliver J. S. and Weiss P. T. (2007). Enhanced sand filtration for storm water phosphorus removal. *Journal of Environmental Engineering*, **133**(5), 485–497.

Erickson A. J., Gulliver J. S. and Weiss P. T. (2012). Capturing dissolved phosphorus with iron enhanced sand filtration. *Water Research*, **46**(9), 6601–6608.

Erickson A. J., Weiss P. T. and Gulliver J. S. (2013). Optimizing Stormwater Treatment Practices: A Handbook of Assessment and Maintenance. Springer Publishing, New York, NY, USA.

Erickson A. J., Arnold W. A., Gulliver J. S., Brekke C. and Bredal M. (2016). Abiotic capture of stormwater nitrates with granular activated carbon. *Environmental Engineering Science*, **33**(5), 354–363.

Galli J. (1992). Analysis of Urban Stormwater BMP Performance and Longevity in Prince George's County, Maryland. Metropolitan Washington Council of Governments, Washington, D.C., USA.

Garcia-Serrana M., Gulliver J. S. and Nieber J. (2017). Infiltration capacity of roadside filter strips with non-uniform overland flow. *Journal of Hydrology*, **45**, 451–462.

Green Roofs for Healthy Cities. (2020). Available from: https://greenroofs.org/about-green-roofs, [Accessed 20 January 2020].

Gurdak J. J. and Qi S. L. (2012). Vulnerability of recently recharged groundwater in principal aquifers of the United States to nitrate contamination. *Environmental Science and Technology*, **46**(11), 6004–6012.

Hatt B. E., Fletcher T. D. and Deletic A. (2008). Hydrologic and pollutant removal performance of stormwater biofiltration systems at the field scale. *Journal of Hydrology*, **365**, 310–321.

Hinman C. and Wulkan B. (2012). Low Impact Development. Technical Guidance Manual for Puget Sound. Publication No. PSP 2012–3. Washington State University, Puyallup Research and Extension Center, Pallyup, WA, USA.

Hoban A. (2019). Water sensitive urban design approaches and their description. In: Approaches to Water Sensitive Urban Design, A. K. Sharma, T. Gardner and D. Begbie (eds), Elsevier, Inc., Amsterdam, Netherlands.

Hogg E. H. and Wein R. W. (1988). Seasonal change in gas content and buoyancy of floating Typha mats. *Journal of Ecology*, **76**(4), 1055–1068.

Holman-Dodds J. K., Bradley A. A. and Potter K. W. (2003). Evaluation of hydrologic benefits of infiltration based urban storm water management. *Journal of the American Water Resources Association*, **39**, 205–215.

Howard A., Mohseni O., Gulliver J. S. and Stefan H. G. (2011). SAFL baffle retrofit for suspended sediment removal in storm sewer sumps. *Water Research*, **45**, 5895–5904.

Hubbard R. K. (2010). Floating vegetated mats for improving surface water quality. In: Emerging Environmental Technologies, V. Shah (ed.), Springer, New York, USA, pp. 211–244.

Hunt W. F., Jarrett A. R., Smith J. T. and Sharkey L. J. (2006). Evaluating bioretention hydrology and nutrient removal at three field sites in North Carolina. *Journal of Irrigation and Drainage Engineering*, **132**, 600–608.

Igielski S., Kjellerup B. V. and Davis A. P. (2019). Understanding urban stormwater denitrification in bioretention internal water storage zones. *Water Environment Research*, **91**(1), 32–44.

Ingvertsen S. T., Cederkvist K., Jensen M. B. and Magid J. (2012). Assessment of existing roadside swales with engineered filter soil, II. Treatment efficiency and in situ mobilization in soil columns. *Journal of Environmental Quality*, **41**, 1970–1981. doi: 10.2134/jeq2012.0116

Johannsen L. L., Cederkvist K., Holm P. E. and Ingvertsen S. T. (2016). Aluminum oxide-coated sand for improved treatment of urban stormwater. *Journal of Environmental Quality*, **45**, 720–727. doi: 10.2134/jeq2015.06.0287

Kang J. H., Weiss P. T., Wilson C. B. and Gulliver J. S. (2008). Maintenance of stormwater BMPs, frequency, effort and cost. *Stormwater*, Nov/Dec, **9**(8), 18–28.

Kim H., Seagren E. A. and Davis A. P. (2003). Engineered bioretention for removal of nitrate from stormwater runoff. Proceedings of the 73rd Annual Water Environment Federation Technical Exposition and Conference (WEFTEC 2000), Anaheim, CA, October 14–18, 2003. Alexandria, VA.

Ko D., Mines P. D., Jakobsen M. H., Yavuz C. T., Hansen H. C. B. and Andersen H. R. (2018). Disulfide polymer grafted porous carbon composites for heavy metal removal from stormwater runoff. *Chemical Engineering Journal*, **348**, 685–692.

Ku H. F. and Simmons D. L. (1986). Effect of Urban Stormwater Runoff on Ground Water Beneath Recharge Basins on Long Island, New York, US Geological Survey Water-Resources Investigations Report 85–4088. Syosset, New York, USA.

LeFevre G. H., Novak P. J. and Hozalski R. M. (2012a). Fate of naphthalene in laboratory-scale bioretention cells, implications for sustainable stormwater Management. *Environmental Science and Technology*, **46**(2), 995–1002.

LeFevre G. H., Hozalski R. M. and Novak P. J. (2012b). The role of biodegradation in limiting the accumulation of petroleum hydrocarbons in raingarden soils. *Water Research*, **46**(20), 6753–6762.

Lenderink G. and van Meijgaard E. (2008). Increase in hourly precipitation extremes beyond expectations from temperature changes. *Nature Geoscience*, **1**, 511–514. doi: 10.1038/ngeo262

Li H. and Davis A. P. (2009). Water quality improvement through reductions of pollutant loads using bioretention. *Journal of Environmental Engineering*, **135**(8), 567–576.

Liu B., Berretta C., Gnecco I. and Ying G. (2009). Control of highway stormwater during event and interevent retention in best management practices. *Transportation Research Record*, **2120**, 115–122. doi: 10.3141/2120-12

Lu L. and Chen B. (2018). Enhanced bisphenol A removal from stormwater in biochar-amended biofilters, combined with batch sorption and fixed-bed column studies. *Environmental Pollution*, **243**, 1539–1549.

Lucas W. C. and Greenway M. (2011). Phosphorus retention by bioretention mecocosms using media formulated for phosphorus sorption, response to accelerated loads. *Journal of Irrigation and Drainage Engineering*, **137**(3), 144–152.

Maestre M. and Pitt R. (2005). A Compilation and Analysis of NPDES Stormwater Monitoring Information. Office of Water, US Environmental Protection Agency, Washington, D.C., USA.

J. Marsalek, E. Watt, E. Zeman and H. Sieker (eds.) (2001). Advances in Urban Stormwater and Agricultural Runoff Source Controls. NATO Earth and Environmental Sciences Series. Kluwer Academic Publishers, Boston, 319 pp.

Mazer G., Booth D. and Ewing K. (2001). Limitations to vegetation establishment and growth in biofiltration swales. *Ecological Engineering*, **17**(4), 429–443.

Moore T. L., Gulliver J. S., Stack L. and Simpson M. H. (2016). Stormwater management and climate change, vulnerability and capacity for adaptation in urban and suburban contexts. *Climatic Change*, **138**(3–4), 491–504.

MPCA. (2019). Minnesota Stormwater Manual. Minnesota Pollution Control Agency, Saint Paul, USA. Available from: https://stormwater.pca.state.mn.us/index.php?title=Main_Page, [Accessed 28 May 2019].

Nocco M. A., Rouse S. E. and Balster N. J. (2016). Vegetation type alters water and nitrogen budgets in a controlled, replicated experiment on residential-sized rain gardens planted with prairie, shrub, and turfgrass. *Urban Ecosystems*, **19**(4), 1665–1691.

Novotny E. V., Murphy D. and Stefan H. G. (2008). Increase of urban lake salinity by road deicing salt. *Science of the Total Environment*, **406**, 131–144.

NYC. (2019). Undated New York City Green Infrastructure Plan, a Sustainable Strategy for Clean Waterways, NYC Department of Environmental Protection and Plan NYC. Available from: http://www.nyc.gov/html/dep/pdf/green_infrastructure/NYCGreenInfrastructurePlan_ExecutiveSummary.pdf, [accessed 8 August 2019].

O'Connor T. P., Muthukrishnan S., Barshatzky K. and Wallace W. (2012). Trace metal accumulation in sediments and benthic macroinvertebrates before and after maintenance of a constructed wetland. *Water Environment Research*, **84**(4), 370–81.

O'Neill S. W. and Davis A. P. (2012a). Water treatment residual as a bioretention amendment for phosphorus. I. evaluation studies. *Journal of Environmental Engineering*, **138**(3), 318–327.

O'Neill S. W. and Davis A. P. (2012b). Water treatment residual as a bioretention amendment for phosphorus. II. long-term column studies. *Journal of Environmental Engineering*, **138**(3), 328–336.

Paus K. H., Morgan J., Gulliver J. S. and Hozalski R. M. (2014). Effects of bioretention media compost volume fraction on toxic metals removal, hydraulic conductivity, and phosphorous release. *Journal of Environmental Engineering*, **140**(10), 04014033. doi: 10.1061/(ASCE)EE.1943-7870.0000846

Pavlineri N., Skoulikidis N. T. and Tsihrintzis V. A. (2017). Constructed floating wetlands, a review of research, design, operation and management aspects, and data meta-analysis. *Chemical Engineering Journal*, **308**, 1120–1132.

Pitt R. (1999). Small storm hydrology and why it is important for the design of stormwater control practices. In: Advances in Modeling the Management of Stormwater Impacts, W. James (ed.), Guelph, Ontario, Vol. **7**.

Pitt R., Field R., Lalor M. and Brown M. (1995). Urban stormwater toxic pollutants, assessment, sources, and treatability. *Water Environment Research*, **67**, 260–275.

Pitt R., Clark S., Parmer K. and Field R. (eds.) (1996). Groundwater Contamination from Stormwater Infiltration. Ann Arbor Press, Ann Arbor, USA, 219 pp.

Pitt R., Clark S. and Field R. (1999). Groundwater contamination potential from stormwater infiltration practices. *Urban Water*, **1**, 217–236.

Pitt R., Chen S. E. and Clark S. (2002). Compacted urban soil effects on infiltration and bioretention stormwater control designs. Global Solutions for Urban Drainage. Proceedings of the Ninth International Conference on Urban Drainage, Portland, Oregon. American Society of Civil Engineers, Washington, DC, USA.

Plaza G. A., Wypych J., Berry C. J. and Brigmon R. (2007). Utilization of monocyclic aromatic hydrocarbons individually and in mixture by bacteria isolated from petroleum-contaminated soil. *World Journal of Microbiology and Biotechnology*, **23** (4), 533–542.

Poor C. J., Conkle K., MacDonald A. and Duncan K. (2019). Water treatment residuals in bioretention planters to reduce phosphorus levels in stormwater. *Environmental Engineering Science*, **36**(3), 265–272.

Qiu F., Zhao S., Zhao D., Wang J. and Fu K. (2019). Enhanced nutrient removal in bioretention systems modified with water treatment residuals and internal water storage zone. *Environmental Science Water Research and Technology*, **5**(5), 993–1003.

Riverkeeper. (2019). Combined Sewage Overflows (CSOs). Available from: www.riverkeeper.org/campaigns/stop-polluters/sewage-contamination/cso/, [Accessed 26 July 2019].

Roman D., Braga A., Shetty N. and Culligan P. (2017). Design and modeling of an adaptively controlled rainwater harvesting system. *Water*, **9**(12), 974.

Roseen R. M., Ballestero T. P., Houle K. M., Heath D. and Houle J. J. (2014). Assessment of winter maintenance of porous asphalt and its function for chloride source control. *Journal of Transportation Engineering*, **140**(2), 1–8.

Scalise C. and Fitzpatrick K. (2012). Chicago deep tunnel design and construction. In: Proceedings of Structures Congress 2012, March 29–31, Chicago, USA, J. Carrato and J. Burns (eds), American Society of Civil Engineers, Washington, DC, USA.

Schneider M. P. W., Scheel T., Mikutta R., van Hees P., Kaiser K. and Kalbitz K. (2010). Sorptive stabilization of organic matter by amorphous Al hydroxide. *Geochimica et Cosmochimica Acta*, **74**, 1606–1619. doi: 10.1016/j.gca.2009.12.017

Shahid M. J., Arslan M., Ali S., Siddique M. and Afzal M. (2018). Floating wetlands, a sustainable tool for wastewater treatment. *Clean-Soil Air Water*, **46**(10), 1800120.

Shamsi U. M. (2018). Green First Approach for Wet Weather Programs. *Journal of Water Management Modeling*, **26**, C439. https://doi.org/10.14796/JWMM.C439

Shutes R. B. E., Revitt D. M. and Mungur Scholes A. S. (1997). The design of wetland systems for the treatment of urban runoff. *Water Science and Technology*, **35**, 19–25.

Steffen J., Jensen M., Pomeroy C. A. and Burian S. J. (2013). Water supply and stormwater management benefits of residential rainwater harvesting in U.S. cities. *Journal of the American Water Resources Association*, **49**(4), 810–824.

Stewart T. W. and Downing J. A. (2008). Macroinvertebrate communities and environmental conditions in recently constructed wetlands. *Wetlands*, **28**(1), 141–150.

Sun X. and Davis A. P. (2007). Heavy metal fates in laboratory bioretention systems. *Chemosphere*, **66**, 1601–1609.

Sun Y., Zhang D. and Wang Z. W. (2017). The potential of using biological nitrogen removal technique for stormwater treatment. *Ecological Engineering*, **106**, 482–495.

Tanner C. C. and Headley T. R. (2011). Components of floating emergent macrophyte treatment wetlands influencing removal of stormwater pollutants. *Ecological Engineering*, **37**(3), 474–486.

Tavakol-Davani H., Goharian E., Hansen C. H., Tavakol-Davani H., Apul D. and Burian S. J. (2016). How does climate change affect combined sewer overflow in a system benefiting from rainwater harvesting systems? *Sustainable Cities and Society*, **27**, 430–438.

Teemusk A. and Mander Ü. (2007). Rainwater runoff quantity and quality performance from a greenroof, the effects of short-term events. *Ecological Engineering*, **30**, 271–277.

TN DOT (Tennessee Department of Transporation). (2019). Open-Graded Friction Course. Available from: www.tn.gov/tdot/maintenance/pavement-office/ogfc.html, [Accessed 27 May 2019].

Traver R. G. and Ebrahimian A. (2017). Dynamic design of green stormwater infrastructure. *Frontiers of Environmental Science and Engineering*, **11**(4), 15.

Tsang D. C. W., Yu I. K. M. and Xiong X. N. (2019). Novel application of biochar in stormwater harvesting. In: Biochar from Biomass and Waste, Fundamentals and Applications, Y. S. Ok, D. Tsang, N. Bolan and J. Novak (eds), Elsevier, Inc., Amsterdam, Netherlands.

Tu M. C. and Traver R. G. (2019). Optimal configuration of an underdrain delivery system for a stormwater infiltration trench. *Journal of Irrigation and Drainage Engineering*, **145** (8), 05019007.

Ulrich B. A., Loehnert M. and Higgins C. P. (2017). Improved contaminant removal in vegetated stormwater biofilters amended with biochar. *Environmental Science – Water Research and Technology*, **3**(4), 726–734.

Unice K. M. and Logan B. E. (2000). The insignificant role of hydrodynamic dispersion on bacterial transport. *Journal Environmental Engineering*, **126**(6), 491.

US EPA. (1994). Combined sewer overflow control policy. *Federal Register*, **59**(75), 18,687–18,698.

US EPA. (2000). Low Impact Development (LID), A Literature Review. Office of Water (4203), EPA-841-B-00-005, Washington, DC, USA.

US EPA. (2005). Stormwater Phase II Final Rule – An Overview. United States Environmental Protection Agency Office of Water, Washington DC, USA. Publication #EPA 833–F-00-001, 2005.

US EPA. (2016). Report to Congress Combined Sewer Overflows into the Great Lakes Basin. U.S. Environmental Protection Agency, Office of Wastewater Management, April 2016, Report EPA 833-R-16-006.

US EPA. (2019). Urban Runoff, Low Impact Development. Available from: www.epa.gov/nps/urban-runoff-low-impact-development, [Accessed 30 May 2019].

US EPA Region 5. (1984). Water Quality. United States Environmental Protection Agency Region 5, Chicago, Illinois, USA. Report #905R84126, August 1984.

Wadzuk B. M., Hickman J., Jr, M and Traver R. G. (2015). Understanding the role of evapotranspiration in bioretention, *Mesocosm study*. Journal of Sustainable Water in the Built Environment, **1**(2), 04014002.

Walker D. J. and Hurl S. (2002). The reduction of heavy metals in a stormwater wetland. *Ecological Engineering*, **18**, 407–414.

Walsh J., Wuebbles D., Hayhoe K., Kossin J., Kunkel K., Stephens G., Thorne P., Vose R., Wehner M., Willis J., Anderson D., Doney S., Feely R., Hennon P., Kharin V., Knutson T., Landerer F., Lenton T., Kennedy J. and Somerville R. (2014). Ch. 2, Our Changing Climate. Climate Change Impacts in the United States. In: The Third National Climate

Assessment, J. M. Melillo, T. C. Richmond and G. W Yohe (eds.), U.S. Global Change Research Program, pp, 19–67. doi: 10.7930/J0KW5CXT

Wilde F. D. (1994). Geochemistry and factors affecting ground-water quality at three storm-water-management sites in Maryland, Report of Investigations 59. Department of Natural Resources, Maryland Geological Survey, Annapolis, MD, USA.

Willard L. L., Wynn-Thompson T., Krometis L. H., Neher T. P. and Badgley B. D. (2017). Does it pay to be mature? Evaluation of bioretention cell performance seven years postconstruction. *Journal of Environmental Engineering*, **143**(9), 04017041.

Winston R. J., Hunt W. F., Kennedy S. G., Merriman L. S., Chandler J. and Brown D. (2013). Evaluation of floating treatment wetlands as retrofits to existing stormwater retention ponds. *Ecological Engineering*, **54**, 254–265.

Wu T. and Sansalone J. (2013a). Phosphorus equilibrium. II, comparing filter media, models, and leaching. *Journal of Environmental Engineering*, **139**(11), 1315–1324.

Wu T. and Sansalone J. (2013b). Phosphorus equilibrium. I, Impact of AlO_x media substrates and aqueous matrices. *Journal of Environmental Engineering*, **139**(11), 1325–1335.

Wu J. S., Holman R. E. and Dorney J. R. (1996). Systematic evaluation of pollutant removal by urban wet detention ponds. *Journal of Environmental Engineering*, **122**(11), 983–988.

Yan Q., James B. R. and Davis A. P. (2018). Bioretention media for enhanced permeability and phosphorus sorption from synthetic urban stormwater. *Journal of Sustainable Water in the Built Environment*, **4**(1), 04017013.

Yue C., Li L. Y. and Johnston C. (2018). Exploratory study on modification of sludge-based activated carbon for nutrient removal from stormwater runoff. *Journal of Environmental Management*, **226**, 37–45.

Chapter 5

Australian case of water sensitive city and its adaptation in China

Kefeng Zhang[1], Miao Wang[1,2], Xiuyu Li[3] and Jiukun Hu[3]

[1]Water Research Centre, School of Civil and Environmental Engineering, University of New South Wales, Sydney, Australia
[2]Department of Civil Engineering, Monash University, Australia
[3]Jiangsu Sponge City Technology Research Institute, Nantong, Jiangsu, China

5.1 INTRODUCTION

Climate change and urbanisation have been challenging our cities, particularly with higher intensity of droughts and floods. For example, the state of New South Wales, Australia experienced intense drought in the winter of 2018, with over $1 billion drought relief measures supported by the state government (NSW Government, 2018). The following spring in the same year brought the worst rain in 45 years, leading to flash flooding that caused property damage, downed trees and loss of human life (BOM, 2018). There is no doubt that with climate change our cities will experience more challenges in the way they manage their water.

Consequently, conventional water management approaches such as pipes and concrete channels are not good enough, and the design of resilient and sustainable water sensitive alternatives are called for. Water Sensitive Urban Design (WSUD) offers alternative solutions to urban water management, for cities to achieve 'water sensitive'. The term WSUD is commonly used to reflect a new paradigm in the planning and design of urban environments that is

Editors: Xiaochang C. Wang and Guangtao Fu
doi: 10.2166/9781789060768_0131

'sensitive' to the issues of water sustainability and environmental protection (Wong *et al.*, 2012). Similar concepts also exist in different parts of the world, e.g. low impact development in US, natural-based solutions in EU, Sustainable Drainage Systems (SuDS) (Fletcher *et al.*, 2015), and more recently Sponge City in China (Nguyen *et al.*, 2019).

Various types of WSUD technologies have been developed over the past decades for sustainable urban stormwater management, such as constructed wetlands, biofilters (or raingardens), swales, ponds, etc. These natural-based systems present significant opportunities for harnessing an alternative water resource for use by cities (Hatt *et al.*, 2006), while simultaneously helping to protect valuable waterways from excessive pollution (Hatt *et al.*, 2009; Malaviya & Singh, 2012; Zhang *et al.*, 2015) and ecosystem degradation (Fletcher *et al.*, 2007), improvement of microclimate (Wong *et al.*, 2013), and amenity values in urban landscapes (Polyakov *et al.*, 2015).

In this chapter we will present two case studies of water sensitive cities that use a number of WSUD technologies for stormwater management – one in Australian and one in China that was developed based on Australian experience. We focus on the technical sides of the WSUD technologies used in these case studies to showcase how these systems manage stormwater, creating a successful water sensitive environment.

5.2 CASE STUDY 1: MONASH CARPARK STORMWATER TREATMENT SYSTEMS

5.2.1 A treatment train that provides both pollution management and landscape value

The Monash Carpark stormwater management system is a typical WSUD project that embraces the treatment train concept, including multiple barriers for stormwater treatment facilities. During rain events, stormwater runs off from the top floor of a multi-level carpark (4500 m^2) and flows into two rainwater tanks for temporary storage; then the stormwater travels via an underground pipe into a sedimentation tank where the pre-treatment of stormwater occurs (e.g. the removal of coarse to medium size sediment). The pre-treated stormwater then enters three different designs of biofilters where the pollutants get further removed via a combination of physical, chemical and biological processes while filtering thought the vegetated media. The treated stormwater finally reaches a recreational pond where the water can be further reused for irrigation of surrounding green spaces.

5.2.2 Key components of the treatment train

5.2.2.1 Rainwater tank

Rainwater tanks (RWT) are widely used in Australia for stormwater retention and fit-for-purpose water supply (Coombes & Kuczera, 2003; Eroksuz & Rahman,

2010). They collect water from impervious surfaces (e.g. roofs, and in this case carpark) during rain events, and thus reduce the amount of stormwater that enters our waterways. The collected water can be used for flushing of toilets, washing of clothes or cars and irrigation of gardens and green space, significantly reducing demand on the potable water supply. Many studies have also shown that RWT can help to mitigate the impacts of urbanisation on stream, as a result of them being able to: (i) restore the natural flow regimes and (ii) restore the water quality closer to natural conditions (Burns *et al.*, 2015; Walsh *et al.*, 2005). It should be noted, however, that to maximise these benefits the RWT should be used frequently, which creates space to capture more water each time when it rains.

5.2.2.2 Sedimentation tank

Sedimentation tanks are often placed upstream of a constructed wetland or biofilters, or during development construction, to reduce sediments loads which is an important component of stormwater quality improvement (Melbourne Water, 2005). They can have two key roles in the stormwater treatment train. The primary function is to remove medium to coarse size sediment (i.e. 125 um or larger) before entering the downstream treatment system. The sedimentation tank can ensure that the vegetation in the biofilter system is not smothered by coarse sediments, and consequently the biofilter is able to target finer particulates, nutrients and other pollutants. Moreover, the sedimentation tanks can also help to attenuate peak flows by providing storage and detention, and consequently protects the vegetation in biofilters against scour during high flows.

In this case, two diversion metal plates were installed inside the sedimentation tank to create longer travel distance (as shown in Figure 5.1). The sedimentation tank is a concrete tank that is also lined to facilitate research purposes (e.g. stormwater mixing and dosing to biofilter). They can also be in different forms,

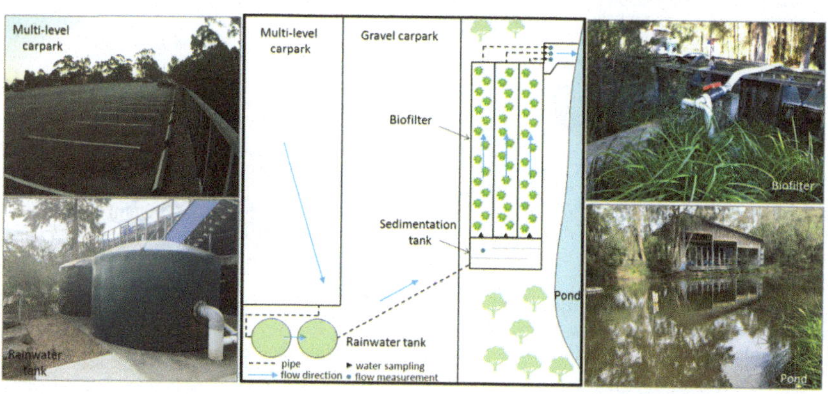

Figure 5.1 The concept of the Monash Carpark stormwater treatment train.

e.g. sedimentation ponds and detention basins that fit better in the natural environment.

5.2.2.3 Stormwater biofilters

5.2.2.3.1 A popular WSUD technology for stormwater treatment

The core treatment barrier for fine sediments and dissolved pollutants in this case are the stormwater biofilters, also named bioretention systems, or raingardens (see Figure 5.2). Biofilters are usually built as trenches or basins filled with carefully engineered, fast-draining filter material and planted with selected plants (Payne *et al.*, 2015), which together are able to remove sediments, metals, nutrients and even pathogens from stormwater. Biofilter plants can tolerate both drained and waterlogged conditions and provide pleasant greenery, so the systems are usually called raingardens. Biofilters are of flexible design and size, and are often installed in parks, along streets and within city squares. These highly popular stormwater treatment systems are therefore both functional and able to increase the amenity of urban space and provide green infrastructure for high density living.

Extensive research and development of stormwater biofiltration technologies have been undertaken in Australia and all over the world over the last decades. This work began with the idea of developing biofilters to better manage the quality and quantity of stormwater for aquatic ecosystem protection (Davis, 2007), but has since been expanded to include application for stormwater harvesting purposes and, even more recently, treatment of multiple water streams (e.g. both stormwater and wastewater) (Barron *et al.*, 2019).

Past studies of biofilter treatment performance generally report high retention of sediment, phosphorus, and heavy metals, but variable removal of nitrogen, particularly nitrate (NO_3^-). Laboratory studies have reported consistently high

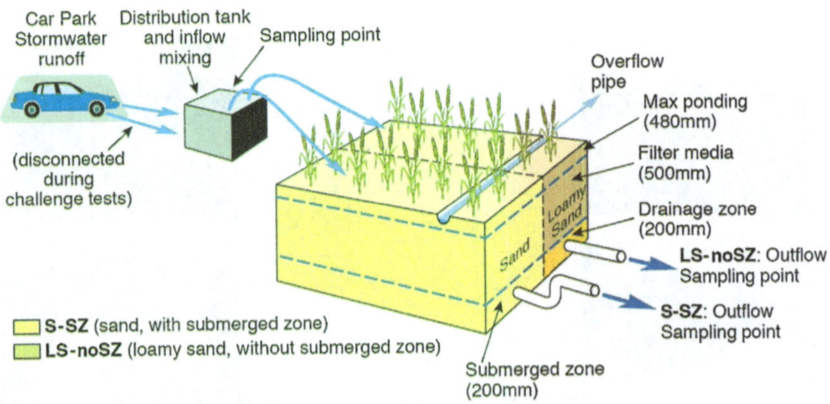

Figure 5.2 The design of the biofilter system (adapted from Zhang *et al.* (2014)).

removal efficiencies (>90%) of total suspended solids (TSS, and see, for example Hsieh and Davis, 2005a, 2005b; Hatt *et al.*, 2007, 2008). Phosphorus removal has generally been moderate to good (Davis *et al.*, 2001, 2006; Hsieh & Davis, 2005a, 2005b; Hsieh *et al.*, 2007) and heavy metal removal very good (>92%) (Davis *et al.*, 2003; Hatt *et al.*, 2008). However, NO_3^- has frequently been shown to leach out of biofiltration systems, thus often resulting in poor total nitrogen (TN) removal (Davis *et al.*, 2001, 2006; Henderson *et al.*, 2007). Further studies have shown that high nitrogen removal rates can be achieved through the careful selection of vegetation (Bratieres *et al.*, 2008; Read *et al.*, 2008b, 2010) and the inclusion of a permanently submerged, carbon-enhanced zone (Kim *et al.*, 2000; Zinger *et al.*, 2007a, 2007b).

5.2.2.3.2 A base of scientific research

This current field biofilter system has been extensively tested to validate the laboratory findings, serving as a very successful base of scientific research. Hatt *et al.* (2009) investigated the hydraulic and pollutant removal performance of this biofilter system from December 2006 to May 2007. They found that the biofilter was able to effectively reduce peak runoff flow rates by at least 80%, and the retention of inflow volume by the media could reduce runoff volume by 33% on average, with a range of 15–98% that is mostly influenced by inflow volume. Vegetation was found to be important for maintaining hydraulic capacity, because root growth and senescence countered compaction and clogging. In terms of pollutant removal, the loads TSS were effectively reduced (>90%) irrespective of the design configuration. Nutrients, however, had very variable removal subject to different designs and operating conditions. The use of filter media with low phosphate content was recommended for biofiltration systems.

The hydraulic performance of the systems was further confirmed by Zhang *et al.* (2014) who challenged the systems with extreme dry and wet operating conditions of the system in 2011 and 2012. Additionally, the authors also found these biofilters were able to remove many organic micropollutants effectively (>80%), including total petroleum hydrocarbons (TPHs), glyphosate, dibutyl phthalate (DBP), bis-(2-ethylhexyl)phthalate (DEHP), pyrene and naphthalene loads even under challenging conditions. Their performance in removing other pollutants, e.g. chloroform, atrazine and simazine, was however, variable. Table 5.1 summarises the average removal efficiencies of heavy metals during these challenge tests, indicating a relatively high and stable removal of most of the metals, especially for the ones that are commonly found in stormwater, e.g. Cu, Zn, Cd, Cd, Mn, and Cd (often >80% removal). Negative removal of some metals (e.g. Ba, Sb) were observed due to their very low inflow concentrations.

Under the same challenge conditions, Chandrasena *et al.* (2016) also found these stormwater biofilters were able to removal an average of ~1.2 log reduction of *Escherichia coli* and over 0.8 log reduction of *Campylobacter*. The results show that the biofilters did not meet the Australian stormwater harvesting guidelines

Table 5.1 Average removal efficiency of metals (%) during the challenge tests.

	Al	Sb	As	Ba	Be	B	Cd	Cr	Co	Cu	Fe	Pb
LS-noSZ (old water)	85.7	5.0	80.0	–8.3	–	67.7	94.0	91.3	82.3	86.3	90.3	97.3
LS-noSZ (new water)	87.7	30.0	70.7	–23.3	–	55.3	93.7	82.7	75.7	81.0	86.3	96.0
S-SZ (old water)	91.3	-20.0	74.0	35.3	–	67.3	97.7	95.3	61.3	85.7	95.0	98.7
S-SZ (new water)	91.0	5.0	59.3	1.7	–	44.0	95.3	83.7	63.3	79.3	88.3	96.0
	Mn	**Hg**	**Mo**	**Ni**	**Se**	**Ag**	**Sr**	**Tl**	**Sn**	**Ti**	**V**	**Zn**
LS-noSZ (old water)	99.3	50.0	33.3	92.7	–	–	35.3	–	–	77.0	71.3	92.0
LS-noSZ (new water)	99.0	50.0	25.0	88.0	–	–	24.3	–	–	80.0	69.0	91.7
S-SZ (old water)	98.3	–	33.3	84.0	–	–	51.0	–	–	91.0	81.7	98.0
S-SZ (new water)	98.3	50.0	0.0	80.0	–	–	29.0	–	–	91.7	77.3	97.7

'–'undetected in all three tests.

for irrigating sports fields and golf course under extreme conditions, and thus further disinfection is needed. However, under normal conditions, the guidelines could be met.

5.2.2.3.3 Recreational stormwater ponds

In this case, the stormwater ponds are placed after the biofilter as a reservoir of the treated water. It not only acts as temporary storage for reuse, but also provides additional water quality improvement (through the emergent vegetation along the borders) and serves as a natural habitat for wildlife (enhancing urban biodiversity) and provides creational and aesthetic values.

5.3 CASE STUDY 2: HOW THIS WAS APPLIED OUTSIDE OF AUSTRALIA

Stormwater treatment technologies (such as biofilters) used in water sensitive cities are often natural-based and direct transfer of these technologies to other climates are not recommended. Therefore, tailoring of the technology to local climate conditions (either outside or inside Australia) is paramount to ensure its performance. In this section, a Water Sensitive City case study, located in the EastHigh Industrial Park, Jurong, Jiangsu Province, China, is introduced. It is an outcome of the project '*Development of Biofiltration Technology for Stormwater Management in Jiangsu Province, China*' by Jiangsu EastHigh Environmental Holdings, Monash University and UNSW Sydney, with funding from the Victoria-Jiangsu Innovation and Technology R&D Fund, as well as the Sino-Australian Centre on Sponge City. This section introduces how the Australian experiences of Water Sensitive Cities and technologies were adopted in China for a successful outcome.

5.3.1 Introduction of EastHigh stormwater treatment systems

This case study also incorporates a treatment train concept as the Monash carpark case. Stormwater is firstly collected from surrounding catchments of about 3000 m^2, including roads and roofs, through the upstream drainage pipe which is connected to the biofilter (Figure 5.3). A mini-sedimentation tank is constructed at the inflow pit for pre-treatment, and the stormwater is then discharged into an 87 m^2 biofilter system which was designed according to the Australian knowledge but with local tailoring research to ensure its performance. The treated stormwater by biofilter is then stored in an 11 m^3 collection pit where the water is used for recycling.

The construction of the project was completed in June 2018. Monitoring of the main treatment system – stormwater biofilters – was also undertaken to validate the hydraulic and treatment performance from November 2018 when the plants had achieved maturity. A picture of the biofilter system is shown in Figure 5.4, taken in September 2018, three months after the completion of the construction.

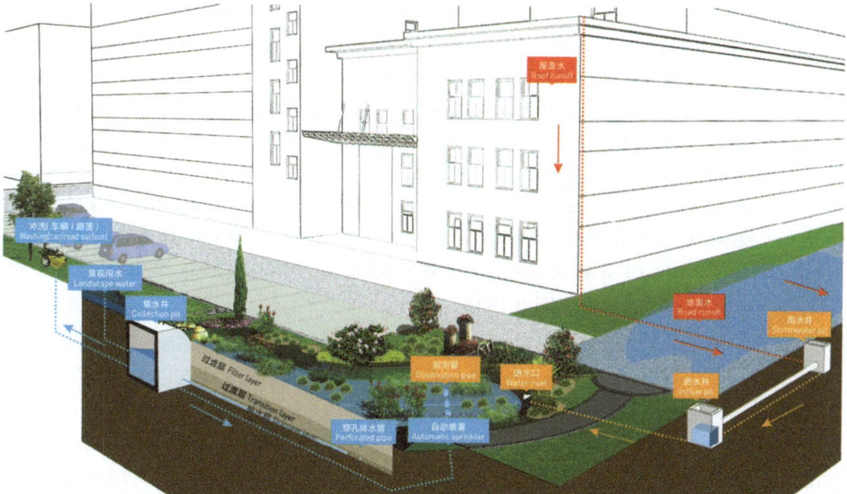

Figure 5.3 The concept of the EastHigh stormwater treatment system.

Figure 5.4 Picture of the biofilter system taken in September 2018 immediately after a rainfall event.

5.3.2 Landscaping

In the design phase of the project, apart from the engineering design to ensure the treatment performance, the landscape designing was also prioritised. In addition to the major treatment system (i.e. biofilter), many local Chinese elements were included, e.g. a small timber arch bridge, a recreational water pond, a small creek and an artificial fountain (Figure 5.5). All the water needed by these elements comes from the treated stormwater by the biofilter; there are lotus and fishes in the recreational pond to enhance the amenity and liveability around the site. In addition, the stormwater biofilter surface is designed as 'leaf' shape with water channels that distribute inflow stormwater to the whole area of the biofilter, i.e. like delivering water to all parts of the leaf.

As such, the whole case study not only serves as a source control facility to manage stormwater pollution and volume, but also offers a liveable place to spend leisure time or a nice short break during or after a busy workday for the people working in the industrial park.

5.3.3 Local tailoring research

Stormwater biofilters are nature-based technologies that use local soils and plants to effectively detain, retain and treat urban runoff, and therefore local research was conducted to tailor this technology for local Jiangsu climate conditions, including plant species and soil media (introduced in the next section).

Figure 5.5 Landscape design concept (top), the recreational pond (waterscape) that receives treated stormwater (bottom left), Arch timber bridge between the pond and biofilter (bottom right, in winter).

5.3.4 The main parts of the biofilter

The main part of the biofilter is a functional facility for stormwater purification. There are several parts needed to be taken into consideration in the design process. The design includes inflow, impermeable layer, media, plants and outflow. All of the key factors of the design follow the criteria in 'Adoption guidelines for stormwater biofiltration systems' (Payne *et al.*, 2015).

5.3.4.1 Inflow pit

An inflow pit is designed to divert the stormwater from the drain (pit A) to the biofilter (see Figure 5.6). It is designed with three chambers: (a) acts like a mini sedimentation tank where the stormwater gets pre-treated through sedimentation,

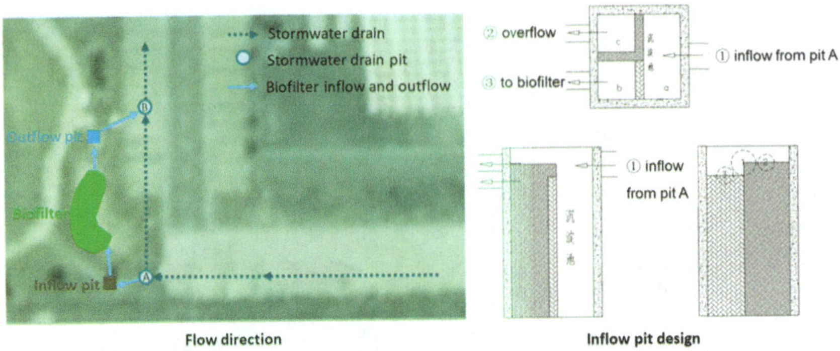

Figure 5.6 Stormwater flow direction for the biofilter (left) and inflow (sediment) pit design (right).

(b) diverts the stormwater into the biofilter system, and (c) is designed to overflow (back to the drain) during a large storm.

5.3.4.2 Media

The filter media is critical to biofilter functioning as reported in many studies (Hsieh & Davis, 2005a; Hunt *et al.*, 2006). For the EastHigh biofilter, there are three layers of media: 500 mm filter layer (fine well-graded sand), 250 mm transition layer (coarse sand), and 150 mm drainage layer (gravel). The submerged zone is set as the drainage and transition layer, i.e. 400 mm. The media is carefully selected based on laboratory experiments by testing a total of about 20 types of media that are locally sourced (Figure 5.7). Three most important criteria according to the Australian Biofiltration Adoption guidelines are (Payne *et al.*, 2015):

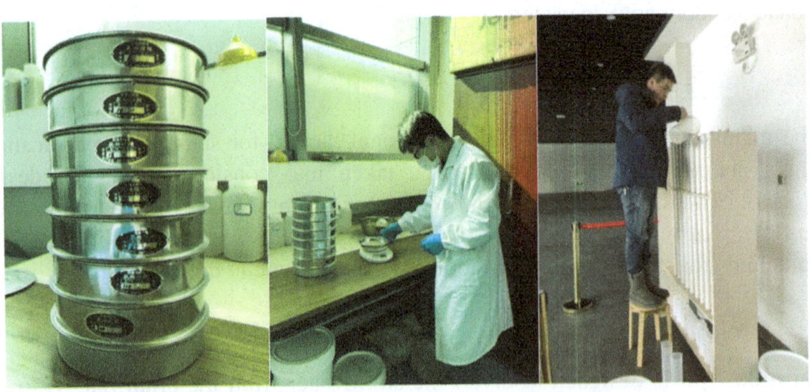

Figure 5.7 Laboratory experiments for selection of media (left – particle size distribution test; right – infiltration tests).

(a) the media is well-graded with their particle size distributions following the adoption guideline (Figure 5.7);
(b) low nutrient content, with total nitrogen (TN) <1000 mg/kg and available phosphate <80 mg/kg, to avoid leaching out of the pollution;
(c) appropriate hydraulic conductivity of between 200 and 500 mm/hr, which is typical for Jiangsu sub-tropical climate conditions (Figure 5.7).

5.3.4.3 Plants

Plants are essential components of functioning biofilters, however biofilter performance is largely impacted by plant species, particularly for nitrogen removal and species tolerance of hydrological conditions (Bratieres *et al.*, 2008; Read *et al.*, 2008a). Thus, it is most important to select effective plants that are locally available and growing Australian native plants is not recommended. Therefore, large laboratory column experiments were set-up to test the performance of a total of 20 Jiangsu local plants in removing pollutants over one year in a greenhouse laboratory, as shown in Figure 5.8.

On the basis of the laboratory experiments results, some ‘effective plants’ were selected to be included in the field biofilter. In addition, ‘effective plants’ often do not have high levels of aesthetic value, e.g. the most effective plant in Australia is *Carexappressa*, which is a sedge that is not regarded as very attractive (Dobbie, 2016). As such, a variety of plants is required to both satisfy the treatment objectives and successful integration of biofilters into the urban landscape, so that the biofilter systems can have higher community acceptance. For this case study, a wide range of effective plants as well as ornamental species are both included. The plants distribution in the biofilter can be seen in Figure 5.9.

Figure 5.8 Experimental set-up of laboratory long-term plant testing (photo taken by Dr Emily Payne, Monash University, with permission for use).

Figure 5.9 Plant distribution in the biofilter. The species highlighted in blue are ornamental species, while the rest are effective plants for pollution reduction.

5.3.4.4 Outflow

In order to reuse the purified stormwater in the outflow pit, several perforated pipes (100 mm diameter) were installed at the bottom of the biofilter for water collection. In addition, to ensure the submerge zone of 400 mm in the biofilter, the outlet of the perforated pipes was set as 400 mm above the bottom of the biofilter. At the downstream of the biofilter, an outflow pit is constructed to store the treated stormwater, where there is also an overflow pipe to allow the water overflow to the downstream stormwater drain. The treated stormwater in the pit is now reused for different purposes, including waterscape, irrigation for the surrounding plants and also for car washing. An auto-dripping irrigation system is also installed for watering surrounding plants.

5.3.4.5 Monitoring

To validate the performance of the stormwater biofilter, the system is equipped with comprehensive monitoring facilities to monitor the quantity and quality of inflow and outflow (Figure 5.10). A rainfall gauge is installed on site to record the rainfall intensity. For the inflow quantity, a flowmeter is installed between the sediment pit and the inlet of the biofilter which is able to obtain the data of water volume and velocity. For the outflow, a water level senor and a V-notch is particularly designed to obtain the outflow rates. A data logger is used for

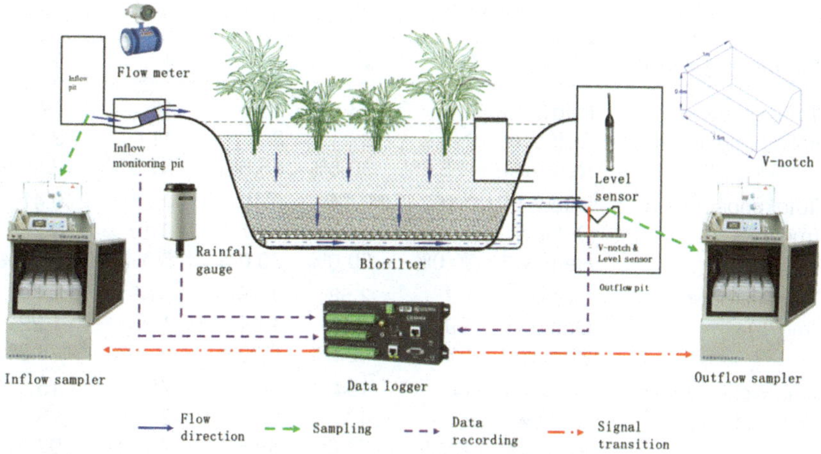

Figure 5.10 Monitoring concept of EastHigh biofilter.

recording all the data from different components, also it is linked to a controller to ensure different components are able to interact with each other successfully. Two auto-samplers are used for taking samples from the inflow pipe and outflow V-notch, triggered by the inflow and outflow rates through the data logger. While this not a standard design for general stormwater treatment projects, it is needed to ensure that the tailored technology is successful.

5.3.4.6 Outflow pollutant concentration

During the monitoring process, five rainfall events were monitored until May 2019. Both the inflow and outflow of the samples were sent for testing, and the outflow of the biofilter indicated a relatively low concentration, the comparison of the inflow and outflow pollutant concentration is shown in Table 5.2.

It can be seen that the reduction rates of different pollutants are varied. The main reason for a low reduction rate is the low inflow pollutant concentration (particularly for TSS and TP). The industrial park is located in a rural area, and the catchment is relatively clean, therefore the road and roof runoff does not contain too many pollutants. The concentrations of nitrogen, COD and BOD5 are, however, within the typical concentration range of stormwater, and the biofilter showed moderate treatment efficiency of these pollutants (~50% reduction). To compare with the 'Environmental quality standards for surface water' (GB 3838-2002) in China, most of the pollutant concentration meets the criteria of category II, which means the biofilter can be considered as a useful stormwater purification facility.

Table 5.2 Pollutant concentration for inflow and outflow and their reduction rate.

Monitoring events		1	2	3	4	5
Total suspend solids (mg/L)	Inflow	7	13	18	16	12
	Outflow	9	9	8	11	9
	Removal	−28.6%	30.8%	55.6%	31.3%	25.0%
Total phosphorus (mg/L)	Inflow	0.15	0.14	0.10	0.20	0.47
	Outflow	0.12	0.07	0.07	0.19	0.29
	Removal	20.0%	50.0%	30.0%	5.0%	38.3%
Total nitrogen (mg/L)	Inflow	1.2	2.58	4.48	3.68	1.7
	Outflow	0.97	0.96	1.44	1.36	1.22
	Removal	19.2%	62.8%	67.9%	63.0%	28.2%
Chemical oxidation demand (mg/L)	Inflow	14	104	67	72	15
	Outflow	8	48	13	21	15
	Removal	42.9%	53.8%	80.6%	70.8%	0.0%
Biological oxidation demand(mg/L)	Inflow	4.2	29	20.4	21.4	4.6
	Outflow	2.3	12.1	3.8	6.5	4.6
	Removal	45.2%	58.3%	81.4%	69.6%	0.0%
Faecal coliforms (unit/L)	Inflow	–	60	2.3×10^6	1.4×10^4	120
	Outflow	–	50	5.3×10^3	30	20
	Removal	–	16.7%	99.8%	99.8%	83.3%

5.4 SUMMARY

This chapter introduced two typical cases of water sensitive cities – one in Australia and the other in China that was designed based on Australian experience with local tailoring research. Both cases use a range of WSUD technologies (from sedimentation tanks, to biofilters and ponds) to form a proper treatment train that ensures the effective management of stormwater in both runoff control and pollution mitigation. Additionally, as successful water sensitive cases, they can also provide amenity values to the environment, with other multiple benefits that can be further investigated, e.g. provision of ecosystem services and mitigation of microclimate. All these collectively can contribute to the transition from current cities into water sensitive cities.

ACKNOWLEDGEMENT

The authors would like to give special thanks to Prof. Ana Deletic (UNSW, Sydney) and A/Prof. David McCarthy (Monash University) for their support, especially with regards to the collection of relevant data at the Monash Car Park case study. The China stormwater treatment system was also supported by Victoria – Jiangsu Innovation and Technology R&D Funding (Vic-Jiangsu TECH).

REFERENCES

Barron N. J., Deletic A., Jung J., Fowdar H., Chen Y. and Hatt B. E. (2019). Dual-mode stormwater-greywater biofilters: the impact of alternating water sources on treatment performance. *Water Research*, **159**, 521–537.

BOM (2018). Monthly Weather Review Australia (August–December 2018). Bureau of Meterology, Melbourne, Australia.

Bratieres K., Fletcher T. D., Deletic A. and Zinger Y. (2008). Nutrient and sediment removal by stormwater biofilters: a large-scale design optimisation study. *Water Research*, **42** (14), 3930–3940.

Burns M. J., Fletcher T. D., Duncan H. P., Hatt B. E., Ladson A. R. and Walsh C. J. (2015). The performance of rainwater tanks for stormwater retention and water supply at the household scale: an empirical study. *Hydrological Processes*, **29**, 152–160.

Chandrasena G. I., Deletic A. and McCarthy D. T. (2016). Biofiltration for stormwater harvesting: comparison of *Campylobacter* spp. and *Escherichia coli* removal under normal and challenging operational conditions. *Journal of Hydrology*, **537**, 248–259.

Coombes P. J. and Kuczera G. (2003). Analysis of the performance of rainwater tanks in Australian capital cities [online]. In: M. J. Boyd, J. E. Ball, M. K. Babister and J. Green (eds). 28th International Hydrology and Water Resources Symposium: About Water; Symposium Proceedings. A.C.T. Institution of Engineers, Barton, Australia, pp. 2.235–2.242.

Davis A. P. (2007). Field performance of bioretention: water quality. *Environmental Engineering Science*, **24**(8), 1048–1064.

Davis A. P., Shokouhian M., Sharma H. and Minami C. (2001). Laboratory study of biological retention for urban stormwater management. *Water Environment Research*, **73**(1), 5–14.

Davis A. P., Shokouhian M., Sharma H., Minami C. and Winogradoff D. (2003). Water quality improvement through bioretention: lead, copper and zinc removal. *Water Environment Research*, **75**(1), 73–82.

Davis A. P., Shokouhian M., Sharma H. and Minami C. (2006). Water quality improvement through bioretention media: Nitrogen and phosphorus removal. *Water Environment Research*, **78**(3), 284–293.

Dobbie M. F. (2016). Designing Raingardens for Community Acceptance. Cooperative Research Centre for Water Sensitive Cities, Melbourne, Australia.

Eroksuz E. and Rahman A. (2010). Rainwater tanks in multi-unit buildings: a case study for three Australian cities. Resources, Conservation and Recycling, **54**(12), 1449–1452.

Fletcher T., Mitchell V., Deletic A., Ladson T. and Seven A. (2007). Is stormwater harvesting beneficial to urban waterway environmental flows? *Water Science and Technology*, **55** (4), 265–272.

Fletcher T. D., Shuster W., Hunt W. F., Ashley R., Butler D., Arthur S., Trowsdale S., Barraud S., Semadeni-Davies A., Bertrand-Krajewski J. L., Mikkelsen P. S., Rivard G., Uhl M., Dagenais D. and Viklander M. (2015). SUDS, LID, BMPs, WSUD and more – the evolution and application of terminology surrounding urban drainage. *Urban Water Journal*, **12**(7), 525–542.

Hatt B. E., Deletic A. and Fletcher T. D. (2006). Integrated treatment and recycling of stormwater: a review of Australian practice. *Journal of Environmental Management*, **79**(1), 102–113.

Hatt B. E., Fletcher T. D. and Deletic A. (2007). Stormwater reuse: designing biofiltration systems for reliable treatment. *Water Science and Technology*, **55**(4), 201–209.

Hatt B. E., Fletcher T. D. and Deletic A. (2008). Hydraulic and pollutant removal performance of fine media stormwater filtration systems. *Environmental Science and Technology*, **42**(7), 2535–2541.

Hatt B. E., Fletcher T. D. and Deletic A. (2009). Hydrologic and pollutant removal performance of stormwater biofiltration systems at the field scale. *Journal of Hydrology*, **365**(3–4), 310–321.

Henderson C., Greenway M. and Phillips I. (2007). Removal of dissolved nitrogen, phosphorus and carbon from stormwater by biofiltration mesocosms. *Water Science and Technology*, **55**(4), 183–191.

Hsieh C. H. and Davis A. P. (2005a). Evaluation and optimization of bioretention media for treatment of urban storm water runoff. *Journal of Environmental Engineering –Asce*, **131**(11), 1521–1531.

Hsieh C. H. and Davis A. P. (2005b). Multiple-event study of bioretention for treatment of urban storm water runoff. *Water Science and Technology*, **51**(3–4), 177–181.

Hsieh C. H., Davis A. P. and Needelman B. A. (2007). Bioretention column studies of phosphorus removal from urban stormwater runoff. *Water Environment Research*, **79** (2), 177–184.

Hunt W. F., Jarrett A. R., Smith J. T. and Sharkey L. J. (2006). Evaluating bioretention hydrology and nutrient removal at three field sites in North Carolina. *Journal of Irrigation and Drainage Engineering*, **132**, 600–608.

Kim H., Seagren E. A. and Davis A. P. (2000). Engineered bioretention for removal of nitrate from stormwater runoff. WEFTEC 2000, Annual Conference & Exposition on Water Quality and Wastewater Treatment, 73rd, Anaheim, CA, United States, Oct. 14–18, 2000, pp. 1349–1358.

Malaviya P. and Singh A. (2012). Constructed Wetlands for Management of Urban Stormwater Runoff. *Critical Reviews in Environmental Science and Technolog*, **42** (20), 2153–2214.

Melbourne Water (2005). WSUD Engineering Procedures: Stormwater. CSIRO Publishing, Clayton, South Australia.

Nguyen T. T., Ngo H. H., Guo W. S., Wang X. C. C., Ren N. Q., Li G. B., Ding J. and Liang H. (2019). Implementation of a specific urban water management – Sponge City. *Science of the Total Environment*, **652**, 147–162.

NSW Government (2018). More than $1 billion in Support to Help Farmers through Worsening Drought. NSW Government, Australia.

Payne E. G. I., Hatt B. E., Deletic A., Dobbie M. F., McCarthy D. T. and Chandrasena G. I. (2015). Adoption Guidelines for Stormwater Biofilter Systems (Version 2). Cooperative Research Centre for Water Sensitive Cities, Melbourne, Australia.

Polyakov M., Iftekhar S., Zhang F. and Fogarty J. (2015). The amenity value of water sensitive urban infrastructure: a case study on rain gardens. Proceedings of 9th IWA Symposium on Systems Analysis and Integrated Assessment, 14–17 June 2015, Gold Coast, Australia.

Read J., Wevill T., Fletcher T. and Deletic A. (2008a). Variation among plant species in pollutant removal from stormwater in biofiltration systems. *Water Research*, **42**(4–5), 893–902.

Read J., Wevill T., Fletcher T. D. and Deletic A. (2008b). Variation among plant species in pollutant removal from stormwater in biofiltration systems. *Water Research*, **42**(4–5), 893–902.

Read J., Fletcher T. D., Wevill P. and Deletic A. (2010). Plant traits that enhance pollutant removal from stormwater in biofiltration systems. *International Journal of Phytoremediation*, **12**(1), 34–53.

Walsh C. J., Roy A. H., Feminella J. W., Cottingham P. D., Groffman P. M. and Morgan R. P. (2005). The urban stream syndrome: current knowledge and the search for a cure. *Journal of the North American Benthological Society*, **24**(3), 706–723.

Wong T. H. F., Allen R., Brown R. R., Deletic A., Fletcher T. D., Gangadharan L., Gernjak W., Jakob C., O'Loan T., Reeder M. and Tapper N. (2012). Stormwater Management in a Water Sensitive City: Blueprint 2012. The Centre for Water Sensitive Cities, Melbourne.

Wong T. H. F., Allen R., Brown R. R., Deletic A., Gangadharan L., Gernjak W., Jakob C., Johnstone P., Reeder M., Tapper N., Vietz G. and Walsh C. (2013). Blueprint 2013 – Stormwater Management in a Water Sensitive City. CRC for Water Sensitive Cities, Melbourne.

Zhang K., Randelovic A., Page D., McCarthy D. T. and Deletic A. (2014). The validation of stormwater biofilters for micropollutant removal using in situ challenge tests. *Ecological Engineering*, **67**, 1–10.

Zhang K., Deletic A., Page D. and McCarthy D. T. (2015). Surrogates for herbicide removal in stormwater biofilters. *Water Research*, **81**(0), 64–71.

Zinger Y., Fletcher T.D. and Deletic A. (2007a). The effect of various intermittent dry-wet cycles on nitrogen removal capacity in biofilters systems. In: P. Coombes and J. Dahlenburg (eds). Proceedings of the Thirteenth International Conference on Rain Water Catchment Systems. Sydney, Australia.

Zinger Y., Fletcher T. D., Deletic A., Blecken G. T. and Viklander M. (2007b). Optimisation of the nitrogen retention capacity of stormwater biofiltration systems, Lyon, France. Paper Presented at the Novatech Conference, Lyon, France, 24–28 June, 2007.

Part II

New Paradigm of Systems Thinking and Technology Advances

Chapter 6

Water cycle management for building water-wise cities

Xiaochang C. Wang[1] *and Li Luo*[2]

[1]*State International S&T Cooperation Centre for Urban Alternative Water Resources Development, Xi'an University of Architecture and Technology, Xi'an, China*

[2]*School of Environmental and Municipal Engineering, Xi'an University of Architecture and Technology, Xi'an, China*

6.1 INTRODUCTION

Water is a chemical substance with a simple but stable molecular structure of H_2O. It cannot be generated or eliminated under normal conditions, so the total mass of water in the whole world is constant. Water itself is transparent, tasteless, odorless, and nearly colorless, but can dissolve various inorganic and organic solutes, and carry fine particulate solids as it flows. All these foreign substances the water carries are called 'impurities' which can enter the water mass and conversely be removed through various natural or artificial actions. Water is essential to human and all living organisms, but any consumption and utilization of water cannot result in any true loss of the water mass. This means that after water use, the water molecules are still there and the only alteration is in their quality, location of existence, and existing state (liquid, gaseous or even solid forms). However, the usability of water may not merely depend on its quantity, but more importantly its quality. Contaminated water cannot be directly supplied for most purposes of water use. In this sense, the problem of water shortage we

Editors: Xiaochang C. Wang and Guangtao Fu
doi: 10.2166/9781789060768_0151

are discussing in several chapters of this book may not only mean a shortage of water mass in most cases, but the shortage of water with quality to meet the requirement of water use.

Fortunately, water is always moving or being moved in different ways and scales. Naturally, the global scale water movement is driven by solar energy and gravitational energy, and brings water mass back to every water body where water can stay. The water movement is also associated with many processes of quality conversion which results in an almost stable water quality in each waterbody under natural conditions.

Human beings are utilizing the abovementioned natural water cycle to obtain source water. To facilitate water use for various purposes, artificial measures are taken for water conveyance and additional quality conversion by consuming energy. The water to be used is from nature and after use it is returned to nature, thus forming various scales of additional artificial water cycles.

Let us come back to the topic of building water-wise cities. We understand that cites are basin connected (the second level action in the IWA Principles for Water Wise Cities), so the water sources provided to us are sustained by the natural water cycle. On the other hand, the water-related infrastructures built for cities are artificial patches to the natural water cycle. In this viewpoint, water cycle management can be proposed as the core concept for formulating strategical schemes for building water-wise cities.

6.2 THINGS TO LEARN FROM THE NATURAL HYDROLOGICAL CYCLE

6.2.1 Natural hydrological cycle

The most important reason for the survival of human beings on the planet Earth through several generations is the presence of plentiful water to support life. With a total surface area of about $5.1 \times 10^8\,km^2$, about 71% of which is ocean surface and 29% land surface, the total volume of water on Earth amounts to about $1.39 \times 10^9\,km^3$ (Agarwal *et al.*, 2019; Nazaroff & Alvarez-Cohen, 2000). If all of Earth's crustal surface were at the same elevation as a smooth sphere, then the resulting water depth covering the Earth would be over 2700 m. This huge amount of water is distributed in a hydrosphere, namely the combined mass of water found on, under, and above the surface of the Earth, including water in liquid and frozen forms in groundwater, oceans, lakes, and streams. Saltwater accounts for almost 97.5% of this amount, whereas freshwater accounts for only about 2.5%. Of the freshwater, 68.9% is in the form of ice and permanent snow cover in the Arctic, Antarctic, and mountain glaciers and 30.8% is in the form of fresh groundwater. Only 0.3% of the freshwater on Earth is in readily accessible lakes, reservoirs, and river systems (Chawla *et al.*, 2020). Table 6.1 shows the distribution of fresh and saline water in the Earth's hydrosphere.

Table 6.1 Distribution of fresh and saline water in the Earth's hydrosphere.

Site	Volume (km^3)	Percent (%)
Ocean	1,350,000,000	97.29
Polar ice caps and glaciers	29,000,000	2.09
Groundwater	8,300,000	0.6
Freshwater lakes	125,000	0.009
Saline lakes and inland seas	104,000	0.0075
Soil and subsoil water	67,000	0.0048
Atmospheric water vapor	13,000	0.00094
Living biomass	3000	0.00022
Stream channels	1000	0.00007
Total	1,390,000,000	100

Source: Nazaroff and Alvarez-Cohen 2000.

Water in the hydrosphere is continuously transformed through various phases in a process called the water cycle or hydrological cycle at global or watershed scales.

6.2.1.1 Global hydrological cycle

Figure 6.1 depicts the major processes in the hydrological cycle on a global scale. As oceans and seas are covering 71% of the Earth's surface, the solar radiation heats the

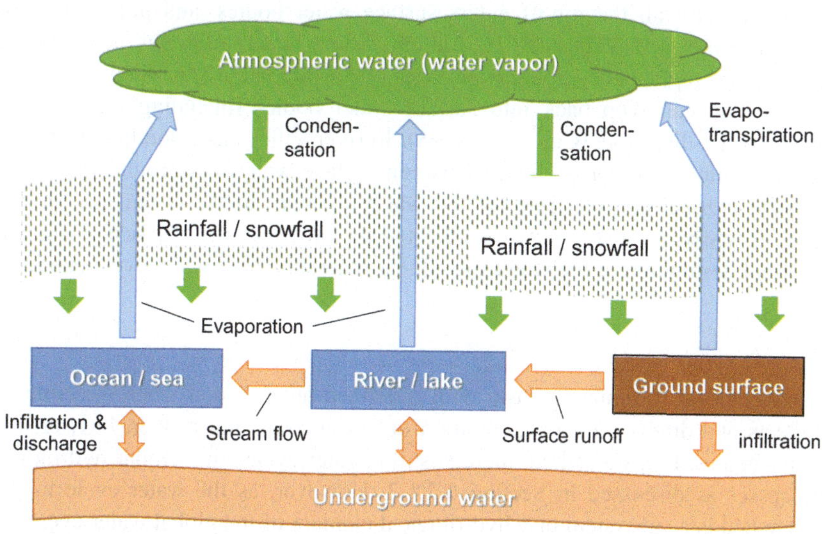

Figure 6.1 Major processes of water movement and transport in the hydrological cycle at the global scale (figure by authors).

seawater and forces it to evaporate as water vapor into the air. A similar process also occurs on other water surfaces, such as rivers, lakes, and other surface water bodies. Water may also be transpired by plants and evaporated from the soil through a process called evapotranspiration. However, the majority of the water vapor from the Earth's surface into the atmosphere is due to seawater evaporation.

As the water molecule has a smaller molecular mass and is less dense than the major components of the atmosphere, such as nitrogen and oxygen gases, the evaporated water vapor is driven by buoyancy and goes upward to a height of 600–1000 m above ground in a humid area or up to 3000 m in a dry area. As altitude increases, air pressure decreases, and the temperature drops. The lower temperature causes water vapor to condense into tiny liquid water droplets which are heavier than the air, and fall unless supported by an updraft. A huge concentration of these droplets over a large space up in the atmosphere becomes visible as a cloud.

Atmospheric circulation moves water vapor around the globe, and cloud particles collide, grow, and fall out of the upper atmospheric layers as precipitation falling to the Earth as rain and/or snow. Most of the precipitation may fall back into the ocean, thus resulting in a shorter cycle of 'seawater evaporation – condensation – precipitation back to the sea.' Moreover, a part of the precipitation may fall onto land, where the water flows over the ground as surface runoff. A portion of the runoff may enter rivers in valleys, with streamflow moving water towards the ocean, resulting in a longer cycle of 'evaporation – condensation – precipitation – surface runoff – streamflow back to the sea.' Runoff may be stored as freshwater in some surface water bodies with ample storage volumes, such as lakes and reservoirs. Not all the runoff enters surface water bodies, and much of it may soak into the ground through infiltration. Some water infiltrates deep into the ground and replenishes groundwater aquifers, while some stay close to the land surface and can seep back into surface water bodies (including the ocean) as groundwater discharge (Cui *et al.*, 2018). In river valleys and floodplains, there is often continuous water exchange between surface water and groundwater in the hyporheic zone, and the time for the water to return to the ocean may be very long. Anyway, no matter how complicated the processes are, the ocean is always the final destination of the water movement where the next water cycle continues.

6.2.1.2 Hydrological cycle of a watershed

A watershed is a drainage basin with a defined area of land where precipitation collects and drains off into a common outlet, such as into a river, bay, or other water bodies. In a watershed, there is still a water cycle with similar hydrological processes as discussed in Section 6.2.1.2. However, as the water cycle in each watershed is a subsystem or a hydrological element of the global water cycle, it is not completely independent of the global water movement. First, the amount of precipitation to the watershed may not be equivalent to the amount of evaporation and evapotranspiration from the watershed area because of the atmospheric movement of water vapor in the broader area. Second, the watershed has an outlet where surplus water flows out of the watershed area (Hester & Little, 2013). Due to these two reasons, the hydrological cycle of a watershed is not a closed water loop but follows a relationship as below:

$$\text{Precipitation} - (\text{Evaporation} + \text{Evapotranspiration}) - \text{Outflow} \\ = \text{Increase in storage} \tag{6.1}$$

where 'Precipitation' is the precipitated water volume within the watershed area, 'Evaporation + Evapotranspiration' is the amount of water evaporated from the watershed area, 'Outflow' is the amount of water that drains off from the outlet of the watershed, and 'Increase in storage' includes the amount of water added to surface water bodies, groundwater aquifers, soil moisture and so on within the watershed area in a given time.

Equation (6.1) is a mass balance relation of water for a watershed. As the precipitation may be affected by weather conditions, the amount of water stored in the watershed may not always be secured. This can explain why droughts occasionally occur in many watersheds worldwide. On the other hand, as the maximum storage volume of water bodies (e.g. lakes and reservoirs) is limited and so is the capacity of outflow through the outlet of the watershed, if the amount of precipitation is abnormally large, the precipitated water may not drain off smoothly and surface flooding may inevitably occur.

6.2.2 Functions of the hydrological cycle

The hydrological cycle has two important functions from the viewpoint of human needs for water resources. One is to secure the quantity of renewable water resources on the earth's surface under a dynamic equilibrium condition, and another is to secure the quality of the water resources.

6.2.2.1 Water quantity secured by the hydrological cycle

It is estimated that the average annual evaporation (including evapotranspiration) amounts to 577,000 km^3, of which 503,000 km^3 evaporates from the oceans and the remaining 74,000 km^3 is from the land by evapotranspiration. As water molecules cannot be naturally created nor destroyed but can only possibly transform from one state to another, the same amount of water (577,000 km^3) returns to earth as a result of precipitation. However, the amount of water that directly precipitates over the oceans is 458,000 km^3 while that over the land is 119,000 km^3. By a simple calculation of the difference between the precipitated amount and the evaporated amount on land, it can be estimated that the net amount of water annually transported from the oceans to the land is 45,000 km^3. This is the principal source of renewable freshwater for all organisms on earth.

Of course, an equivalent amount of water eventually discharges to the oceans through surface flows (rivers) and subsurface flows (groundwater outflow) to maintain the mass balance of water circulation in the global hydrological cycle, as discussed in Section 6.2.1.2.

According to the Food and Agriculture Organization of the United Nations (FAO, 2003), the total water resources of the world are estimated at 43,764 km^3 per year, which is roughly the amount of water annually transported from the oceans to the land (45,000 km^3 as calculated above). This amount of water is called the annual renewable water resource, which, by definition, is the water replenishable to replace the portion of water depleted by usage and consumption for various purposes. If the current world population of 7.7 billion is considered, then the average per capita water resource is about 5700 m^3 per year. However, water resources are unevenly distributed throughout the world. America (including North, South, and Central) possesses about 45.3% of the world water resources, but its population accounts for only 13.5% of the world population. In contrast, Asia has the largest population (about 60% of the world population) while it possesses only about 28.5% of the world's water resources.

As the annual precipitation amount significantly influences the amount of annual renewable water resources, there is a fluctuation in the amount of world water resources. We do not have reliable worldwide data to show such kinds of fluctuation but can take China as an example to see how the amount of water resources varies in the territory of this large country. Figure 6.2 is depicted using the data from the National Bureau of Statistics of China (NBSC, 2020) from

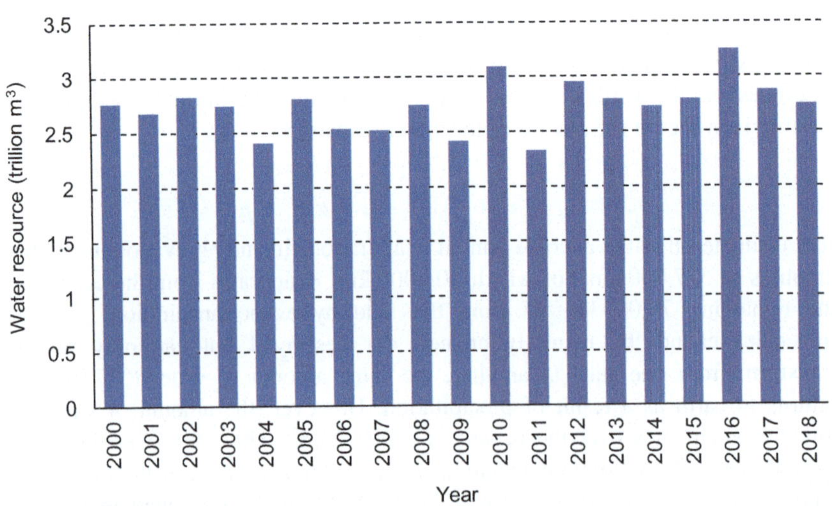

Figure 6.2 Annual water resource in China from 2000 to 2018 (data source: NBSC, 2020).

2000 to 2018. Historical data show that the long-term average of total water resources in China amounts to 2829.6 km^3 (2.8296 trillion m^3) per year. However, the short-term average of total water resources in the past 19 years is calculated as 2.7380 trillion m^3 per year, indicating an apparent decline in water resources compared with the long-term average. Very low amounts of total water resources occurred in 2004, 2009, and 2011, at about 2.41, 2.42, and 2.33 trillion m^3, respectively, which are 14.8, 14.5, and 17.7% below the long-term average. These years are known as extremely dry years in the past two decades. Nonetheless, in wet years, such as 2010 and 2016, the amount of total water resources shows increases over the long-term average.

The renewable water resources for the whole world and any country or region include surface water resources (from rivers, lakes, reservoirs, etc.) and groundwater resources (from exploitable groundwater aquifers). Water in all these water bodies is replenished through the processes in the hydrological cycle as discussed in Section 6.2.1.

6.2.2.2 Water quality secured by the hydrological cycle

The turnover of water from the oceans to land through evaporation and precipitation is accompanied by the conversion of saline to freshwater, which is the most important process of 'water treatment' for removing salts and all impurities to obtain pure H_2O molecules. Driven by solar radiation, the cleanest natural energy, the water purified amounts to 45,000 km^3 per year, or about 123.3 km^3 per day, on average, and provides high-quality water to replenish surface and subsurface water bodies. This is the primary step of water purification through the hydrological cycle.

After reaching the land, natural elements, of which many are minerals essential to the health of human beings and other living organisms, are dissolved from soils, rocks, and other mineral media into the precipitated water during surface runoff, streamflow, and underground movement in groundwater aquifers. This is a secondary step of water quality adjustment through the hydrological cycle.

During flow in waterways, e.g. streams and rivers, and storage in water realms without apparent flow, e.g. lakes and reservoirs, impurities in the water can be removed and/or assimilated under a series of natural physical (sedimentation, entrapment, etc.), physicochemical (natural coagulation, complexation, and precipitation, filtration, adsorption, ion-exchange, etc.), chemical (oxidation, etc.), and biological (decomposition, degradation, etc.) actions (Korenaga *et al.*, 2017; Oki & Kanae, 2006). Nowadays, we use the terminology of 'self-purification' to explain the function of water quality conversion within waterbodies. The most classic work was carried out by Streeter and Phelps (Long, 2020) who considered both organic matter and dissolved oxygen in a stream and developed partial differential equations for modeling organic decay during streamflow. Various studies have also been conducted for characterizing natural processes in streams, lakes, and groundwater aquifers to remove organic matter, nutrients, and other pollutants (Dai, *et al.*, 2020; Tong *et al.*, 2019). Natural purification is the subsidiary function of water quality stabilization through the hydrological cycle.

6.2.3 Thermodynamic characteristics of the hydrological cycle

Let us consider again the global hydrological cycle regarding its thermodynamic characteristics. First, water movement through the hydrological cycle is driven by solar energy and gravitational potential energy. The former transforms water from its liquid state to gaseous state (the water vapor) and drives it up to the atmosphere. The latter forces the liquid droplets condensed from the water vapor to fall as rainfall or snowfall, and then forces the precipitated water to move along the ground slope, forming surface runoff and streamflow. Water infiltration down to subsurface and groundwater movement are also due to gravitational force. Both solar energy and gravitational potential energy belong to green energy or sustainable energy, which, by definition, can never be exhausted so that it always meets the needs of the present without compromising the ability of future generations to meet their own needs (Stassen *et al.*, 2019).

Second, as water in various states (liquid, vapor, or even ice) are all confined in what we call the hydrosphere, the combined mass of water found on, under, and above the surface of the Earth, there is not any inflow to or outflow from the hydrosphere during any process in the hydrological cycle. Therefore, taking the out layer of the hydrosphere (several thousand meters above the Earth's surface) as the boundary for mass balance analysis of water molecules, the referred system (a spherical volume with the Earth in it) is a closed system with no exchange of water molecules without space, and the total mass of water remains constant. Therefore, the movement of water through the hydrological cycle is reversible, namely a reversible cycle with a cyclical reversible process in which the system and its surroundings will be returned to their original states.

With the above characteristics, it can be concluded that the hydrological cycle can spontaneously maintain a dynamic equilibrium condition without any unnatural interference. It is virtually a thermodynamically sound system with each hydrological element (e.g. each surface or subsurface water body) in it also in a dynamic equilibrium state.

6.2.4 Human disturbance of the hydrological cycle

The discussions above are based on a primitive assumption of the state of the hydrological cycle with unnatural interferences. From ancient times, human beings have depended on water of relatively stable quantity and quality from various water bodies to support their domestic and productive water use. However, according to the scale of water use, human activities inevitably impose impacts or disturbances on the hydrological cycle.

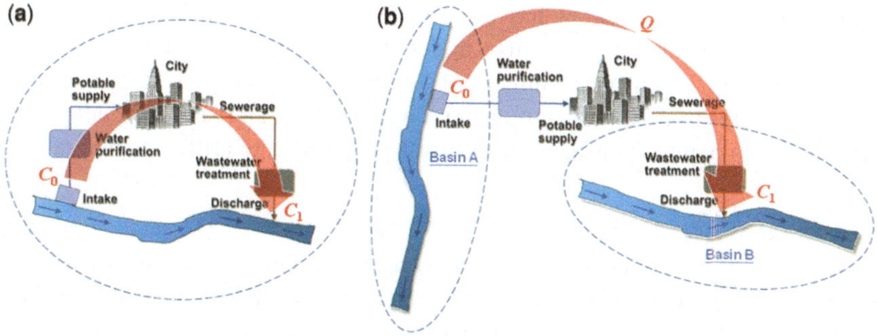

Figure 6.3 Disturbance of natural streamflow due to urban water supply and sewage discharge (figure by authors). (**a**) Intra-basin disturbance. (**b**) Inter-basin disturbance.

Figure 6.3 shows the possible disturbance of natural streamflow due to urban development under an assumption that a city depends on a river for water supply from its upstream and discharges the used water to its downstream (Figure 6.3 (a)), or a city takes water from a river in basin A for urban supply and discharges the used water to another river in basin B (Figure 6.3(b)).

In the case of Figure 6.3(a), the major disturbance to the natural streamflow of the river is the change of water quality from C_0 at the upstream section where water is withdrawn to C_1 at the downstream section where the used water (treated or untreated effluent) is discharged. The stream flowrate may not change so much for the river as a whole. This is a typical case of 'intra-basin disturbance', usually for a riverside city. According to the scale of water supply and the remaining pollutant loading in the discharged flow into the river, the water quality downstream may be severely deteriorated.

In the case of Figure 6.3(b), the disturbance to the natural streamflow of the river in basin A is the loss of flowrate Q due to water supply to the city, while that to the natural streamflow of the river in basin B is an increase of flowrate Q at the section of discharge and an increase of pollutant loading to the stream. This is a typical case of 'inter-basin disturbance.' It brings about significant changes in the hydrological conditions of the two rivers in different basins.

6.3 URBAN WATER CYCLE

In Section 6.2, we discussed the hydrological cycle at both the global and watershed scales. We learnt that all the natural processes involved in the hydrological cycle perform functions either to ensure water quantity and/or to ensure water quality in all freshwater bodies. Such a natural water cycle, on the whole, is a thermodynamically sound system under a dynamical equilibrium state. Human beings, as well as other living organisms in the world, are enjoying the grace of nature for obtaining water to support their lives. However, our large-scale utilization of water resources has more or less disturbed the hydrological cycle, especially after people began to live in densely populated settlements – the cities. As indicated in Figure 6.3, human consumption of freshwater from nature will not cause water to disappear, but create a water bypass and alter its quality before returning it to nature. This also forms a water cycle due to artificial elaboration.

6.3.1 Characteristics of the urban water cycle

Figure 6.4 is a conceptual depiction of the process of human utilization of water, the engineered system to facilitate water use, and how this engineered system is connected with natural waters. The basic process of human utilization of water consists of four sub-processes, namely water intake, water supply, water use, and used water disposal. In prehistoric times or even nowadays in remote areas, all these are conducted by using manpower, such as fetching water from a stream or well with a bucket, carrying the bucket back home, using the water at home, and finally discarding the used water randomly. Conversely, for modern cities, an engineered system with the four subsystems shown in Figure 6.4 is provided. Such an engineered system connects with natural waters at least at two locations – one at the start of the water resource subsystem where freshwater is withdrawn, and another at the end of the wastewater and drainage subsystem where used water is discharged back to natural waters. This engineered system and the connected natural waters form a water loop is named as an 'urban water cycle'.

We take the engineered system as the artificial (engineering) component of the urban water cycle because water is mechanically forced to flow through artificially built facilities to facilitate human water use. The function of each component (subsystem) of the engineered system is summarized as follows: (1)

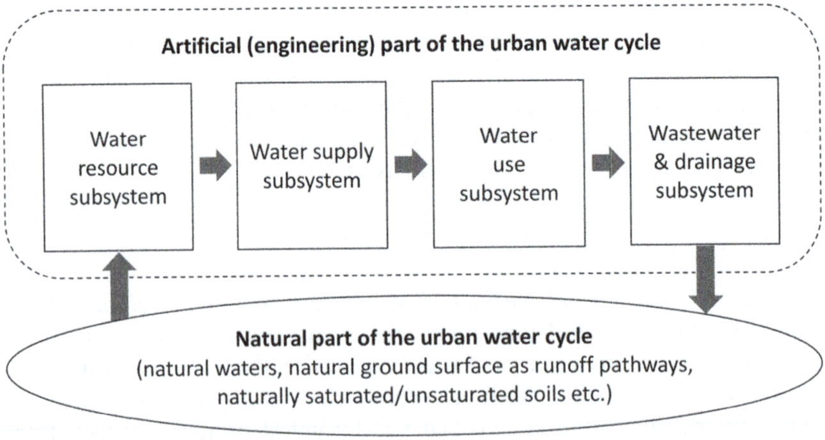

Figure 6.4 Urban water cycle consisting of artificial (engineering) and natural components (figure by authors).

the water resource subsystem provides sufficient raw water to a city by booster pumps and conveyance pipelines, (2) the water supply subsystem consists of water purification plants for treating the raw water to meet the quality requirement for human use (usually for drinking), and a distribution network to supply the treated water (potable water) to users, (3) the water use subsystem usually covers the whole city with various apparatus to facilitate water use for all purposes, and (4) the wastewater and drainage subsystem consists of sewage networks for collecting the used water (conventionally called wastewater) and send it to wastewater treatment plants where pollutants are separated from the wastewater and the treated effluent is finally discharged. This subsystem also includes facilities for urban drainage during rainy days through separated and/or combined pipelines for smooth drainage of surface runoff (Bach *et al.*, 2014).

As shown in Figure 6.4, the engineered system discussed above is connected with natural waters at least at two locations, namely the waterbody that provides source water and the waterbody that receives the used water (treated wastewater) and urban drainage. In fact, in the urban area, the natural ground surface provides surface runoff pathways and the naturally saturated and/or unsaturated soil provides pathways for water infiltration into groundwater aquifers. The source for water supply to the city and the urban discharge/drainage receiver may be the same waterbody as that depicted in Figure 6.3(a) or water bodies in neighboring watersheds as that depicted in Figure 6.3(b). In either case, if the human disturbance has not brought about a significant variation of the dynamic equilibrium state of the local hydrological system (with related waters in it), this natural component of the urban water cycle can be capable of continuous provision of source water and accommodation of urban discharge/drainage. On the other hand, with the rational design of the artificial component of the urban water cycle (the engineered system with application of up to date technologies), the human disturbance can also be reduced to a permissible extent so that the whole urban water cycle can be maintained in a healthy state, or in other words, a dynamic equilibrium state, in terms of both water quantity and quality.

6.3.2 Conventional modern urban water system

We use the expression of 'conventional modern' to characterize the urban water system widely built over the world from the late 1800s or early 1900s till nowadays, where 'conventional' means the manner widely accepted or standardized, and 'modern' differentiates the past time of more than one century from earlier times. The conventional modern urban water system, formulated as a result of the industrial revolution, is characterized by centralized potable water supply to provide water of sufficient quality to meet various demands for water use, and centralized sewerage for collecting and transporting human wastes swiftly out of the urban area (Tambo, 2004). In addition to the provision of large water distribution and used water collection networks, the introduction of water purification facilities (first by slow sand filtration and later by rapid sand filtration with disinfection) and wastewater treatment (symbolized by the activated sludge process) are important innovations in water and sanitation for urban societies (Li *et al.*, 2018). As the main objective of the system design was to satisfy the human desire for comfort and better sanitation, the maximized utilization of available water sources and the use of water bodies for the assimilation of discharged wastes has been commonly practiced everywhere. However, the shortcomings become more and more obvious with the increasing world population and growth in the number and scale of cities. These include but are not limited to the following:

(1) Indiscriminate use of high-quality water: In almost all cities, the total quantity of water supplied to users is of potable quality, but less than half of it is used for drinking, cooking, or other purposes that really require high quality.

(2) Endless requirements for quality improvement: In the late 1800s or early 1900s when the conventional urban water system was put into operation, the processes for drinking water purification and wastewater treatment were 'conventional' with applications of basic technologies. This is not because of the lack of advanced technologies at that time, but mainly due to the easiness to obtain good quality source water and sufficient self-purification capacity of the waters to receive wastewater discharge. However, as water pollution became increasingly severe in many countries and regions, especially in the vicinity of large cities, more and more sophisticated treatment was required for the provision of safe drinking water and the reduction of pollutant loadings to receiving waters. High costs of water and wastewater treatment have increased economic difficulties in many cities.

(3) Continuous or limitless system expansion: As coverage of the whole service area is the task for a centralized urban water system, it always needs expansion and/or upgrading to meet increasing demands due to the enlargement of urban areas and population growth. Even the maintenance and rehabilitation of the massive water and wastewater networks are heavy tasks in many cities.

(4) Difficulties in practicing water reuse and resource recovery: The conventional urban water system was designed following an 'end-of-the-pipe' model, characterized by a sequence of production-utilization-wastage because water reuse and resource recovery were not topics at all for all the cities developed decades ago. However, when the water resource is no longer plentiful, and its reclamation and reuse become necessary, the conventional system is found to be unsuitable to meet this new requirement. On the other hand, although we have realized for a long time that water, fertilizers, and other resources from the wastewater

stream can be reclaimed for various uses because most wastewater treatment plants are located outside the city, the reclaimed water and resources have to be sent back to the city area by long-distance transportation, which needs additional energy and/or cost input.

(5) Lack of harmonic relation with natural waters: In urban water system design, natural waters are usually taken as source providers and waste receivers but not important components of the urban water cycle. The harmonic relation between engineered facilities and nature was seldom taken into consideration in planning the whole system.

6.3.3 Urban water system toward a new paradigm

From a historical view, human utilization of water has so far roughly undergone two distinctive paradigms. The first is almost nature-dependent, while the second is largely engineering-dependent. The nature-dependent paradigm is the paradigm of very basic water supply by fetching water directly from surface waters and/or wells, with the use of streets and simple street drainage for stormwater and wastewater conveyance. Water quality at that time was generally excellent due to sufficient natural purification capacity of streams and/or groundwater aquifers to assimilate pollutants. The timeline for this paradigm is not very clear but can date back to the period from B.C. to the Middle Ages. It can still be found in remote areas nowadays. The engineering-dependent paradigm is the paradigm of water supply by engineered systems. It also has not a clear start time in history but gradually evolved from nature-dependent to more engineering-dependent (decrease in nature dependence and increase in engineering dependence), as depicted in Figure 6.5, with rough periods and typical engineered facility types for water supply and sewerage.

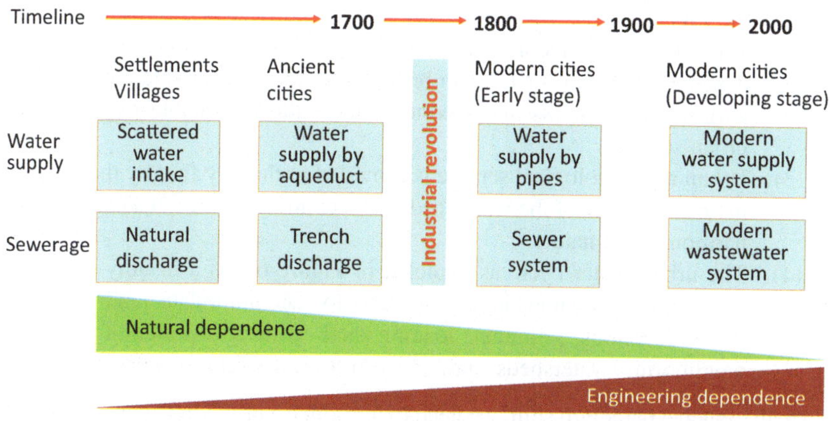

Figure 6.5 Evolution of urban water and wastewater systems with a decrease of nature dependence and increase of engineering dependence (figure by authors).

Now we come back to continue our discussion of the urban water cycle, as depicted in Figure 6.4, regarding the functions of the natural and artificial components. Since the natural component belongs to the hydrological cycle, as discussed in Section 6.2.2, its function is, first, to secure a sufficient quantity of water for the city. The second function is to secure the water quality of both the source water and the receiving water. In contrast, the function of the artificial or engineering component is to supply water for various uses and bring the used water back to nature. In terms of water quantity, a balance between the withdrawable water source and the demand for water supply should be reached. Conversely, in terms of water quality, the pollutant loading of the discharged wastewater and urban drainage should not exceed the carrying capacity of the receiving waterbody. The installation of wastewater treatment facilities is, in principle, for the protection of the receiving waterbody. However, if the qualitative balance between the natural and artificial components is broken, the urban water cycle will enter a vicious circle: pollution of natural water results in deteriorated source water quality so that more sophisticated drinking water treatment is required, and to protect natural waters, wastewater treatment facilities have to be upgraded to achieve higher pollutant removal goals. Such a condition is currently encountered by many cities.

It is widely recognized that we need a paradigm shift in building our urban water systems. The tendency of decreasing nature dependence and increasing engineering dependence, as shown in Figure 6.5, has to be changed so that a new paradigm toward the future may take the form of 'engineering in nature' with the following characteristics:

(1) A city is located in a watershed within the natural hydrological cycle and the watershed determines the hydrological boundary of the city.
(2) Any engineered facility provided for utilization of water in the city is designed in a way first not to bring about any irreversible damage to the hydrological process and second to meet the reasonable requirement of urban service.
(3) Urban aqua-ecological service is provided to the city first by the favorable natural property of the watershed and second by an artificial elaboration in harmony with nature.
(4) The urban water cycle as depicted in Figure 6.4 is a healthy water loop with sufficient metabolic capacity to accommodate and assimilate its endogenous pollutant loading and with no pollution export to neighboring watersheds and/or hydrological cycle at a larger scale.

To realize such a paradigm shift, we need to introduce a brand-new concept of water cycle management.

6.4 CONCEPTUAL SCHEME OF WATER CYCLE MANAGEMENT

Table 6.2 shows a conceptual scheme of water cycle management (WCM) for cities. It includes the management of water sources, water quality, water use, and waste discharge, each with specific objectives, with the overall management aiming at urban water sustainability.

6.4.1 Resource management

The primary water source for urban water supply is the natural waters within the watershed where a city is located. In cases where the city and the source water are hydrologically located in adjacent watersheds, we would like to put all of them into the scope of the urban water cycle for the discussion. The primary source of water is usually the available freshwater suitable for drinking water supply and other applications. In addition to such types of 'conventional' water source, there are also 'alternative' water sources that can be made applicable for cities, such as the harvestable rainwater, reclaimed used water (alternative terminology of 'wastewater'), and even water from saline or brackish sources.

The utilization of conventional water sources is usually prioritized because of their favorable quality. However, every freshwater body has its limit of water withdrawal beyond which its aquatic environmental condition will be deteriorated

Table 6.2 Conceptual scheme of WCM for cities.

Item	Coverage	Objectives
Resource management	Conventional water source Alternative water source	Meet the reasonable requirement of urban service without irreversible damage on natural waters
Quality management	Quality for water use Quality for ecological health	Meet the quality requirement of water for various water uses and ecological safety
Water use management	Potable water use Non-potable water use	Fit-for-purpose water use and minimization of freshwater consumption
Discharge management	Urban floods Pollutant loading Recycle & reuse	Urban flooding control, minimization of pollutants discharged to natural waters and maximization of resource recovery
Overall management		Sustainable water service and urban water environment protection

(Schornikov *et al.*, 2014). Therefore, the restriction of overexploitation of source water is the main point of conventional water resources management. In case the conventional source water is insufficient for urban water supply, the utilization of alternative water sources should become a realistic option.

Alternative water sources development encompasses the utilization of the hydrological cycle and/or urban water cycle. For example, rainwater harvesting collects water directly in the processes of precipitation and/or surface runoff (Figure 6.1), while water reclamation from the discharged used water stream is through an interception of the outflow from the artificial component to the natural component of the urban water cycle (Figure 6.4). As rainwater is a seasonally obtainable source, its potential for utilization much depends on the natural and artificial storage capacity in the local area. In contrast, the reclaimed water can be viewed as a stable water source because an almost fixed amount of the used water is collectible daily. As far as economically and technologically permissible, there is no limit to water withdrawals from these alternative sources (Hiratsuka & Wakae, 2019).

6.4.2 Quality management

Quality management of water for meeting the requirement of water use has so far been a prioritized task. Of the various purposes of urban water use, drinking water requires a quality that should not cause any negative effect on human health after daily oral intake (Goncharuk *et al.*, 2018). Historically this was not difficult because water from natural sources was usually clean and after removing the naturally originating turbid and colored substances (usually by physicochemical processes of coagulation, sedimentation, and filtration when surface water was the source) and carrying out disinfection to inactivate pathogenic bacteria (usually by chlorination), the water could be made drinkable for tap water supply (Han *et al.*, 2020). Although additional treatment, such as carbon adsorption and more sophisticated processes including advanced oxidation, membrane filtration, and so on, have been required in many water purification plants due to source water contamination, the conventional coagulation-sedimentation-filtration treatment is still the basic process of water purification. For safe drinking water supply there is always a dilemma between source water protection and upgrading of drinking water treatment processes. From the viewpoint of water cycle management, the upmost principle should be to prevent foreign pollutants from entering the urban water cycle, so source water protection should be a wise choice.

The target of water quality management should also be placed on urban ecological health. As an urban planning term, ecological health refers much to the 'greenness' of cities for which the visibility of water is a very important factor (He *et al.*, 2020).

In fact, the quality for water use and that for ecological health are closely interrelated because when we put a city into its related watershed and consider them as a whole, without an ecological healthy condition for the watershed we may not obtain good source water to ensure good quality for various water uses.

6.4.3 Water use management

The conventional manner of urban water use is characterized by using tap water of drinkable quality for most or almost all domestic and municipal purposes. Although many cities have been practicing water-saving to reduce per capita and total water consumption, a large amount of high-quality tap water is still wasted because it is not consumed for potable use. In Section 6.4.1.1 we added alternative waters into the water resources for management. In most cases, these alternative waters are of inferior quality compared to freshwater. Unless necessary, it may not be an apt option to convert the quality of the harvested rainwater, or that of the reclaimed water, to drinkable level through sophisticated treatment processes and with high energy consumption. Therefore, 'fit-for-purpose' water use should be the principle of urban water use management.

Fit-for-purpose is a common term to describe the ideal level of quality for a given use (Coonrod *et al.*, 2020). Regarding water use, it first stresses the importance that various source waters should be rationally utilized. Freshwater from nature should be preferentially used for potable purposes while alternative waters, usually as supplemental sources, should mainly be used for non-potable water supply. Taking domestic (household) water supply as an example, of the per capita water demand ranging from several ten liters to several hundred liters per day in different countries and regions of the world (Kuski *et al.*, 2020), the amount actually used for drinking (including cooking) is no more than 20 liters. If the water for in-house washing, bathing, and toilet flushing is also accounted for, the required amount will be up to 50–80 liters (Tamura and Ogawa, 2012). The considerably higher per capita water demand or consumption than this required amount in many countries is an indication of overconsumption of tap water for other purposes, such as gardening, etc. which may not really need high-quality tap water. Dual-pipe water supply, one for potable and another for non-potable, can be an option for minimizing freshwater source utilization and facilitating alternative water use (Hambly *et al.*, 2015).

On the other hand, fit-for-purpose water use also implies that water quality conversion (treatment) before being supplied for a given use should meet the given quality criteria but not require over-processing. Taking water reclamation from discharged wastewater as an example, technologically it is possible to transform domestic wastewater into a potable quality level, such as what is done in Singapore for NEWater production (Schnoor, 2009). However, in most cases, the reclaimed water is used for gardening and urban irrigation, replenishing urban lakes/streams, and other environmental purposes. Therefore, the required treatment processes should be the most appropriate ones but not the most sophisticated ones, as long as water reuse may not result in negative environmental impacts (Wang *et al.*, 2018).

Water use management for minimizing freshwater consumption is an important task especially for cities facing water shortage problems.

6.4.4 Discharge management

In the urban water cycle depicted in Figure 6.4, the water to be discharged back to natural water bodies includes the used water and stormwater from the urban area, and related watershed. The task of discharge management to be discussed here deals with the quantitative and qualitative aspects.

Quantitatively, the used water discharged from various users is with an almost constant average flowrate, on a daily basis, and the urban sewage system, including sewer network, transfer pipelines, and treatment facilities, is designed with a capacity for the smooth conveyance of the sewage flow under ordinary conditions (e.g. in the dry weather). Parallel to the sewage system, an urban drainage system is also to be provided for discharging surface runoff on rainy days. The design of the drainage system is usually based on a given rainfall intensity corresponding to a prescribed return period (Hou & Ning, 2007). There are also cases of combined sewers that are designed to simultaneously collect surface runoff and sewage water in a shared system, especially in many older cities (Kamei-Ishikawa *et al.*, 2016). During dry weather or small storms, all flows are handled by the treatment facilities, while during large storms, some of the combined stormwater and sewage is allowed to be discharged untreated to the receiving waterbody. For diverting flows in excess of the peak design flow of the sewage treatment facilities, relief structures, which are called stormwater regulators or combined sewer overflows, are constructed in combined sewer systems. When constructed, combined sewer systems are typically sized according to a prescribed interception ratio (the capacity to carry a mixed flow of sewage and surface runoff over the average dry weather sewage flow). For either the separate or combined systems, smooth transmission of the sewage and surface runoff flows to prevent urban flooding is the main objective of discharge management in the quantitative aspect.

As the discharged flows eventually enter the receiving water bodies which are usually important water environmental elements of a city, their quality management is also required. For most cities, the reduction of pollutant loading of the sewage flow by proper wastewater treatment is the main measure to be taken. Many nations have put forward strict regulations on effluent discharge from treatment facilities for water environmental protection. Another trend is to incorporate water reclamation and useful materials recovery into the treatment scheme (Kog, 2020). On the other hand, surface runoff during a storm may carry pollutants from nonpoint sources to result in pollution of the receiving waters as well. Therefore, various measures are also taken within the scheme of low impact development (LID) and others for reducing pollutants (Eckart *et al.*, 2018). In the case of combined sewers, serious water pollution can be caused during combined sewer overflow (CSO) events when combined sewage and surface runoff flows exceed the capacity of the sewage treatment plant, or of the maximum flow rate of the system which transmits the combined sources (Rathnayake & Faisal, 2019). CSO management, thus, becomes very important for some older cities where combined sewers are still in service.

6.4.5 Overall management

Under the WCM principle, we should also stress the systematic management of the whole urban water cycle, targeting sustainable water services and urban water environmental protection. Each of the subsystems shown in Figure 6.4 is linked with others, as well as the natural elements of the watershed. Water from each of the utilizable sources is with its characteristic quality which meets the requirement of a specific purpose of water use after a minimum or simplest quality conversion process. The treated effluent from sewage treatment facilities and harvestable stormwater are also alternative sources for water use. Therefore, the discussions in Sections 6.4.1–6.4.4 for resource, quality, water use, and discharge management should be put into an overall management scheme.

6.5 WCM CONCEPT APPLICATION FOR WATER SOURCE ENLARGEMENT TO RESTORE A WATER CITY

6.5.1 Background

As an example of the WCM concept application, we introduce a case of urban water system planning in Xi'an, a megacity in northwestern China, to solve the problem of water shortage and the restoration of an aquatic city.

Xi'an was the ancient capital city of China. In ancient times, there was plentiful water running down from the nearby Qinling Mountains, feeding many rivers passing near the city and forming the ancient beauty of 'Eight Rivers Surrounding the Capital.' Over time, climate change, hydrogeological variation, rapid industrialization and urbanization, overuse, and improper management of the water resources all resulted in the disappearance of the ancient water quality and abundance. Located in the middle of the Yellow River basin, the annual precipitation in Xi'an is about 550 mm, yet the evaporation amount far exceeds this amount of rainfall. Although the Wei-River passing through the northern suburb of the city is the largest tributary of the Yellow River, due to the overconsumption of the river water in the upstream area, it is almost impossible, nowadays, to withdraw surface water for water supply to Xi'an. Groundwater used to be important source water, but for the prevention of ground subsidence, its development has to be strictly prohibited. Since the 1990s, more than 70% of the water supply to the central urban area has depended on water transfer from a set of dams about 140 km away, built on rivers originating from the Qinling Mountains.

Within the recent urban development plan, a major governmental investment project has been implemented. Aiming at a restoration of the ancient water city, the so-called 'Eight-Rivers Regeneration' project includes the construction of water supply lines, restoration of seven wetlands, rehabilitation of eight river channels, and reconstruction of 28 lakes and ponds, including some surrounding parks designed to imitate Tang Dynasty (AD 618–907, the most prosperous time of Xi'an as the ancient capital city) landscapes known from ancient illustrations. The project also includes the expansion of urban green spaces to make the city more resilient and environmentally sustainable. Figure 6.6 shows the general plan of the project.

However, one of the bottlenecks to this large project is the supply of sufficient water for maintaining a healthy urban water system, especially the 28 lakes/ponds that need water replenishment regularly to maintain their favorable environmental and landscape values. Under the current conditions, domestic, municipal, and industrial water demands are continuously increasing in this megacity of circa 10 million population and only a limited amount of water from natural resources can be allocated for environmental uses. Associated with the

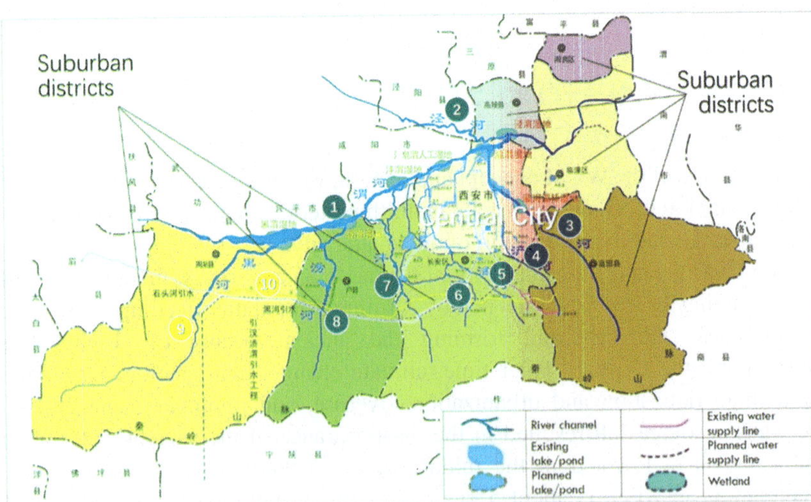

Figure 6.6 General plan of the 'Eight-Rivers Regeneration' project. For the river channels, Nos. 1–8 are the historical 'Eight Rivers' of Wei (a tributary of the Yellow River), Jing, Ba, Chan, Jue, Hao, Feng, and Lao; No. 9 is the source water river for water supply to Xi'an, and No. 10 is the water transfer channel (adapted from Xi'an Water Authority, 2013).

implementation of the project, a water source enlargement plan has been formulated following the WCM concept.

6.5.2 Water source enlargement plan

6.5.2.1 Requirement of source enlargement

For the 'Eight-Rivers Regeneration' project, water source enlargement solely targets replenishing the 28 lakes/ponds. Table 6.3 summarizes the results of water budget analysis for evaluating the required amount of water to be supplemented.

The average available water resources for Xi'an amounts to 2347 million m^3/yr (Xi'an Water Authority, 2019). If the current total population of 10.2 million is considered (Xi'an Municipal Bureau of Statistics and NBS Survey Office in Xi'an, 2020), the per capita water resource is only about 230 m^3/yr, indicating a severe water shortage condition for this city. As water resources should be prioritized for domestic and production water supply, the amount of water allocated for the environment is very limited. As shown in Table 6.3, the currently available water source of 141 million m^3/yr, mainly from natural stream flows, for replenishing these lakes and ponds is far below the total demand of 304 million m^3/yr, which is calculated based on the requirement to maintain a favorable landscape for each lake or pond.

6.5.2.2 Source enlargement measures

There are generally two categories of measures to be taken for the enlargement of water sources for replenishing urban lakes/ponds, namely, alternative water resource development and increasing water use efficiency.

Table 6.3 Results of water budget analysis for lakes/ponds replenishment.

No	Item	Unit	Value	Remarks
1	Water surface area	ha	2060	
2	Storage volume	million m^3	65	
3	Annual net evaporation loss	million m^3	5.4	Evaporation – Precipitation
4	Annual leakage loss	million m^3	4.6	Local experiential data
5	Replenishment plan	time/yr	4~6	Lake/pond specific
6	Annual replenishment amount	million m^3/yr	294	Sum of all lakes/ponds
7	Total water demand	million m^3/yr	304	Sum of items 3, 4 and 6
8	Currently available water source	million m^3/yr	141	Mainly from streamflow
9	Annual water deficiency	million m^3/yr	163	Deference of items 7 and 8

6.5.2.2.1 Alternative water resource development

As a megacity, there are currently 22 centralized domestic wastewater treatment plants in operation and the volume of wastewater treated amounts to about 2.2 million m^3/d. About 0.385 million m^3/d of this wastewater is produced as reclaimed water with quality meeting the standard for environmental reuse including recreational water replenishment (Xi'an Municipal Bureau of Statistics and NBS Survey Office in Xi'an, 2020). The potential of reclaimed water use for non-potable supply is about 140 million m^3/yr at present and will increase to about 215 million m^3/yr in 2030. This is the most promising alternative water resource for the project.

Another alternative water resource can be from rainwater harvesting. Although most of the lakes/ponds do not have considerable catchment areas for receiving sufficient amounts of natural runoff, as they are mostly built on low-lying lands, it is possible to implement measures that connect their inlet structures with local stormwater regulation and drainage facilities. For effective harvesting of rainwater or stormwater runoff, the provision of sufficient storage volume is usually required. Fortunately, these lakes and ponds have their own storage volume to meet this requirement.

6.5.2.2.2 Increasing water use efficiency

One important characteristic of landscape water use is that there is no actual water loss other than surface evaporation and subsurface leakage. The main purpose of water replenishment is to renew the lake water so as to prevent water quality deterioration due to long-term stagnation. In the project area, the hydraulic retention time (HRT) for small-scale urban lakes or ponds is preferably two months, whereas that for a larger lake can be up to three months, from past experiences (Chang *et al.*, 2020). As shown in Table 6.3, the required amount for replenishment is far larger than that for evaporation and leakage. As long as the water quality can be secured, water may be used more than once for replenishment purposes. Cycling flow may be a measure for individual lakes or ponds, but it may be more feasible to practice cascading water use between lakes if elevation difference can be utilized for gravity flow. On the other hand, water use efficiency may also be increased by flow regulation between streams connecting lakes and ponds.

6.5.2.3 Formulation of a quasi-natural water cycle for water source enlargement

Putting all the available natural and alternative sources into an integrated application scheme, a quasi-natural water cycle is conceptually formulated as shown in Figure 6.7.

This water cycle covers the watersheds where Xi'an city is located (approximately the whole area shown in Figure 6.6). The natural component of

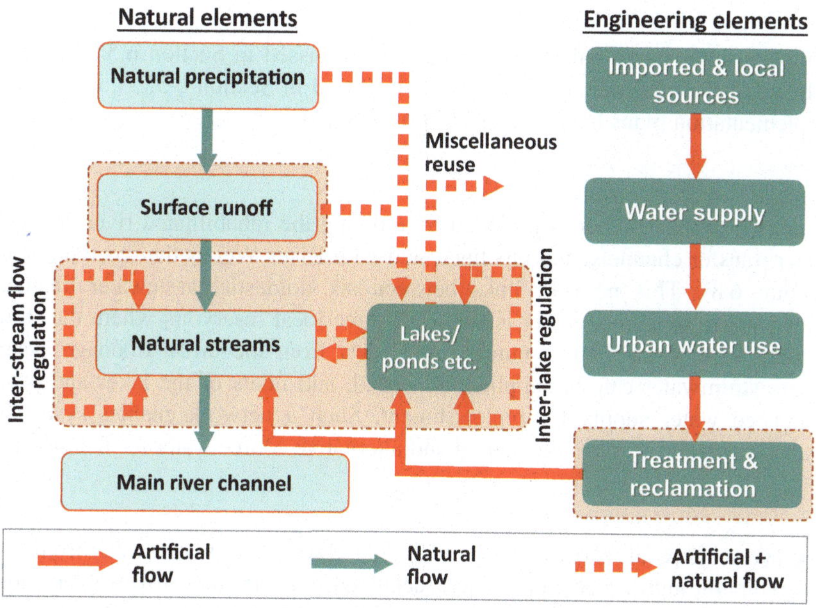

Figure 6.7 A quasi-natural water cycle for integrated water management to enlarge water sources for the 'Eight-Rivers Regeneration' project. A, B, and C are the three measures of water source enlargement as water reuse, rainwater harvesting, and cascading water use, respectively (figure by authors).

the water cycle includes all the hydrological elements, such as atmospheric precipitation, surface runoff, and natural streams shown in Figure 6.6. The Wei River (No. 1 in Figure 6.6) is the main river channel receiving all stream flows. The artificial or engineering component of the water cycle includes all the engineering facilities for water sources, water supply, water use, and wastewater treatment, and final disposal. For this project, as special attention is paid to the lakes and ponds which need water replenishment, these water bodies become hydraulic nodes between the natural and engineering components. On one hand, as open water bodies, they may receive natural precipitation, stormwater runoff, and/or stream water replenishment. On the other hand, the reclaimed water from domestic wastewater treatment facilities can become an important source for water replenishment.

Within such a quasi-natural water cycle, source enlargement can be fulfilled by three measures, namely water reclamation from domestic wastewater (A in Figure 6.6) which is a source with almost constant flow, rainwater harvesting from surface runoff (B in Figure 6.6) which is a seasonal water source, and cascading water use (C in Figure 6.6) including inter-stream and inter-lake flow regulations. The lakes and ponds, as well as the river channels, also provide a water storage volume buffering between water supply and uses.

6.5.2.4 Implementation plan

Following the source enlargement measures discussed in Section 6.5.2.2 and the scheme of the quasi-natural water cycle discussed in Section 6.5.2.3, engineering implementation plans have further been formulated.

6.5.2.4.1 Water supply network

A water supply network is provided by utilizing the rehabilitated river channels, water transfer channels, and pipelines, and with River Wei as the receiving water (Figure 6.8). This network links between six domestic wastewater treatment plants where reclaimed water is supplied, three local reservoirs where the stored natural flow is supplied, a number of outlets from the urban drainage system where stormwater can be supplied as needed, and inlets of the lakes and ponds that need water supply for replenishment. Such a network enables systematic management of the available natural and alternative source water for the project.

6.5.2.4.2 Source water distribution

The lakes and ponds shown in Figure 6.8 are fed with river water, water from local reservoirs through water supply lines, reclaimed water from nearby WWTPs, local rainwater harvesting facilities (not shown in Figure 6.8 in detail), or multiple sources depending on their locations and source availability.

For lakes and ponds adjacent to the rehabilitated river channels, direct use of the river water is preferable from both the quantity and quality aspects. In addition to

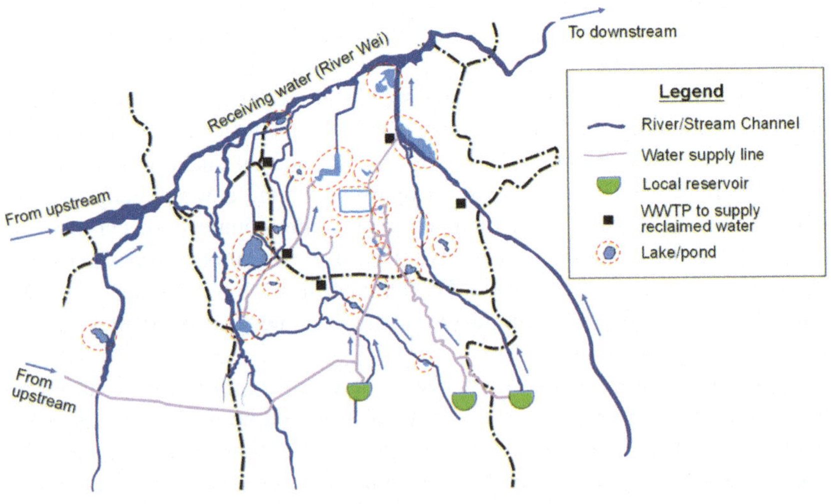

Figure 6.8 Water supply network with river/stream channels linking between local reservoirs, WWTPs, and lake/pond inlets and outlets for natural/alternative water supply and receiving outflows (adapted from Xi'an Water Authority, 2013).

these natural rivers, several drainage channels in the central urban area are also utilized for source water distribution. In rainy seasons their main functions are still stormwater drainage while in dry seasons they become water channels for transferring reclaimed water from WWTPs and/or harvested rainwater from storage facilities to lakes/ponds as alternative sources for water replenishment.

6.5.2.4.3 Realization of cascading water use

Topographically, Xi'an has a ground slope from the mountain foot to its south to the main channel of River Wei to its north. For a group of lakes with their elevations of water surface descending along the ground slope, cascading water use can be made possible between them by using the outflow from a lake at the upstream side as the inflow to another lake at the downstream side. This is mainly for the case of a series of lakes supplied with natural water from the same local reservoir for a reduction of the total demand for such types of high-quality source water.

With the water supply network shown in Figure 6.8, water regulation between different river/stream channels for mitigating the imbalanced demand-supply relationship can also be made possible for more reasonable and efficient use of the limited water resource.

6.5.2.4.4 Water quality protection

The formulation of the water network shown in Figure 6.8 has well realized the basic consideration of a quasi-natural water cycle of Figure 6.7. It fully utilizes part of the urban watershed where a number of natural streams flow toward a common receiving water. All artificial water structures, including the water lines, lakes, and ponds, are harmoniously superimposed onto the natural flow system and add its new hydrological elements without altering the hydrological characteristics of the original watershed. Such a water system not only enables the smooth distribution of water to each of the lakes and ponds for replenishment but also assists water quality protection and even improvement because the good flow and circulation of water in the whole system significantly increase the self-purification capacity.

Water from natural streams and local reservoirs are of good quality in terms of organic content (COD or BOD) and nutrients (P and N) as the most favorable water source for lake replenishment. The harvested rainwater, if the initial surface runoff with high pollutants concentration is not included, is also good for lake replenishment. In contrast, the water reclaimed from domestic wastewater, though well treated in WWTPs, is usually with higher concentrations of organics and nutrients. This is, in many cases, an obstacle for using reclaimed water as the sole source for replenishing landscape lakes due to the possible occurrence of water eutrophication. Through the water network shown in Figure 6.8, water quality deterioration is prevented by: (1) allocation of larger amounts of flow to the lakes that are solely replenished by reclaimed water so as to shorten their hydraulic retention time for the prevention of algae growth; (2) dilution of the reclaimed water with water from other sources during water distribution; and (3) artificial flow circulation within the water body or between water bodies.

6.5.3 Effects of water source enlargement

The implementation of the water source enlargement plan through the quasi-natural water system discussed above has mitigated the problem of water shortage for the replenishment of lakes and ponds restored and/or built within the 'Eight-Rivers Regeneration' project. Table 6.4 summarizes the overall effects. The conventional source includes water directly from natural streams and local reservoirs, which amounts to 141 million m^3/yr, as indicated in Table 6.4. The alternative sources include the amount of water supplied to some of the lakes through cascading water use (amounting to 56 million m^3/yr), water harvested from surface runoff in rainy seasons (amounting to 24 million m^3/yr), and reclaimed water from WWTPs (amounting to 86 million m^3/yr). The total amount of water from alternative sources reaches 166 million m^3/yr and has covered the annual water deficiency of 163 million m^3/yr indicated in Table 6.4. This indicates that by the management of the quasi-natural water cycle shown in Figure 6.7, the available water source can be enlarged to meet the needs for replenishing the 28 lakes and ponds. Of the total amount of 307 million m^3/yr to be supplied for lakes/ponds replenishment, that from the conventional water sources, namely stream flows, takes 45.9%, while that from alternative sources takes 54.1%.

It is notable that cascading water use virtually covers 18.3% of the water supply for some of the lakes and ponds. Although this is not truly an amount of water adding to the water source, an equivalent amount of high-quality stream flow is effectively saved due to the improvement of water use efficiency.

For environmental water use in a megacity in water-deficient regions, reclaimed water is an important alternative source. For this project, reclaimed water covers 28% of the water supply for lakes/ponds replenishment. Through the water

Table 6.4 Water supply for lakes/ponds replenishment by various water sources.

Water source		Annual supply (million m^3/yr)	Percent of supply (%)
Conventional source	Streamflow[a]	141	45.9
Alternative source	Cascading water use[b]	56	18.3
	Rainwater harvesting[c]	24	7.8
	Reclaimed water[d]	86	28.0
Total		307	100.0

[a]From natural streams and local reservoirs.
[b]For lakes/ponds in series.
[c]Combining with utilization of urban drainage system.
[d]Supplied from domestic wastewater treatment plants.

supply network shown in Figure 6.8, reclaimed water use is practiced either by using it as the sole source or mixing it with water from other sources. With the increased percentage of reclaimed water use, more frequent replenishment is required for maintaining a favorable landscape condition (Ao *et al.*, 2018).

REFERENCES

Agarwal R., Chhabra P., Goyal A. P. and Srivastava S. (2019). Predictive modelling to assess groundwater pollution and integration with water quality index. *International Journal of Engineering and Advanced Technology*, **8**(5), 1076–1084.

Ao D., Luo L., Mawuli D., Chen R., Xue T. and Wang X. C. (2018). Replenishment of landscape water with reclaimed water: optimization of supply scheme using transparency as an indicator. *Ecological Indicators*, **88**, 503–511.

Bach P. M., Rauch W., Mikkelsen P. S., McCarthy D. T. and Deletic A. (2014). A critical review of integrated urban water modelling – Urban drainage and beyond. *Environmental Modelling & Software*, **54**, 88–107.

Chang N., Luo L., Wang X. C., Song J., Han J. and Ao D. (2020). A novel index for assessing the water quality of urban landscape lakes based on water transparency. *Science of the Total Environment*, **735**, 139351.

Chawla I., Karthikeyan L. and Mishra A. K. (2020). A review of remote sensing applications for water security: quantity, quality and extremes. *Journal of Hydrology*, **585**, 124826.

Coonrod C. L., Ben Yin Y., Hanna T., Atkinson A. J., Alvarez P. J. J., Tekavec T. N., Reynolds M. A. and Wong M. S. (2020). Fit-for-purpose treatment goals for produced waters in shale oil and gas fields. *Water Research*, **173**, 115467.

Cui Y., Chen X., Gao J., Yan B., Tang G. and Hong Y. (2018). Global water cycle and remote sensing big data: overview, challenge, and opportunities. *Big Earth Data*, **2**(3), 282–297.

Dai Z., Zhang J., She R., Hu N., Xia S., Ma G., Han R. and Ming R. (2020). Numerical investigation on re-oxygenation efficiency of stepped overflow weir in urban stream. *Journal of Cleaner Production*, **258**, 120583.

Eckart K., McPhee Z. and Bolisetti T. (2018). Multiobjective optimization of low impact development stormwater controls. *Journal of Hydrology*, **562**, 564–576.

FAO (Food and Agriculture Organization of the United Nations). (2003). Review of World Water Resources by Country: Water Report 23. Food and Agriculture Organization of the United Nations, Rome.

Goncharuk V. V., Pshinko G. N., Rudenko A. V., Pleteneva T. V., Syroeshkin A. V., Uspenskaya E. V., Saprykina M. N. and Zlatskiy I. A. (2018). Genetically safe drinking water. requirements and methods of its quality control. *Journal of Water Chemistry and Technology*, **40**(1), 16–20.

Hambly A. C., Henderson R. K., Baker A., Stuetz R. M. and Khan S. J. (2015). Application of portable fluorescence spectrophotometry for integrity testing of recycled water dual distribution systems. *Applied Spectroscopy*, **69**(1), 124–129.

Han Z. M., An W., Yang M. and Zhang Y. (2020). Assessing the impact of source water on tap water bacterial communities in 46 drinking water supply systems in China. *Water Research*, **172**, 115469.

He J., Wu X., Zhang Y., Zheng B., Meng D., Zhou H., Lu L., Deng W., Shao Z. and Qin Y. (2020). Management of water quality targets based on river-lake water quality response relationships for lake basins –A case study of Dianchi Lake. *Environmental Research*, **186**, 109479.

Hester E. T. and Little J. C. (2013). Measuring environmental sustainability of water in watersheds. *Environmental Science & Technology*, **47**(15), 8083–8090.

Hiratsuka A. and Wakae K. (2019). Study on sustainable rainwater resource utilization-towards deepening of homo environments. *Journal of Water Resource and Protection*, **11**, 491–528.

Hou S. B. and Ning Q. (2007). Comment and proposals on the lead sealing management of ship engine room sewage discharge systems. *Journal of Maritime Law and Commerce*, **38**(1), 67–73.

Kamei-Ishikawa N., Yoshida D., Ito A. and Umita T. (2016). Cesium and strontium loads into a combined sewer system from rainwater runoff. *Journal of Environmental Management*, **183**, 1041–1049.

Kog Y. C. (2020). Water reclamation and reuse in Singapore. *Journal of Environmental Engineering*, **146**(4), 03120001.

Korenaga J., Planavsky N. J. and Evans D. A. D. (2017). Global water cycle and the coevolution of the Earth's interior and surface environment. *Philosophical Transactions of the Royal Society A* **375**, 20150393.

Kuski L., Maia E., Moura P., Caetano N. and Felgueiras C. (2020). Development of a decentralized monitoring system of domestic water consumption. *Energy Reports*, **6**, 856–861.

Li C., Zhao Y., Ouyang J., Wei D., Wei L. and Chang C. -C. (2018). Activated sludge and other aerobic suspended culture processes. *Water Environment Research*, **90**(10), 1439–1457.

Long B. T. (2020). Inverse algorithm for Streeter–Phelps equation in water pollution control problem. *Mathematics and Computers in Simulation*, **171**, 119–126.

Nazaroff W. W. and Alvarez-Cohen L. (2000). Environmental Engineering Science. John Wiley & Sons Inc, New York, London.

NBSC (National Bureau of Statistics of China), (2020) China Statistical Yearbook-2019 (in Chinese). China Statistics Press, National Bureau of Statistics of China, Beijing.

Oki T. and Kanae S. (2006) Global hydrological cycles and world water resources (Review). *Science* **313**(5790), 1068–1072.

Rathnayake U. and Faisal A. H. M. (2019). Dynamic control of urban sewer systems to reduce combined sewer overflows and their adverse impacts. *Journal of Hydrology*, **579**, 124150.

Schnoor J. L. (2009). NEWater Future? *Environmental Science & Technology*, **43**(17), 6441–6442.

Schornikov E. I., Zenina M. A. and Ivanova E. V. (2014). Ostracods as indicators of the aquatic environmental conditions on the northeastern Black Sea shelf over the past 70 years. *Russian Journal of Marine Biology*, **40**(6), 455–464.

Stassen C., Dommenget D. and Loveday N. (2019). A hydrological cycle model for the Globally Resolved Energy Balance (GREB) model v1.0. *Geoscientific Model Development*, **12**, 425–440.

Tambo N. (2004). Urban metabolic system of water for the 21st century. *Water Science & Technology: Water Supply* **4**(1), 1–5.

Tamura M. and Ogawa Y. (2012). Visualization of the coated layer at the surface of rice grain cooked with varying amounts of cooking water. *Journal of Cereal Science*, **56**(2), 404–409.

Tong Y., Li J., Qi M., Zhang X., Wang M., Liu X., Zhang W., Wang X., Lu Y. and Lin Y. (2019). Impacts of water residence time on nitrogen budget of lakes and reservoirs. *Science of the Total Environment*, **646**, 75–83.

Wang S., Liu F., Wu W., Hu Y., Liao R., Chen G., Wang J. and Li J. (2018). Migration and health risks of nonylphenol and bisphenol a in soil-winter wheat systems with long-term reclaimed water irrigation. *Ecotoxicology and Environmental Safety*, **158**, 28–36.

Xi'an Municipal Bureau of Statistics and NBS Survey office in Xi'an (2020). Xi'an Statistical Yearbook-2019 (in Chinese). China Statistics Press, Beijing.

Xi'an Water Authority (2013). Eight-Rivers Project Plan (in Chinese). Xi'an Water Authority, Xi'an, China.

Xi'an Water Authority (2019). Xi'an Water Resources Bulletin 2018 (in Chinese). Xi'an Water Authority, Xi'an, China.

Chapter 7

Resilient infrastructures for reducing urban flooding risks

Maria Pregnolato[1], Dawei Han[1], Anna Lo Jacomo[1], Raffaele De Risi[1], Jitendra Agarwal[1] and Jing Huang[2]

[1]Department of Civil Engineering, University of Bristol, Bristol, BS8 1TR, UK
[2]Business School, Hohai University, No. 8 Focheng Road, Nanjing, 211100, P.R. China

7.1 INTRODUCTION

The world's population living in urban areas is expected to increase from 55 to 68% by 2050 and the number of megacities with >10 million inhabitants from 33 to 43, with a faster pace of urbanisation in developing countries (United Nations, 2019). As the world continues to urbanize, sustainable development increasingly relies on the successful planning and management of urban growth, especially in hazard-prone regions. Natural hazards have heavily affected cities in recent years, for example, Hurricane Florence in 2018 and Harvey, Irma and Maria in the USA in 2017, the 2011 earthquake and tsunami in Japan, Cyclone Nargis in Sri Lanka in 2008, the Indian Ocean tsunami in 2004. The frequency and intensity of these phenomena seem also to be increasing due to climatic changes, with significant environmental, social and economic impacts (Stewart & Deng, 2014).

Natural catastrophes cause losses to people, properties and infrastructure according to their exposure and vulnerability. In particular, infrastructure represents a determining factor in limiting the impact of the events (Arrighi *et al.*, 2019; Garschagen *et al.*, 2016). Roads, for example, can provide access to

Editors: Xiaochang C. Wang and Guangtao Fu
doi: 10.2166/9781789060768_0181

emergency operation and evacuation, while if destroyed, entire areas can be isolated from support and aid (Arrighi *et al.*, 2020). In addition, infrastructure typically comprises various geographically extensive and interdependent systems (Chang, 2016); this interlinked nature results in cascading effects, i.e. disruptions in one system affect one or more other systems. For instance, the power supply system provides essential input (i.e. electricity) to transportation systems (e.g. to run electric trains), or water supply system (e.g. to run water pumps) (Pregnolato *et al.*, 2020). Consequently, the impact of natural catastrophes is often disproportionately large.

Modern cities are evidently complex and vulnerable environments, and at the same time a concentration of resources and wealth. When taking a long-term view, a resilient city is a system which includes the capability to withstand and bounce back from adverse events, and resilience is necessary for sustainable urban growth (Elmqvist *et al.*, 2019). As infrastructure is a core component of disaster risk reduction, the current challenge is to manage the resilient city's transformation process based on resilient infrastructure, thus enabling the city to provide services to its inhabitants even under adverse conditions (Pregnolato *et al.*, 2020).

In the context of highly vulnerable urban systems to hazards, adapting to reduce the harm is recognised as a primary need of the modern society (Aerts *et al.*, 2013). As adaptation is still to be completely defined and developed, strategies currently consist in 'learning by doing' and include all available options due to the uncertainties related to future climatic and socio-economic conditions. The implementation of adaptation measures involves decision-making and financing (Pregnolato & Dawson, 2018). At the stage of planning, various measures should be taken into account, alongside a range of decision time horizons (i.e. short-term, medium-term, and long-term) and uncertainties. By estimating the benefits from adaptation, innovative interventions related to infrastructure and urban planning could be seen as opportunities by investors and planners (Dawson *et al.*, 2015).

In recent years, a wide bulk of research has challenged practice and ways of thinking for the transformation of existing cities into adaptive and resilient environments; readers can refer to comprehensive works and reviews in published literature (e.g. Batty, 2013; Birkmann & Mechler, 2015; DEFRA, 2016). This chapter aims to discuss the role of infrastructure in resilient cities with a focus on adaptation strategies; it will review the main notions and concepts, and discuss a case study as proof of concept. It is intended to delineate a flood-wise city from an infrastructural point of view, illustrating advances in contemporary practice.

7.1.1 Definition of main terms

In literature, words like 'risk' and 'resilience' are becoming increasingly popular, with different interpretations. This chapter refers to the definitions of terms given below.

- Hazard: a process, phenomenon or human activity that may cause loss of life, injury or other health impacts, property damage, social and economic disruption or environmental degradation (UNDRR, 2019).
- (National) Infrastructure: the fundamental facilities and systems serving a country, city, or other area, including the services and facilities necessary for its economy to function (O'Sullivan & Sheffrin, 2003).
- Reliability: a measure of the margin between demand and capacity, expressed in terms of probability of failure (UNDRR, 2019).
- Risk: the product of the probability of a hazard and the consequential damage, summed over all possible events, which is often quoted in terms of an expected annual damage (Hall *et al.*, 2003).
- Resilience: the ability of assets, networks and systems to anticipate, absorb, adapt to and/or rapidly recover from a disruptive event (Cabinet Office, 2011).
- Adaptation: adjustments in natural or human systems in response to actual or expected climatic stimuli or their effects, which moderate harm or exploit beneficial opportunities (IPCC, 2014).

7.2 REVIEW OF THE CONTEXT

7.2.1 Flooding hazard

The overflow of water that submerges land that is usually dry is a flood. Flooding is one of the most frequent and costliest hazards in many countries worldwide. This phenomenon can be caused by a range of triggers (Table 7.1), namely: (i) rivers, canals, mountain streams, or periodic water sources (generally, riverine flooding) – due to water exceeding the capacity of the water system; (ii) heavy rain (flash floods) – due to intense and sudden rainfall that overwhelms drainage and does not allow the soil to absorb the runoff; (iii) groundwater – due to prolonged rainfall that saturates the soil, often associated with high levels of surface water; (iv) sea and ocean (coastal floods) – due to sea water floods from estuaries and coastal lakes, usually associated with high tide levels, strong winds and high waves (storm surge); (v) drain and sewer – due to a blockage or failure within the drainage system, not necessarily attributed to weather; (vi) snowmelt – due to surface runoff associated with melting snow and ice; (vii) infrastructure – due to accidental failure of flood defense infrastructure (e.g. dams).

Flood risk combines the probability of flooding and the consequential damage (Hall *et al.*, 2003). Thus, flood risk depends upon the characteristic of the hazard trigger (usually represented by one or more intensity measures such as flood depth), the characteristics of the exposure (land use, assets value) and the vulnerability of the exposed elements to the hazard. Flood risk maps usually help to identify locations where there is potential of significant flood risk. These maps can be produced by means of simulation models. First, for a flood event of

Table 7.1 Different types of flood.

Flood Type	Cause	Duration	Damage
1. Riverine	Water exceeding the capacity of a body of water	Weeks to months	High
2. Flash flood	Intense sudden rainfall	Days to weeks	Medium-high
3. Groundwater	Prolonged rainfall	Weeks to months	Medium-high
4. Coastal	Storm surge	Weeks to months	High
5. Urban	Drainage system overwhelmed	Days to weeks	Medium-high
6. Snowmelt	Snow and ice melting	Weeks to months	Medium-high
7. Infrastructure failure	Accidental failure of e.g. dams	Days to weeks	High

reference, the runoff per catchment area is calculated, accounting for topographic features, by implementing a hydrologic model that converts precipitation to runoff. Next, a detailed hydraulic model is used in conjunction with the hydrologic model output to define a flow versus depth relationship for flood inundation extent (Merz *et al.*, 2010). There are a wide variety of models that account for varying degrees of physical complexity and offer subtly different solutions to a given problem (e.g. Neal *et al.*, 2012).

The damage estimation consists of evaluating costs and losses caused by floods to assets (e.g. buildings, infrastructure, environment) and human lives and health. Possible climatic changes could affect flood seasonality and intensity, e.g. cause changes in rainfall, snow accumulation, and snowmelt; the consequences of flooding could be also exacerbated by urbanization (e.g. increase of impermeable surfaces) and land-use (e.g. buildings on the floodplain). To reduce flood losses within current and future risks, communities need to increase their resilience to flood events, by enhancing the robustness of critical infrastructure (see Section 7.2.2) and developing cost-effective intervention strategies (see Section 7.2.3).

7.2.2 Infrastructure resilience from a system perspective

An infrastructure consists of a network of man-made systems and processes that function cooperatively and synergistically to produce and distribute essential goods or services. Modern infrastructure has evolved from collections of discrete physical components such as buildings and bridges, roads or emergency services into a tightly interconnected and interdependent physical, cyber and human components. Critical infrastructures are 'those elements of national infrastructure the loss or compromise of which would result in major detrimental impact on the availability, delivery or integrity of essential services, leading to severe economic or social consequences or to loss of life' (CPNI, n.d.). The categorization of infrastructure varies but it typically includes the following sectors: communications, energy, transport, water, emergency services, financial services, government, food and health.

All infrastructures are subject to disruption due to different actions which could be internal to the system or external. Natural hazards such as floods, cyclones and earthquakes typically affect several infrastructure systems at the same time, resulting in damage to the infrastructure on a large scale. Man-made or technological hazards such as sabotage, explosions, fire and component failures alone usually do not result in widespread failure. Increasing complexity and interdependency of infrastructure systems can increase the risk of failure by propagating disruptions, so that the actual scale of the impact goes hugely beyond the area of the hazard. There are many examples (e.g. hurricane Katrina in USA 2005, tsunami in Indian Ocean region 2004, Christchurch earthquake 2010, Tohoku earthquake 2011) where their infrastructures were severely damaged resulting in wide-spread disruption to the societal functioning. In other cases (e.g. Eyjafjallajökull volcano 2011, heavy snow in the UK in early 2009, blackout in the Northeast USA in 2003) there was limited physical damage to the infrastructure system, but their functioning was disrupted, affecting the societal operations.

Predicting and managing the response of physical and human components of the infrastructure to the shocks and stresses is central to the functioning of society. Much of the infrastructure in Europe and in the USA is aged and in need of improvement as Tables 7.2 and 7.3 show; the grades are based on expert views of resilience, economic and social aspects, condition and capacity, leadership and other qualitative evidence.

The state of infrastructure in many other countries is not very different; some developing countries do not yet have adequate infrastructure while others regularly suffer from natural hazards including floods.

Table 7.2 State of the UK infrastructure (ICE, 2014).

Infrastructure Sector	Grade	Infrastructure Sector	Grade
Water and wastewater	B	Local transport	D–
Flood risk management	C–	Strategic transport networks (highways, air, ports)	B
Waste and resource management	C+	Energy	C–

Key: A – Fit for the future; B – Adequate for now; C – Requires attention; D – At risk; E – Unfit for purpose.

Table 7.3 State of the infrastructure in the USA (ASCE, 2017).

Infrastructure Sector	Grade	Infrastructure Sector	Grade
Drinking water	D	Ports	C+
Dams	D	Inland waterways	D
Levees	D	Roads	D
Wastewater	D+	Rail	B
Solid waste	C+	Bridges	C+
Hazardous waste	D+	Energy	D+

Key: A: Exceptional – fit for the future, B: Good – adequate for now, C: Mediocre – requires attention, D: Poor – at risk, F: Failing/critical – unfit for purpose.

7.2.2.1 Infrastructure risk and resilience

Typically, consequences of infrastructure failure can be grouped under three headings: (i) human – fatalities, injuries and psychological damage; (ii) economic – repair, replacement and compensation costs, traffic delay, re-routing and management costs, loss of business, reputation and share value; (iii) environmental – CO_2 emissions and pollutant release, energy costs. The UK summer floods in 2007 (Cabinet Office, 2008) are exemplary to demonstrate the scale of disruption to infrastructure and potential consequences: five water treatment works and over three hundred sewage treatment works were affected; the Mythe water treatment works in Gloucester alone resulted in cutting off the water supply to 350,000 people for 17 days; Walham substation came very close to failure which could have affected the electricity supply to nearly half a million people; Ulley Reservoir came close to being breached which could have resulted in the loss of life, damage to an important motorway, a major electricity substation and the gas network to Sheffield.

While each asset or component of infrastructure (e.g. power station or water pumping station) is designed to meet its performance requirements, predicting the knock-on consequences is not straightforward due to spatial distribution and interdependency of assets. The tools of scientific knowledge are well-established to model the demand and capacity of individual components or systems and arrive at probabilities of failure, i.e. reliability. To study the behaviour of geographically-distributed infrastructure, graph-theoretic tools are being increasingly used (Galvan & Agarwal, 2018). In a graph model of an infrastructure system, the nodes represent the components where the service is generated or delivered, the edges represent the connections between the components. The size of these networks can vary greatly depending upon the level of model required, e.g. national or city level. The effect of disruptions can be modelled by the loss of nodes or edges. A range of metrics are used to assess the consequences (Ouyang, 2014). For example, information centrality quantifies the consequences of removing a node in terms of the change in the efficiency of the network. The objective of such models is to identify the elements that are critical from the whole-system perspective, e.g. the failure of a few elements due to a localised hazard can result in loss of functionality over a much larger region. Such vulnerable elements may either be redesigned/strengthened, or protection and recovery measures may be put in place for rapidly restoring the infrastructure functionality. While such models are useful for planning and high-level decision-making, they are not intended to substitute physical models (e.g. hydraulic analysis of pipe networks or electric circuit analysis).

7.2.3 Adaptation strategies and adaptation benefits

Flood risk mitigation strategies can traditionally be classified into two main categories: structural and non-structural (Thampapillai & Musgrave, 1985). Structural measures are physical constructions and techniques aiming at reducing the flooding hazard; structural strategies modify the streamflow of rivers and channels leading to the reduction of the frequency and intensity of floods. Structural measures are further sub-classified into active and passive measures. Active structural measures modify the hydrograph involving mechanical or electrical systems (e.g. pumping), reducing and delaying the maximum peak discharge (e.g. on- and off-stream floodplain storages). The passive structural measures mitigate flooding by modifying the riverbed and its surroundings without involving mechanical or electrical systems (e.g. dams, levees, cleaning of the riverbed section from sediment, hydraulic bypass). Non-structural measures are procedures that do not require physical constructions; they consist of actions that lead to promoting knowledge, enforcing best practices, raising awareness and implementing strategic policies (e.g. flood early warning systems, land use regulations, flood insurance).

Over the last decade policy makers and stakeholders have been moving from the classical flood protection paradigm to the new concept of flood risk management. Specifically, urban drainage management has evolved significantly from a conventional 'rapid disposal' approach to a more integrated and sustainable 'design with nature' approach. Examples of this paradigm include new trending approaches worldwide, such as: Integrated Urban Water Management (IUWM), Water Sensitive Urban Design (WSUD), Sustainable Urban Drainage Systems (SuDS), Sponge Cities and Low Impact Development (LID) (De Risi *et al.*, 2018).

7.2.3.1 Monetary and non-monetary benefits from adaptation

To assess which adaptation strategy needs to be applied, it is fundamental to be able to select among different alternatives. This selection requires the identification of the losses (i.e. the risk) associated with flood damages (Table 7.4). Damages caused by floods are generally classified into tangibles and intangibles (Nadal *et al.*, 2009).

Table 7.4 Classification of direct/indirect, tangible/intangible damages.

	Tangible Damages (Market Losses)	**Intangible Damages (Non-Market Losses)**
Direct	E.g. repair costs, replacement, cleaning costs, debris removal	E.g. casualties, injuries
Indirect	E.g. business interruption, rerouting	E.g. increase of inequalities

Analogously, losses due to flooding can be categorised into market versus non-market and direct versus indirect losses (De Risi *et al.*, 2018).

Direct market losses are the negative impacts of the disaster itself on goods and services and are generally determined using observable data (e.g. repair costs). Direct non-market losses are costs that are caused by the disaster itself but whose economic value cannot be readily quantified because they are not themselves traded on markets (e.g. anxiety, mental suffering, environment degradation). Indirect losses are not caused by the immediate disaster itself but rather by secondary effects. For example, damage to an infrastructure may cause business interruptions that continue far beyond the duration of the actual flooding itself. Therefore, flooding may cause indirect losses on economic activity outside of the flooded area as well, e.g. losses through supply chains during the 2011 flooding in Thailand impacted the electronics industry.

Engineers quantify risk in economic terms; therefore, they mainly focus on direct market losses. This quantification is conventionally performed convoluting the hazard, vulnerability and exposure models (De Risi *et al.*, 2013), and it is often referred to in terms of an Expected Annual Loss (EAL) (Hall *et al.*, 2003). Such integration in fact leads to the assessment of the EAL, which is the average annual loss expected for the asset at stake. The EAL is a key element for the selection of the best mitigation strategy.

A conventional tool adopted to select the best mitigation alternative is the Cost-Benefit Analysis (CBA). CBA is a commonly used method to compare the cost and benefit of different risk mitigation strategies over an investigated time interval (Dong & Frangopol, 2017). Engineers can quantify the cost of the mitigation strategy adopting any quantitative survey technique, and the benefit consisting of the reduction of EAL due to avoided costs after the application of the mitigation strategy. A CBA can be performed in many different ways. It has been recently demonstrated that Life Cycle Cost (LCC) and Return on Investment (ROI) are efficient decision variables for evaluating the financial feasibility and economic performance, respectively, of a set of flood mitigation strategies over time (De Risi *et al.*, 2018). In the same study it has also been demonstrated that LCC and ROI analyses yield identical rankings of mitigation alternatives only if the public policy or program being evaluated produces benefits only in the form of avoided costs. The presence of other types of benefits breaks the equivalence of LCC and ROI.

7.3 FLOOD-WISE USE OF URBAN INFRASTRUCTURE

Urban development represents an opportunity to link resilience and sustainability. Jingdezhen is an example of urbanised area and mid-size industrial city that is rapidly growing in the Jiangxi province in China (Figure 7.1). Jingdezhen lies at the interface between the Huangshan mountains in the north-west, and Poyang Lake, China's largest freshwater lake, in the south-east. The city is built in a low-lying area at the confluence of the Changjiang river with its two tributaries, the Nanhe and Xihe rivers. The Changjiang river drains an area of 6222 km^2 and has an average flow rate of 89 m^3/s, and high inter-annual and inter-seasonal variability (Hohai University, 2015). The city has a sub-tropical monsoon

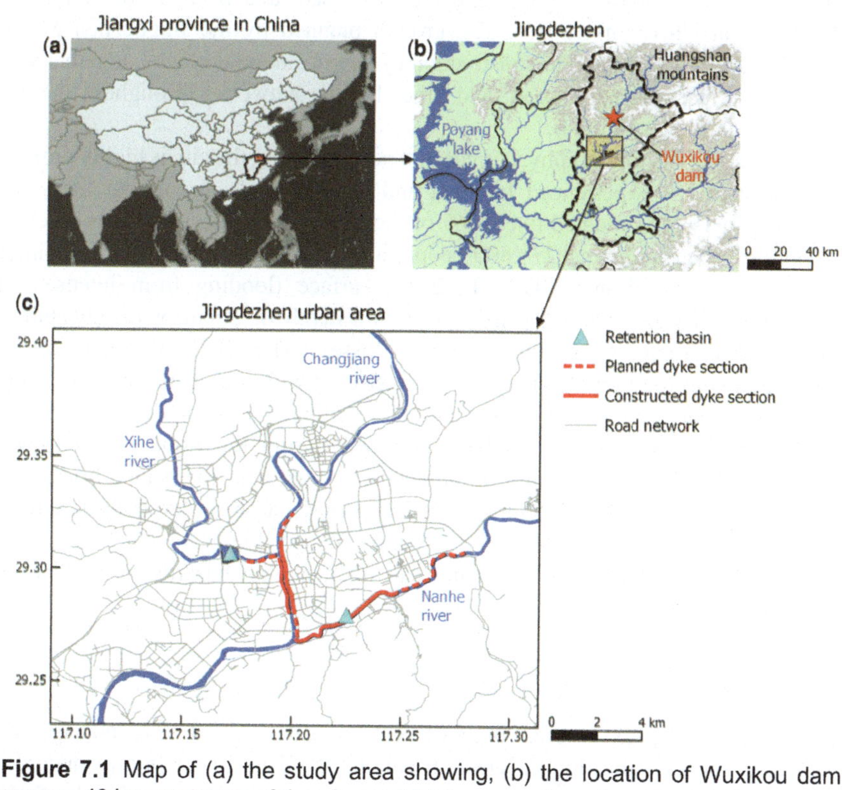

Figure 7.1 Map of (a) the study area showing, (b) the location of Wuxikou dam, approx. 40 km upstream of the city, and (c) levee sections to protect the city from a 1-in-20 year flood and two retention basins to store water from the Xihe and Nanhe tributaries. The road network is obtained from Open Street Map.

climate, characterised by warm temperatures and abundant rain. The average annual rainfall is 1800 mm, with half of the rainfall concentrated in the period between April and June (Hohai University, 2015). Between July and August, the city frequently experiences extreme rainfall events due to typhoons. The urban drainage infrastructure of the city consists of 26 km of pipes, but only 19% of it meets the standard for a one-year return period event. Issues such as bottleneck pipe sections, mild pipe gradients and insufficient outlets also contribute to the poor drainage capacity, and the rapid urbanisation is further increasing the pressure on the network (Wang *et al.*, 2018).

The city's population was approximately 480,000 in 2013 and it is predicted to increase to 1.2 million by 2050 (World Bank, 2013); this growing population is putting pressure on infrastructure, e.g. the water supply demand is forecasted to increase from 455,000 to 550,000 m^3/day between 2020 and 2030 (Artelia International, 2012). Its economy is growing at a rate of over 8% per year (World Bank, 2013) with regional GDP in 2017 corresponding to 87.8 billion Yuan, and ongoing urban growth would extend the urbanised area from 33 to 78 km^2 by 2030 (Hohai University, 2015). The growth planned by the Jingdezhen Master Plan 2012–2030 will include construction of new residential blocks, together with expanded water, drainage, road, and other lifeline networks throughout the city (Figures 7.2, 7.3(a) and (b)).

As a result of its topography, climate and infrastructure, the city is highly prone to flooding and has experienced frequent and severe floods throughout the years. Flooding in 1998 inundated over 90% of the urban area, in some places reaching a depth of 10 m, affecting 271,800 people and causing losses of over 2.3 billion Yuan (2.6% of regional GDP). In 2010, surface flooding from intense rain flooded an area of 9.1 km^2 up to a maximum depth of 2.8 m, with 2.96 billion (3.4% of regional GDP) in economic losses. The 2010 flooding caused interruptions to the power supply which interfered with the operation of drainage pumps that were being used to alleviate the flooding. The same occurred in 2016 when pumping stations had to be shut down during the flood due to power outages and the risk of collapse of an electricity pole near the Nanhe bridge that was still carrying traffic. A total of 119,700 people had to be evacuated, and the city sustained economic losses of 1.9 billion Yuan (Wang *et al.*, 2018). These events demonstrate how infrastructure can interact with hazards to amplify the consequences in urban areas.

7.3.1 Flood risk management in Jingdezhen

In 1998, the Jiangxi region authorities launched a flood risk management project to increase the flood protection beyond the 1-in-5-year flood standard at that time. The objective is to provide 1-in-100 level flood protection by 2050, with expected annual losses limited to 0.5% of GDP, and to ensure no fatalities due to floods. Structural measures were designed to protect up to a 1-in-50 year

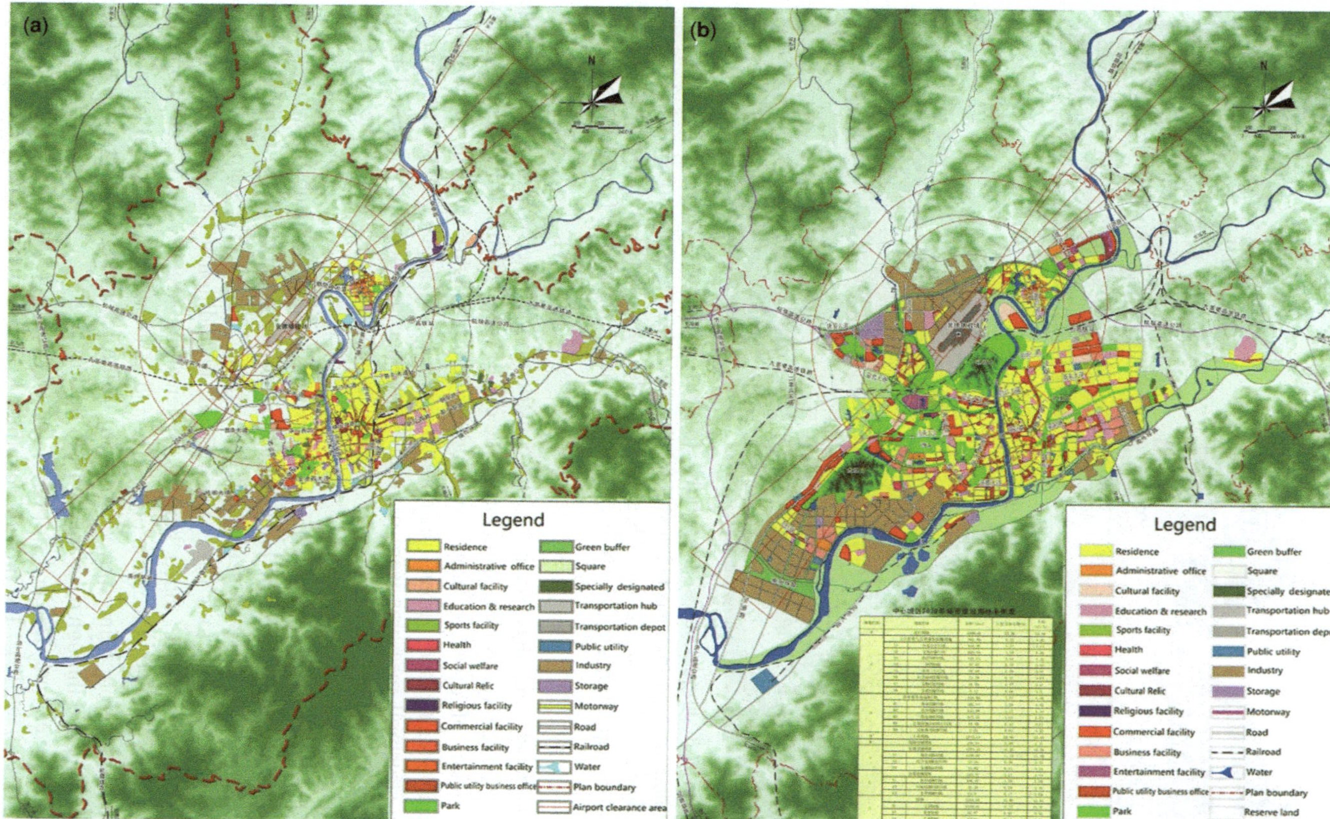

Figure 7.2 Land use in (a) 2010 and land use planning in (b) 2030 for the city of Jingdezhen, according to Jingdezhen Master Plan (2012–2030).

Figure 7.3 Jingdezhen's snapshots: (a) the Old Town, with the characteristic architecture and low-rise buildings; (b) the new development consists in high-rise buildings; (c) example of SUDs (permeable pavement) in parking areas; (d) greenery and pedestrian walkway, as part of the construction of the levee section.

flood level, and these include: (a) construction of the Wuxikou dam on the Changjiang river, 40 km upstream of the city (Figure 7.1(b)); (b) the construction of 58 km of dikes to protect the urban centre (Figure 7.1(c)); (c) drainage improvements and urban retention basins to manage stormwater. Further non-structural measures were planned to manage the flooding up to a 1-in-100 flood, including enabling the city to better evacuate and recover in the event of a flood (Figure 7.3(c) and (d)).

The Wuxikou dam on the Changjiang river is located 40 km upstream of the city. The dam (46 m high, 538 m long) has a total storage capacity of 427 million m^3 and an installed hydropower generation capacity of 32 MW. The dam is designed to protect the city up to a 1-in-50-year flood. In addition to flood mitigation, it is expected to provide added benefits by ensuring security of water supply, to satisfy the forecasted increase of the demand. The current abstraction point on the river is therefore being moved upstream to the reservoir, from which water will be channelled through a separate pipe to the city.

The levees are designed to protect the city up to a 1-in-20-year flood (Hohai University, 2015). To provide continued access to the river by the city inhabitants, the levees include stairs that lead to a green area along the river, with pedestrian walkways, play areas, and diverse vegetation (Figure 7.3(d)). The construction of the levees also includes 16 pumping stations that drain the flood waters and limit the extent of damage in case of flooding.

Drainage improvements include the construction of separated stormwater and foul water sewers in newly developed parts of the city, and an upgrade of the network in the old part of the city to have the capacity to drain a 1-in-20 year, 24-hour-duration, rainfall event. Urban stormwater management capacity has also been increased by constructing two large water retention areas, the Changnan Lake and Laonanhe retention basin in the low-lying areas of the city, to increase storage capacity from the two tributaries. Laonanhe retention basin has a total surface area of over 61,200 m^2 and a total storage depth of 4.5 m. It will include leisure space with playgrounds and a pleasant waterfront area, paved with permeable bricks and stone to enable infiltration of rainwater. Vegetated ditches will contribute to cleaning the rainwater that flows to the lake.

Non-structural measures are also being carried out to improve flood management capability. This includes training local decision makers, conducting visits to observe and learn from best-practice in other cities in China and abroad, conducting an education campaign to raise awareness within the local population, updating the forecasting system to include the influence of the operation of the Wuxikou dam, and including risks associated with dam or levee break into the existing emergency response plan.

7.3.2 Costs and benefits from adaptation measures

The total cost of the flood protection project is 513.7 million US$, which corresponds to approximately 4% of the regional GDP in 2017. Of this, 114.9 million (22%) is used for constructing the dam, 384.6 million (75%) for implementing the resettlement action plan, and 9.4 million (2%) for additional non-structural mitigation measures. Resettlement costs thus represent most of the project cost: the construction of the levees means displacing over 2000 homes and over 300 businesses from the flood prone riverbanks and resettling them to other parts of the city, while building the dam involves resettling 9800 people from villages within the reservoir area (Artelia International, 2012). Funding for the project comes from a partnership funding involving the local government (41.1%), supplemented by the regional (9.4%) and national government (30%), together with a loan from the World Bank (19.5%).

For the Wuxikou dam, eight scenarios were modelled to assess the dam effect in terms of flood alleviation in the future (2030), as compared to the present day (2010) (Figure 7.4). Four maps (Figure 7.4(a), (c), (e) and (g)) simulate various return periods, land use, drainage network, and embankment for the present day (2010). The other four maps (Figure 7.4(b), (d), (f) and (h)) are for a 2030 land use plan, while the drainage network and embankment do not change with respect to 2010.

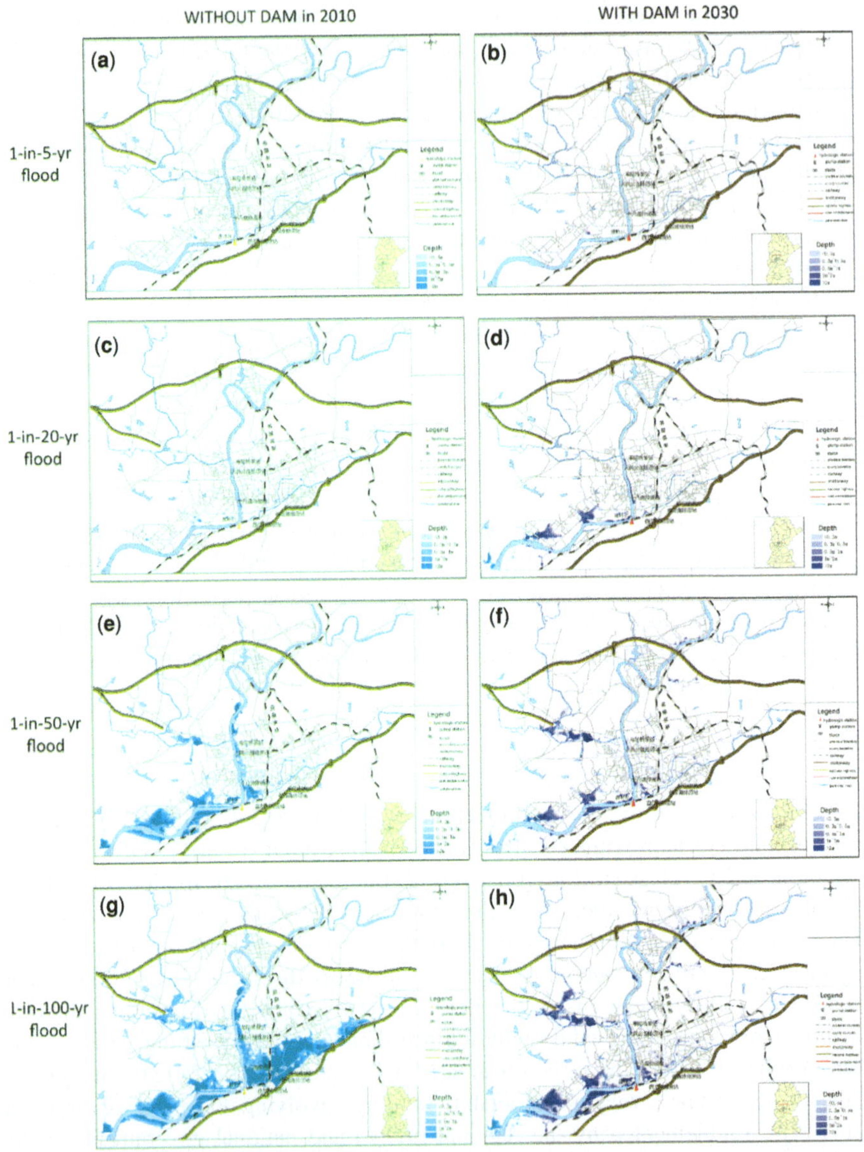

Figure 7.4 Modelled flood maps for present (2010) and future (2030) scenarios: (a/b) 1-in-a-five-year flood event without/with Wuxikou dam; (c/d) 1-in-a-20-year flood event without/with Wuxikou dam; (e/f) 1-in-a-50-year flood event without/with Wuxikou dam; (g/h) 1-in-a-100-year flood event without/with Wuxikou dam. 2030 scenarios include urban development according to land use planning.

With respect to the 1-in-a-five-year flood event, the flood footprint increased by +67% from the current and with respect to the 1-in-a-20-year flood event, the flood footprint increased by +69%. On the contrary, with respect to the 1-in-a-50-year flood event, the flood footprint decreased by 42% and with respect to the 1-in-a-100-year flood event, the flood footprint reduced by –36% (Figure 7.5 and Table 7.5).

The interventions are also expected to bring other improvements in the quality of life in the city. Green areas along the levees and urban storage areas will provide public leisure space and access to nature which improves wellbeing. The construction of the dam will provide a reliable source of water to mitigate shortages. Buildings along the riverfront that were exposed to risks of erosion of the riverbanks were resettled to new areas away from the river, providing safer living conditions to the residents. The key objectives that underly flood management efforts are indeed to safeguard social stability and to promote stable economic development.

The embankment of the urban reach of Changjiang River is designed to withstand the shocks of 1-in-50-year flood with the Wuxikou Dam; several sections of the embankment should be raised if using the flood protection standard of 1-in-100-years. However, raising the existing embankment would lead to high costs due to the engineering quantity and urban planning (e.g. resettling the population), while non-structural measures were proven to reduce the total loss if supplemented to relatively low protection standards. Therefore, the combination of structural (levees, dam, drainage upgrades and retention basins) and non-structural (evacuation, insurance, raising awareness) measures was found to be the most cost-effective way of limiting future damage.

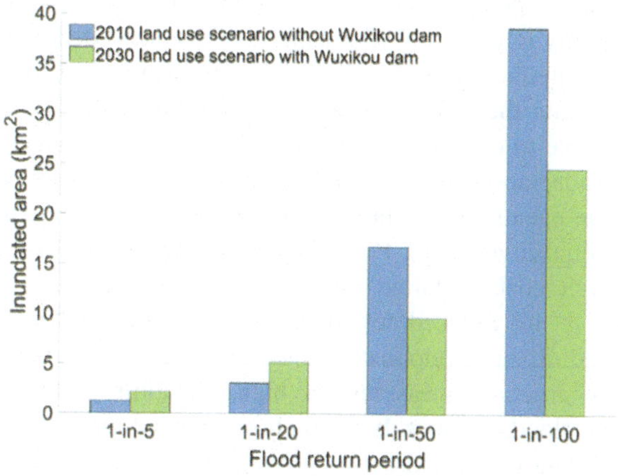

Figure 7.5 Inundated area (km^2) per flood return period and scenario.

Table 7.5 Modelled flood depths and footprints for the eight scenarios (2010 without dam/2030 with dam).

Return Period	Area in Different Depth (km^2)					Total Flood Area (km^2)
	<0.5 m	0.5–1 m	1–2 m	2–3 m	>3 m	
Flood area in 2010 scenarios						
1-in-5-year	1.070	0.182	0.056	0.001	–	1.309
1-in-20-year	2.419	0.367	0.223	0.037	0.041	3.087
1-in-50-year	4.569	2.544	3.692	2.698	3.241	16.744
1-in-100-year	6.369	4.514	8.235	6.683	12.917	38.718
Flood area in 2030 scenarios						
1-in-5-year	1.777	0.163	0.142	0.070	0.046	2.198
1-in-20-year	3.653	0.802	0.444	0.180	0.147	5.226
1-in-50-year	4.826	1.668	1.326	1.380	0.486	9.686
1-in-100-year	7.289	5.240	5.572	3.746	2.831	24.678

7.4 DISCUSSION

Flood losses are expected to increase due to climatic changes and urbanization. The case study of Jingdezhen presented an example of an ongoing flood risk management project for enhancing urban resilience and reducing the impacts of flooding in cities through infrastructure. The combined approach using structural and non-structural measures reflects the recent shift in thinking and practice, which has moved away from a flood control approach, towards increased flood risk management.

Strategies that are being implemented combine increasing flood protection with an improved ability to cope with flooding. As the reservoir, levees, and retention lakes reduce the frequency of flooding, other measures (e.g. education) will be essential to maintain the awareness and preparedness of the population to ensure that the city is able to cope when a flood does occur. In a context of remarkable urban growth, adaptation (structural) measures that target low-probability high-impact flooding events (i.e. 1-in-50/1-in-100-year events) guarantee a higher return of the initial investment. For less extreme events (i.e. 1-in-5/1-in-20-year events), the high costs required are less justified because, for example, the inundated area is not reduced for the 2030 period. However, future scenarios included planned urban development which increases urban runoff; this runoff is tackled by non-structural measures for high-probability events. This approach recognises that floods cannot be entirely prevented and aims to promote a philosophy of living with floods.

7.4.1 Next frontier of research

The combination of structural and non-structural measures promotes cost-effective planning for enhancing urban resilience. These measures target flood alleviation as the main benefit but encompass a wider range of positive effects. For example, the Laonanhe retention basin improved the urban quality of Jingdezhen by including a leisure space with playgrounds and a public waterfront area. However, no method exists in current practice and research to account for such co-benefits. Future research could investigate about how to include co-benefits from adaptation measures into existing economic appraisal (EAL, ROI, etc.).

There is increasing research effort on identifying interdependencies between different infrastructure systems. This is expected to lead to scenarios that have the potential to be critical from a system-of-systems perspective. These can then be used for designing appropriate protection measures from floods and/or preparing rapid recovery plans. Infrastructure improvements could indeed form part of a city-wide resilience strategy, to ensure that existing and newly constructed lifelines are able to cope in the event of a flood. For example, isolating and waterproofing the electricity supply to ensure continued operation during floods; identifying priority road sections and junctions that could be strengthened to maintain connectivity of the city during and after flooding; or better understanding the criticality of the regional rail, road, and power transmission grids, so that new construction can be combined with reducing the vulnerability of the networks as a whole.

7.5 CONCLUSION

Flooding risk to cities has to be managed through resilient and sustainable planning, especially in fast-developing areas. The planning of a flood-wise city requires understanding of the potential consequences from a hazard, designing for structural measures to reduce these consequences and preparing the community to withstand impact. The city of Jingdezhen showed how structural (e.g. dikes, levees) and non-structural measures (e.g. preparedness) could be successfully combined for reducing flooding risk for future scenarios. Future research could follow up on how to integrate adaptation co-benefits into current economic appraisal of adaptation measures and how to address resilience from a system-of-system perspective.

ACKNOWLEDGEMENTS

This work was supported by the Engineering and Physical Sciences Research Council (ESPRC) LWEC (Living With Environmental Change) fellowship (EP/R00742X/1 and 2); EPSRC WISE (Water Informatics: Science and Engineering) grant (EP/L016214/1); Jiangxi Wuxikou Integrated Flood

Management Project Management Office (Consulting Services for: Development of Master Plan for Jingdezhen City Integrated Flood Risk Management (no. JDZ-WXK-ZX-09); National Natural Science Foundation of China (NSFC) grant (no. 71601070); Resilient Economy and Society by Integrated SysTems modelling (RESIST), Newton Fund via Natural Environment Research Council (NERC) and Economic and Social Research Council (ESRC) (NE/N012143/1).

REFERENCES

Aerts J., Botzen W., Bowman M., Dircke P. and Ward P. (2013). Climate Adaptation and Flood Risk in Coastal Cities. Taylor and Francis, London. http://ncl.eblib.com/patron/FullRecord.aspx?p=1576080, (accessed 20 April 2018).

Arrighi C., Pregnolato M., Dawson R. J. and Castelli F. (2019). Preparedness against mobility disruption by floods. *Science of the Total Environment*, **654**, 1010–1022.

Arrighi C., Pregnolato M. and Castelli F. (2020). Indirect flood impacts and cascade risk across interdependent linear infrastructures. *Natural Hazards and Earth System Sciences Discussion*, in review. https://doi.org/10.5194/nhess-2020-371

Artelia International (2012). Jiangxi Wuxikou Integrated Flood Management Project – Environment Impact Assessment Executive Summary. World Bank Group, Washington, DC. http://documents.worldbank.org/curated/en/178691468243566032/Environmental-impact-assessment-executive-summary, (accessed 13/11/2020).

ASCE (2017). 2017 Report Card for America's Infrastructure. American Society of Civil Engineers, Virginia, USA. http://www.infrastructurereportcard.org

Batty M. (2013). The New Science of Cities. MIT Press, Cambridge, MA.

Birkmann J. and Mechler R. (2015). Advancing climate adaptation and risk management. New insights, concepts and approaches: what have we learned from the SREX and the AR5 processes? *Climatic Change*, **133**(1), 1–6.

Cabinet Office (2008). The Pitt Review: Lessons Learned from the 2007 Floods. Crown Copyright, London.

Cabinet Office (2011) Keeping the country running: natural hazards and infrastructure. http://www.gov.uk/government/publications/keeping-the-country-running-natural-hazards-and-infrastructure, (accessed 20 April 2017).

Chang S. E. (2016). Socioeconomic Impacts of Infrastructure Disruptions. Oxford Research Encyclopaedia of Natural Hazard Science, Oxford, UK. doi: 10.1093/acrefore/9780199389407.013.66. naturalhazardscience.oxfordre.com, (accessed 20 April 2018).

CPNI (n.d.). Critical National Infrastructure. Centre for Protection of National Infrastructure, London. http://www.cpni.gov.uk/critical-national-infrastructure-0

Dawson R. J., Affleck A., Blythe P., Campbell D., Fu G., Gibbon J., Glendinning S., Garrod G., Heidrich O., Kafouros M., Kerr N., Knoeri C., Passarella M., Purnell P., Robertson M., Roelich K., Sandham R., Spencer D. and Steinberger J. (2015). Are You Being Served? Alternative Infrastructure Business Models to Improve Economic Growth and Well-being. Centre for Earth Systems Engineering Research, Newcastle. ISBN: 978-09928437-1-7

DEFRA (2016). Climate Change Risk Assessment 2017. DEFRA, London, UK. www.gov.uk/government/publications, (accessed 21 August 2019). ISBN: 9781474137416

De Risi R., Jalayer F., De Paola F., Iervolino I., Giugni M. and Topa M. E. (2013). Flood risk assessment for informal settlements. *Natural Hazards*, **69**(1), 1003–1032.

De Risi R., De Paola F., Turpie J. and Kroeger T. (2018). Life cycle cost and return on investment as complementary decision variables for urban flood risk management in developing countries. *International Journal of Disaster Risk Reduction*, **28**, 88–106.

Dong Y. and Frangopol D. M. (2017). Probabilistic life-cycle cost-benefit analysis of portfolios of buildings under flood hazard. *Engineering Structures*, **142**, 290–299.

Elmqvist T., Andersson E., Frantzeskaki N., McPhearson T., Olsson P., Gaffney O., Takeuchi K. and Folke C. (2019). Sustainability and resilience for transformation in the urban century. *Nature Sustainability*, **2**(4), 267–273. http://dx.doi.org/10.1038/s41893-019-0250-1

Galvan G. and Agarwal J. (2018). Community detection in action: identification of critical elements in infrastructural networks. *ASCE Journal of Infrastructure Systems*, **24**(1), 04017046.

Garschagen M., Hagenlocher M., Comes M., Dubbert M., Sabelfeld R., Yew Jin L., Grunewald L., Lanzendörfer M., Mucke P., Neuschäfer O., Pott S., Post J., Schramm S., Schumann-Bölsche D., Vandemeulebroecke B., Welle T. and Birkmann J. (2016). World Risk Report 2016. Bündnis Entwicklung Hilft and UNU-EHS, Berlin. ISBN: 9783946785026

Hall J. W., Dawson R. J., Sayers P. B., Rosu C., Chatterton J. B. and Deakin R. (2003). A methodology for national-scale flood risk assessment. *Proceedings of the ICE-Water and Maritime Engineering*, **156**(3), 235–247.

Hohai University (2015). Consulting Services for: Development of Master Plan for Jingdezhen City Integrated Flood Risk Management. Hohai University, Nanjing, China.

ICE (2014). The State of the Nation: Infrastructure 2014. Institution of Civil Engineers, London. http://www.ice.org.uk/media-and-policy/policy/state-of-the-nation-infrastructure-2014

IPCC (2014). Climate Change 2014: Impacts, Adaptation, and Vulnerability. The Intergovernmental Panel on Climate Change, Geneva, Switzerland. http://ipcc-wg2.gov/AR5/, (accessed 20 April 2019).

Merz B., Kreibich H., Schwarze R. and Thieken A. (2010). Assessment of economic flood damage. *Natural Hazards and Earth System Sciences*, **11**(6), 1697–1724.

Nadal N. C., Zapata R. E., Pagán I., López R. and Agudelo J. (2009). Building damage due to riverine and coastal floods.*Journal of Water Resources Planning and Manageement – ASCE*, **136**(3), 327–336.

Neal J., Schumann G. and Bates P. (2012). A subgrid channel model for simulating river hydraulics and floodplain inundation over large and data sparse areas. *Water Resources Research*, **48**(11), 1–16.

O'Sullivan A. and Sheffrin S. M. (2003). Economics: Principles in Action. Pearson Prentice Hall, Upper Saddle River, NJ, p. 474. ISBN 978-0-13-063085-8

Ouyang M. (2014). Review on modelling and simulation of interdependent critical infrastructure systems. *Reliability Engineering and Systems Safety*, **121**, 43–60.

Pregnolato M. and Dawson D. A. (2018). Adaptation investments for transport resilience: Trends and recommendations. *International Journal of Safety and Security Engineering*, **8**(4), 515–527.

Pregnolato M., Jaroszweski D., Ford A. and Dawson R. J. (2020). Climate extremes and their implications for impact modeling in transport. In: Climate Extremes and Their Implications for Impact and Risk Assessment, J. Sillmann, S. Sippel and S. Russo (eds.), Elsevier, Amsterdam, pp. 195–216. https://doi.org/10.1016/B978-0-12-814895-2.00011-2

Stewart M. and Deng X. (2014). Climate impact risks and climate adaptation engineering for built infrastructure. *Journal of Risk and Uncertainty in Engineering Systems, Part A: Civil Engineering*, **1**(1), 1–12.

Thampapillai D. J. and Musgrave W. F. (1985). Flood damage mitigation: a review of structural and non-structural measures and alternative decision frameworks. *Water Resources Research*, **21**(4), 411–424.

UNDRR (2019). Terminology. United Nations Office for Disaster Risk Reduction, Geneva, Switzerland. http://www.unisdr.org/we/inform/terminology#letter-h

United Nations (2019). World Urbanization Prospects: The 2018 Revision (ST/ESA/SER. A/420). United Nations, Department of Economic and Social Affairs, Population Division, New York. https://population.un.org/wup/Publications/Files/WUP2018-Report.pdf, (accessed 20 August 2019).

Wang Z., Wang H., Huang J., Kang J. and Han D. (2018). Analysis of the public flood risk perception in a flood-prone city: The case of Jingdezhen city in China. *Water*, **10**(11), 1577. doi: 10.3390/w10111577

World Bank (2013). International Bank for Reconstruction and Development Project Appraisal Document – Jiangxi Wuxikou Integrated Flood Management Project. World Bank Group, Washington, DC. http://www-wds.worldbank.org/external/default/WDSContentServer/WDSP/IB/2014/06/04/000470435_20140604100811/Rendered/PDF/PAD7220PAD0P13010Box385222B00OUO090.pdf

Chapter 8

Building resilience in water supply infrastructure in the face of future uncertainties: Insight from Cape Town

Upeshika Heenetigala, Leon Kapetas and Richard Fenner

Department of Engineering, University of Cambridge, CB2 1PZ, Cambridge, UK

8.1 INTRODUCTION

Water is a renewable but finite resource that is subject to enormous temporal and spatial variability which has an influence on how people and ecosystems value its importance (Postel, 2000; Rijsberman, 2006). Sustainable Development Goal 6 reaffirms that access to water and sanitation is essential for human and environmental health, and economic prosperity (United Nations, 2017). Water scarcity, which is perceived to be a growing systemic risk (Mekonnen & Hoekstra, 2016), is defined as an excess of demand over available supply under prevailing institutional arrangements and/or prices. In terms of potential impact, the World Economic Forum (2018) has consistently listed water crises as one of the top five global risks since 2015. Approximately two-thirds of the global population experience severe water scarcity for at least one month of the year and half a billion people face it all year round (Mekonnen & Hoekstra, 2016).

The effects of water scarcity are particularly sensitive in the rapidly growing cities of the Global South as climate change highlights existing vulnerabilities in areas with high proportions of urban poor (Araos *et al.*, 2017). These areas have

Editors: Xiaochang C. Wang and Guangtao Fu
doi: 10.2166/9781789060768_0201

limited capacity to cope and recover from loss whilst being highly exposed to water-related risks, exacerbated by rapidly expanding populations, urbanisation and chaotic economic growth (MacAlister & Subramanyam, 2018). The lack of access to safe drinking water and sanitation results in a vicious cycle of poor hygiene, malnutrition, health and poverty which has adverse effects on people's wellbeing and livelihoods (Rijsberman, 2006). It also destabilises food production and security, ecosystem health, sustainable economic growth, infrastructure development and political and social stability (MacAlister & Subramanyam, 2018; Postel, 2000).

Making long-term water infrastructure decisions is challenging when faced with the climatic and socio-economic changes that an uncertain future might bring. It is therefore critical to test whether integrated, systems-based strategies are more resilient and adaptive to change than traditional assessment approaches to select water supply options. This chapter develops an options evaluation framework for water supply infrastructure options to assist in the early stages of long-term planning decisions. While many studies have been conducted to develop and test such frameworks, few consider the concept of resilience as a standalone, risk-based criterion. There are several definitions of 'resilience' which distinguish between the community, ecological and engineering sense of the term. While Birkland and Waterman (2008) have suggested community resilience is based on damage prevention, speedy recovery and preservation of community functionality, Holling (1996) stresses that engineering resilience focuses on efficiency, constancy and predictability while ecological resilience focuses on persistence, change and unpredictability. The broader concept of resilience can be interpreted as involving the interaction of the resource, the physical infrastructure, the management and financing arrangements involved in service provision, and the behaviour of the water consumers, especially as resources become limited. In the context of water supply provision, this chapter will focus on one aspect of this conceptualisation, namely the physical infrastructure, whereby 'resilience' will be used to evaluate the ability of water supply sources to withstand climate change induced trends (Bichai *et al.*, 2015). This is in keeping with 'engineering resilience', aptly defined as 'the degree to which the system minimises level of service failure magnitude and duration over its design life when subject to exceptional conditions' by Butler *et al.* (2014) in the Safe and SuRe Framework.

To develop a holistic framework, it is crucial to understand which driving forces contribute to the problem of water scarcity to be able to create boundaries for the outcomes of the framework. The method chosen to test this framework is by means of a case study, namely the crippling drought that was experienced in Cape Town in South Africa from 2015 to 2018. The three main components of the options evaluation framework are listed below and illustrated in Figure 8.1.

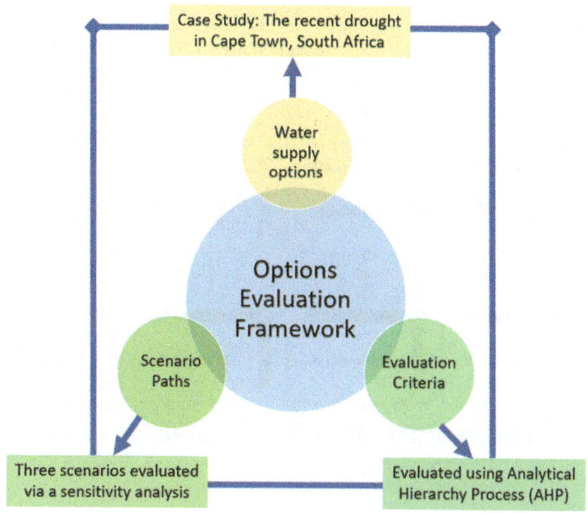

Figure 8.1 Options evaluation framework.

The components of the framework are:

(1) Water Supply Options: The focus will be on four water supply options that were considered by local officials in Cape Town during the drought to test the framework.

(2) Evaluation Criteria: Six key evaluation criteria will be used to assess the options using the Analytical Hierarchy Process (AHP) method. The AHP is an additive MCDA method that generates a single score for each alternative which enables options to be directly comparable to each other (Ainger & Fenner, 2014). This approach has been widely utilised for urban water system analysis as it considers multiple criteria in a decision-making environment which improves the analytical rigor of the process (Hajkowicz & Collins, 2007; Lai *et al.*, 2008). A complex decision problem is expressed as a hierarchy, as shown in Figure 8.2, whereby the overall objective of the decision lies at the top of the hierarchy and the criteria affecting the decision and the alternatives lie on each descending level (Fong & Choi, 2000).

(3) AHP is based on the pairwise comparison of criteria/alternatives with respect to criteria whereby decision-makers are asked to express preference for one option over another using the Fundamental Scale developed by Saaty (1987).

(4) Scenario Paths: Rather than establishing real stakeholder preferences from primary field work that would provide context to the situation on the ground in Cape Town, a sensitivity analysis is conducted to assess how the results

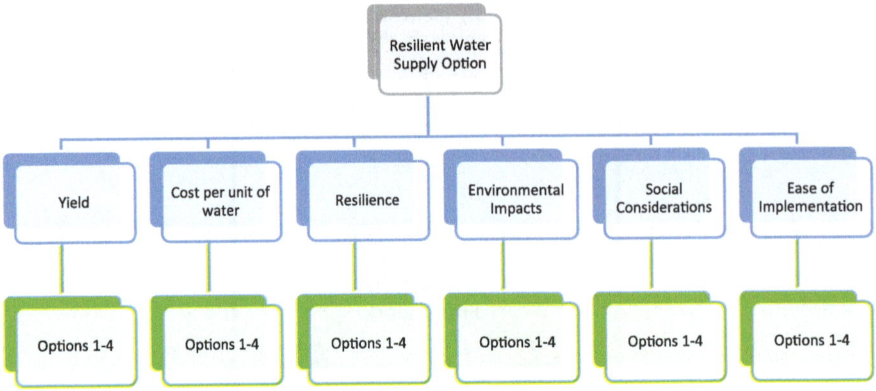

Figure 8.2 Hierarchical structure of the AHP (adapted from Saaty, 1987).

of the AHP change when the weights of the criteria are altered. This is conducted through the lenses of three 'perspectives' derived from the cultural theory of Thompson (1988) and adopted by Hoekstra (2000) for the field of water management. Key themes from the Hierarchist, Egalitarian and Individualist perspectives will be explored and adapted for the area of water supply management.

8.2 THE DROUGHT IN CAPE TOWN

2017 was one of the driest years on record in recent decades in Cape Town, following two successive dry winters in 2015 and 2016, in what scientists are referring to cumulatively as a 1-in-300-year event (Wolski, 2018). In January 2018 'Day Zero', the point when water levels in the reservoir system reach 13.5% of capacity and taps are turned off, was originally calculated to fall on the 12th of April (DWS, 2018). However, following a rise in reservoir levels due to well overdue rainfall, Cape Town averted a crisis that could have made them the first major city in the modern era to run out of water (McKenzie & Swails, 2018).

8.2.1 Water resources

Cape Town is serviced by the Western Cape Water Supply System (WCWSS), which is an integrated system of 14 dams, pump stations, pipelines, canals and tunnels (Basholo, 2016). Approximately 88% of the system's water comes from surface runoff from mountain catchments with the remaining 12% coming from various augmentation schemes such as the Atlantis groundwater aquifer recharge scheme (DWS, 2018). The majority of rain falls during the winter months between May and September and therefore it is crucial to store water to meet the demands of the whole year, particularly during the drier summer months (Mauck,

2017). The total dam storage capacity is approximately 900 Mm^3, with the six major dams – the Theewaterskloof, Voëlvlei, Berg River, Wemmershoek and the Steenbras Upper and Lower dams – storing 99.6% of the systems raw water (Mauck, 2017). The WCWSS has an unconstrained annual yield of approximately 570 Mm^3, of which Cape Town is allocated a maximum of 399 Mm^3a^{-1} (Currie *et al.*, 2017). Twenty-nine and 7% of the remaining annual yield is allocated for agricultural use and to other municipal areas respectively (DWS, 2018).

Due to the heavy reliance of the WCWSS on surface water storage for its supply, dam levels have been closely monitored for many years. Historically, for the first few months of the year during the summer dam levels drop and then increase mid-year during the rainy season. According to research conducted by the DWS, at the beginning of 2018 dam levels were 15.5 and 24.4% lower than at the same point in 2017 and 2016 respectively. In March 2018 total dam storage levels were at 22.9%, however by the middle of May 2018 the gap had closed and the current dam storage levels, as at July 2018, were 56.4% (City of Cape Town, 2018).

8.2.2 Water system vulnerabilities

The following section will assess the changing patterns of risk accumulation which influenced aspects of the risk equation and invariably contributed to the severity of the water crisis.

8.2.2.1 Climate variability

Cape Town has a Mediterranean climate with maximum summer temperatures of 26.9°C and winter temperatures ranging from 9.1 to 17.7°C (Tadross & Johnston, 2012). Mean annual rainfall is relatively low at approximately 700 mm/y and has become erratic over the last few years with the city experiencing irregular episodes of hydrological drought, including between 1986 and 1988, 2000 and 2001, 2004 and 2005 and more recently 2015 and 2018 (Mukheibir & Ziervogel, 2007).

Predicting rainfall for the Western Cape has proven to be a difficult task, with premier forecast institutions such as the European Centre for Medium-Range Weather Forecasts and the International Research Institute for Climate and Society failing to provide accurate forecasts (Wolski *et al.*, 2017). This is mainly because of the relationship between El-Niño, which brings the drought to South Africa, and La Niña episodes, which brings wet conditions, and the Western Cape's winter rainfall patterns is weak and inconsistent (Wolski *et al.*, 2017). 'The well-below-average rainfall of 2016 and 2017 occurred during a weak La Niña and a weak El-Niño respectively. This does not reflect the expected El-Niño-rainfall relationship' (Wolski *et al.*, 2017). Research conducted by the University of Cape Town suggests that future climatic projections show a shift towards a drier, more drought-prone climate which means that the possibility of extreme drought events are also increasing (Wolski *et al.*, 2017).

8.2.2.2 Population growth and urbanisation

Cape Town's population has grown by an estimated 3% since 2011 and 52% since 1996, when the population was just over 2.5 million people (Currie *et al.*, 2017). The increasing demand for water has put pressure on the current water supply system, whose capacity has not increased to match the growing population, and made the city increasingly vulnerable to water scarcity (DWAF, 2009).

8.2.2.3 Water supply and demand management

Based on low rainfall levels in 2004–2005, the DWS issued the city with a warning in 2007 stating that new water sources would be required by 2015 (Olivier, 2017). The CCT implemented several successful demand management initiatives to conserve water and reduce inefficiencies in the system (Sousa-Alves, 2015). These included plumbing leak detection and meter repairs, pipe replacement, pressure reduction, water restrictions or load shedding, stepped tariffs and user education programmes (City of Cape Town, 2017b). A review of these strategies found that water loss reduced from 23.7% in 2009/2010 to 14.7% in 2013/2014 (Sousa-Alves, 2015).

However, the only significant addition to the WCWSS since 1995 has been the Berg River dam, which increased storage capacity by 14% by adding 130 Mm^3 of the total 900 Mm^3 capacity (Bohatch, 2017). It could therefore be deduced that the success of these demand management strategies effectively delayed the need to develop more storage capacity in the form of dams, which was to the city's detriment in the face of a 1-in-300-year drought.

8.2.2.4 Water pricing and social inequality

Post-apartheid, the new South African constitution in 1994 guaranteed water as a human right (RSA, 1996). The three-step block tariff structure introduced in Cape Town in the 1960's was replaced by a five-step block tariff in 1998 (Smith, 2004). The motivation for block tariffs was 'to meet the constitutional obligation of ensuring a pro-poor tariff by cross-subsidizing low-end users through high rates to high-level users' (Smith, 2004). Further motivated by social equity concerns, the Free Basic Water policy was introduced in 2001 to provide 6 kilolitres of water per month to every household regardless of income level or size at no cost – one of a handful of such policies in the world (Szabó, 2015). Together, the water tariff subsidies the first 6 kilolitres by increasing prices in the remaining steps of the tariff. Based on an estimation of 50 litres per capita daily and assuming a nuclear family structure of four people, mostly prevalent among affluent communities, the downside of the pro-poor policy is that it increases the price of water for households of more than five people, which is the norm in many informal settlements (Smith, 2004). So much so that although the informal settlements are densely populated, they only consume about 5% of the city's total water supply (Dawson, 2018).

Furthermore, the provision of affordable water to households requires not only determining the price of water but also developing the infrastructure for piped water and sanitation. Reports suggest that 99.8% of households have 'access to piped water inside the dwelling or yard or within 200 metres from the yard' (City of Cape Town, 2017a). Due to personal safety concerns, issues of overcrowding and malfunctioning water points, access to water may be reduced for people who are required to leave their dwellings (Currie *et al.*, 2017). This delineation suggests that 87.3% of households in the city have access to water inside the dwelling or yard (Currie *et al.*, 2017).

The drought has had polarising impacts on the cities affluent and poor. Whilst the poor are accustomed to queuing for free water at scarce communal taps within their settlements, the affluent have resorted to drilling boreholes to create their own private supplies of water, thereby going off-grid (Sieff, 2018). The city estimates that approximately an additional 20 000 boreholes have been drilled during the drought, and has little knowledge on what environmental impacts the often unregulated, decentralised drilling will have on groundwater levels (Cheslow, 2018). This will have dire knock-on effects on the provision of free water to those that depend on it the most. Removing the households that consume larger volumes of water from the equation means that it will be harder to generate enough revenue to subsidise water, further marginalising the poor (Cotterill, 2018). Cape Town bears the legacy of apartheid through high levels of inequality in access to water. Inequity plays out in water access very obviously, and highlights that water and social justice are intrinsically linked. The drought has demonstrated that the crisis is not only about a physical water scarce city but also that it is a product of water apartheid (Dawson, 2018).

8.2.2.5 Invasive alien plants species

These are a great threat to Cape Town's water security and climate resilience. It is estimated that invasions by alien trees, such as pines, wattles and eucalyptus, are diminishing water resources by using water yields of 38 Mm^3 per annum (104 MLD) which substantially impact the WCWSS (Singels *et al.*, 2018; Slingsby and Botha, 2018). Aside from depleting the water supply, they also intensify wildfires, reduce agricultural productivity and threaten globally significant biodiversity (Singels *et al.*, 2018).

8.2.3 Demand management

'Day Zero' was not avoided by chance. There had been a steady decrease in water demand since 2011 amidst long-standing drought fears and subsequent successful demand management initiatives. As a result, Cape Town consumed 315 $Mm^3\ a^{-1}$ of its allocation in 2014, down 2.18% from 320 $Mm^3\ a^{-1}$ in 2013 (Currie *et al.*, 2017). The severity of the drought prompted the City of Cape Town (CCT) to develop a two-phase disaster management plan in the event that dam levels were to drop to an extent that they could no longer sufficiently provide water to the metropolitan area (DWS, 2018).

Restrictions were gradually enforced from December 2015 till February 2018, when the city implemented Phase 1 of the plan whereby Level 6B restrictions were put in place which required 45% of savings for urban users (DWS, 2018). Individuals were restricted to 50 litres per capita per day, resulting in a total restricted allocation of 450 MLD for the entire system, which was a vast contrast from the peak summer consumption of 1200 MLD in 2015 (DWS, 2018). From a social perspective, 50 litres of water per day entails less water for personal hygiene and basic amenities such as cooking and cleaning. The use of grey water from a household's baths, showers and washing machine as an ancillary source of water for toilet flushing and gardening was encouraged (Western Cape Government, 2018).

Aside from restricting demand which saw 'one of the most drastic civic water conservation campaigns ever conceived' (Cotterill, 2018), once allocations were reached in January, DWS (2018) stopped releases to irrigation boards thus reducing drawdown from the system. Furthermore, a sizeable transfer of 10 billion litres of water in February 2018 from the Groenland Water Users' Association in the adjacent Elgin and Grabouw catchment region helped increase dam levels (eNCA, 2018).

8.2.4 Long-term solutions – supply augmentation

The city has accepted that in order to ensure the resilience of the WCWSS against future drought crises, it needs to invest in diversified supply sources which reduces its reliance on rainfall runoff and dam storage (DWS, 2018). Cost considerations factor strongly in the equation whereby an appropriate balance between the cost of water and the assurance of supply will have to be found (DWS, 2018). The opportunity cost of investing in infrastructure that is not used is high in South Africa (Cotterill, 2018). The supply augmentation options evaluated by local authorities in Cape Town 2018 which form the focus of this chapter are: (i) desalination, (ii) groundwater augmentation, (iii) wastewater reuse and (iv) surface water transfer.

8.3 OPTION CHARACTERISATION ANALYSIS

By responding to uncertainties related to the impacts of climate change and an increasing demand, well considered water resource strategies have the opportunity to set the policy direction for managing supply and demand into the future. This work tests the options evaluation framework set out in Figure 8.1 using the key resilience-building criteria defined in Figure 8.3, which includes both qualitative and quantitative criteria, to analyse the four water supply options being considered by the CCT.

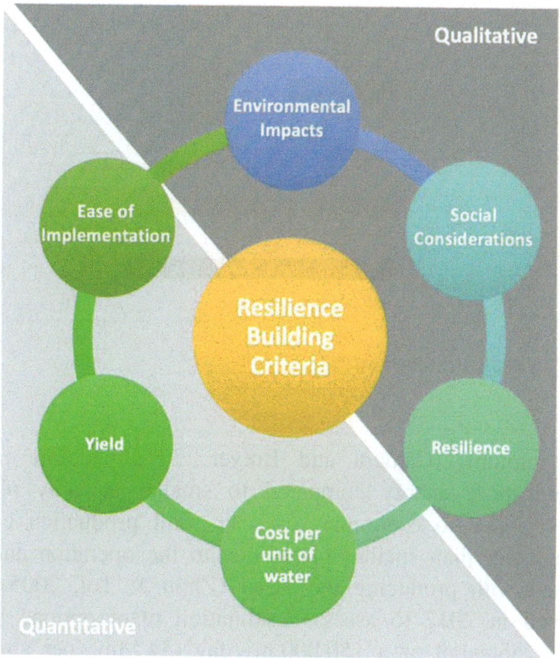

Figure 8.3 Key evaluation criteria.

The six key evaluation criteria were selected via a rigorous process which narrowed them down from an exhaustive list of criteria. This process was based on two important factors. First, the aim of the framework is to provide resilient and adaptive water-supply solutions to enhance the security of water systems. Second, by carefully considering the context in Cape Town, it was important to select criteria that were believed to be critical to the decision-making process. The criteria selected were yield, cost per unit of water, resilience, environmental impacts, social considerations and ease of implementation. The following sections define and analyse each of the six criteria against the four options.

8.3.1 Criteria 1 (C1): yield (m^3/day)

Determining the functional yield of each of the options is based on a range of different factors including economies of scale and safe environmental extraction limits, as discussed below and summarised in Figure 8.4.

8.3.1.1 Option 1: desalination plant

Desalination plant capacity is a major cost factor that needs to be considered. Large capacity plants, which are those that have capacity greater than 50 000 m^3/day

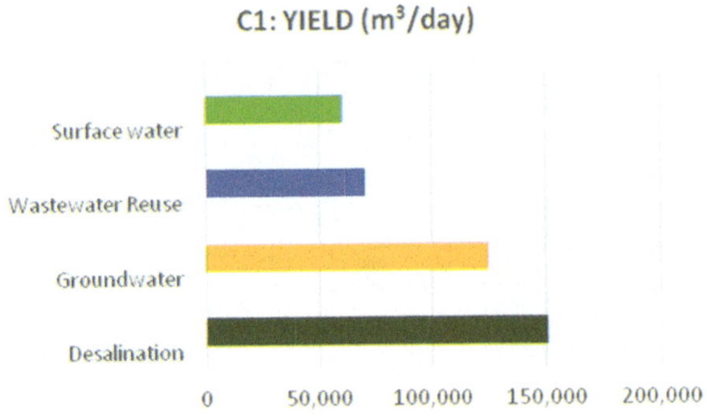

Figure 8.4 C1: yield (m^3/day).

(18 Mm^3 per annum) (Caldera and Breyer, 2017), require a high capital investment and more energy compared to smaller capacity plants (Younos, 2005). However, due to economies of scale, unit production costs for larger plants can be lower than smaller plants due to the operation and management costs associated with producing the water (Zhou & Tol, 2005). A feasibility study conducted in 2017 to assess desalination plant options in Cape Town highlighted the potential of a 150 000 m^3/day (54 Mm^3 per annum) Seawater Reverse Osmosis (SWRO) desalination plant (Bosman, 2017). It is important to note that while desalination is scalable, this may not be the case within the duration of a drought.

8.3.1.2 Option 2: groundwater augmentation scheme

Groundwater is seen as a valuable resource in the Western Cape due to its vast potential for storage, however it is largely underutilised for bulk water supply, making up only 1.7% of the water supplied to the CCT (Mauck, 2017). The city's groundwater resources include the Atlantis Aquifer, the Cape Flats Aquifer (CFA) and the Table Mountain Group (TMG) Aquifer. Considering the combined extraction volumes of the three separate schemes, the total scale of the Groundwater Augmentation Scheme is in the range of 125 000 m^3/day (45 Mm^3 per annum), which is discussed in more detail below.

(1) Atlantis Aquifer

The vast majority of the groundwater contributing to the bulk water supply comes from the Atlantis Aquifer which supplies approximately 12 000 m^3/day (4.3 Mm^3 per annum) through an innovative Managed Aquifer Recharge (MAR) scheme developed in 1979 (DWS, 2018; Luker, 2017). The New Water Program is investigating the possibility of sustainably extracting a further 20 000 m^3/day (7.2 Mm^3 per annum) with recharge from treated wastewater (DWS, 2018).

(2) Cape Flats Aquifer (CFA)

Cape Flats Aquifer has above sea-level storage capacity of more than 600 Mm^3 (DWS, 2018). Studies have been conducted to assess the storage potential and feasibility of MAR at Philippi and Mitchells Plain, two sites south of the Cape Flats. It has been found that the total sustainable yield of these schemes is approximately 18 Mm^3 per annum which equates to 50 000 m^3/day without risking seawater intrusion (Mauck, 2017).

(3) Table Mountain Group Aquifer (TMG)

Table Mountain Group Aquifer has approximately 1000 Mm^3 of storage capacity (DWS, 2018). Investigations are ongoing to determine the optimal locations for abstractions and input into the WCWSS (DWS, 2018). It is expected that the sustainable yield will be more than 20 Mm^3 per annum with recharge, which equates to 55 000 m^3/day (DWAF, 2007a).

8.3.1.3 Option 3: wastewater reuse treatment plant

The CCT is considering treating effluent to potable standards for supply from the existing Faure Water Treatment plant (DWS, 2018). Several factors were taken into consideration when sizing the scheme such as the source yield, the existing infrastructure capacities and the winter demands in the recipient zones (DWAF, 2007b). The CCT is considering a scheme which yields 26 Mm^3 per annum, translating to approximately 70 000 m^3 per day (DWS, 2018).

8.3.1.4 Option 4: surface water transfer scheme

Feasibility studies have been undertaken to assess options to pump winter water from the Berg River to the Voëlvlei Dam in the Berg River-Voëlvlei Augmentation Scheme (BRVAS), once the ecological water requirements of the river and the estuary have been met (DWAF, 2012). Of the options considered, a 4 m^3/s pump station with a step-pump operating appeared to be more easily implemented and operated, and offered a slightly higher resulting yield of 23 Mm^3 per annum, which results in a yield of approximately 60 000 m^3 per day (DWAF, 2012).

8.3.2 Criteria 2 (C2): cost per unit of water

Infrastructure spending in South Africa has to be weighed up carefully with regard to other pressing societal needs such as housing, healthcare and education. The cost criterion is therefore very crucial in the decision-making process. The normalised cost per cubic meter ($/$m^3$ in USD) of water is calculated to provide an indication of the relative cost for each option. It is based on a simplified method whereby the sum of the annual capital expenditure (CAPEX) over the loan repayment period and operational and maintenance costs (OPEX) is divided by the annual yield of the scheme as per the equation below.

$$\text{Cost per Unit of Water} = \frac{(\text{CAPEX}/20\ \text{years}) + \text{OPEX}}{\text{Annual Yield}}$$

Several assumptions are made in this calculation:

- the aim is not to make a profit on the sale of the water, rather to recover the full cost to produce it;
- the goal is to have a balanced budget; and
- to ensure the provision of basic services for all.

The following sections describe the key components which have been considered in the CAPEX and OPEX costs for each of the options and Table 8.1 summarises the results.

8.3.2.1 Option 1: desalination plant

The cost per unit of desalinated water depends on a range of factors including size and type of plant, plant location and the source and quality of incoming feed water (Caldera & Breyer, 2017). The components of CAPEX consist of direct and indirect costs including but not limited to construction, equipment, engineering, land, buildings, insurance, project financing and contingency costs – which is generally about 12% of the total direct costs (CMI, 2017; Younos, 2005). OPEX consists of fixed costs including insurance and amortization costs and variable costs including energy, chemicals, membrane replacement and labour costs (CMI, 2017).

8.3.2.2 Option 2: groundwater augmentation scheme

Costs associated with groundwater extraction are related to borehole depths, yield and location and are sensitive to water quality and associated treatment costs (DWS, 2018). The components of CAPEX include the cost of drilling and

Table 8.1 C2: cost per unit of water ($\$/m^3$).

Options	CAPEX (USD)	OPEX (USD/annum)	Yield (m^3/day)	Cost per unit ($\$/m^3$)
Desalination	350 900 000[1]	30 000 000[1]	150 000	0.87
Groundwater	72 000 000[2]	12 000 000[2]	125 000	0.34
Wastewater Reuse	12 857 008[3]	1 066 375[3]	70 000	0.07
Surface water	21 100 000[4]	643 000[4]	60 000	0.08

[1]Bosman, 2017; [2]DWS, 2018; [3]Department of Water Affairs and Forestry (DWAF, 2007a, 2007b); [d]Department of Water Affairs and Forestry (DWAF, 2012).

infrastructure connections. The OPEX budget includes pumping, electricity and treatment costs.

8.3.2.3 Option 3: wastewater reuse treatment plant

The CAPEX components considered for the required infrastructure components include pipelines and pump stations for the treated effluent and potable water, waste water treatment works and contingency costs (DWAF, 2007b). The OPEX costs are related to energy consumption, maintenance and replacement of all membranes and electro-mechanical plant every 10 years (DWAF, 2007b).

8.3.2.4 Option 4: surface water transfer scheme

The CAPEX components considered are related to the mechanical, electrical and civil costs of building the pump station and the outlet structure, contingency costs and professional and property fees (DWAF, 2012). The OPEX costs considered are related to civil, electrical and mechanical maintenance costs and energy costs.

8.3.2.5 Summary of cost per unit of water analysis

Table 8.1 summarises the CAPEX and OPEX costs and the yields used to calculate the cost per unit of water for each option.

8.3.3 Criteria 3 (C3): resilience

In 2015, the Department for the Environment, Food and Rural Affairs (Defra) and the Environment Agency (UK) commissioned the 'Strategic Water Infrastructure and Resilience' project to provide future guidance on water resource, supply and environmental resilience in the face of hazards, including extreme weather events in the UK. This chapter has adapted and built on the findings of the project, focusing on seven resilience characteristics which are more specific to the options considered in Cape Town. The means of measuring each of the options against the resilience characteristics are illustrated in the 'Resilience Scale Scoring Mechanism' tool in Figure 8.5. While the scales vary for each of the characteristics (e.g., the scale for 'Resilience to temperature extremes' is between 0 and 2 whereas the scale for 'Reliant on rainfall' is between 0 and 1), all of the scales have been standardised to fall between the range of −1 and 2, thereby enabling easy comparison between characteristics.

The tool developed acknowledges that water systems need to be resilient to water supply shortages and surpluses, catering to both ends of the hydrological spectrum and providing a buffer to both extremes – droughts and floods. The adaptive capacity of the system is also considered via the 'storage' characteristic. This acknowledges the dynamic nature of uncertainty and assesses whether the capacity of a system within itself is flexible enough to adapt to stresses and shocks, whilst taking advantage of opportunities.

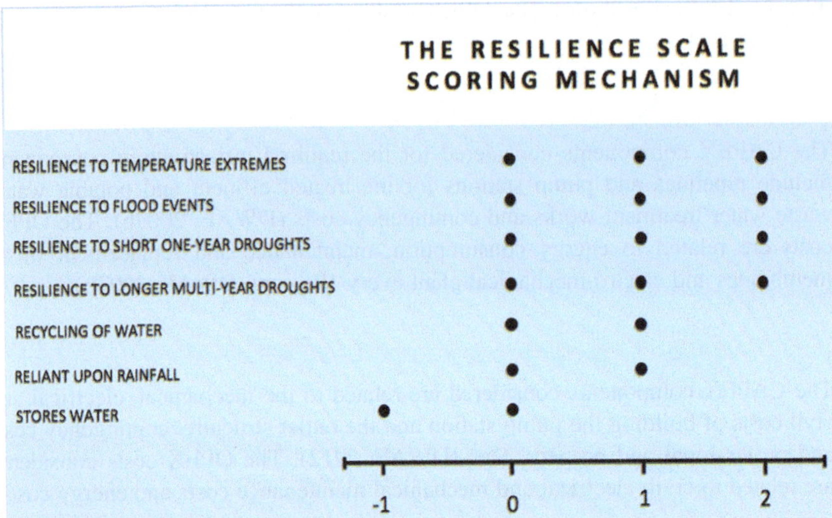

Figure 8.5 Resilience scale scoring mechanism tool.

The following sections detail the analysis of each of the options against the seven resilience characteristics and the results of this are summarised in Table 8.2 and Figure 8.6.

8.3.3.1 Option 1: desalination plant

Desalination adds an alternate source of water to the supply that is independent of rainfall and resilient to temperature extremes brought on by climate change. It is therefore considered to provide high resilience to environmental flows, drought and flood events (Defra, 2015). However, as the plant capacity is technically locked-in, it is unable to produce more water if demand increases beyond its capacity.

8.3.3.2 Option 2: groundwater augmentation scheme

Cape Town has abundant aquifer storage, more so than the combined storage of the dam system (DWS, 2018). Groundwater abstraction generally provides higher resilience to shorter drought events and temperature extremes because the storage capacity of the aquifers means that they are less susceptible to changes in rainfall patterns (Defra, 2015). However, they are less resilient to multi-year droughts as aquifer storage becomes depleted and safe environmental abstraction limits are reached. That being said, in the event that demand increases temporarily, it may be possible to increase abstractions provided that there is recharge. From a flood risk perspective, provided there is sufficient protection to the source, groundwater

Table 8.2 C3: summary of resilience characteristics analysis.

Alternatives	Resilience characteristics							
	Resilience rating scale (0–2)				Recycling of water Scale (0–1)	Reliant upon rainfall Scale (−1–0)	Stores water Scale (0–1)	Total score −1 to 10
	Short one-year drought	Longer multi-year drought	Temperature extremes	Flood events scale				
Desalination	2	2	2	2	0	0	0	8
Groundwater	2	1	2	2	0	−1	1	7
Wastewater reuse	2	2	2	2	1	0	0	9
Surface water	1	0	1	1	0	−1	1	3

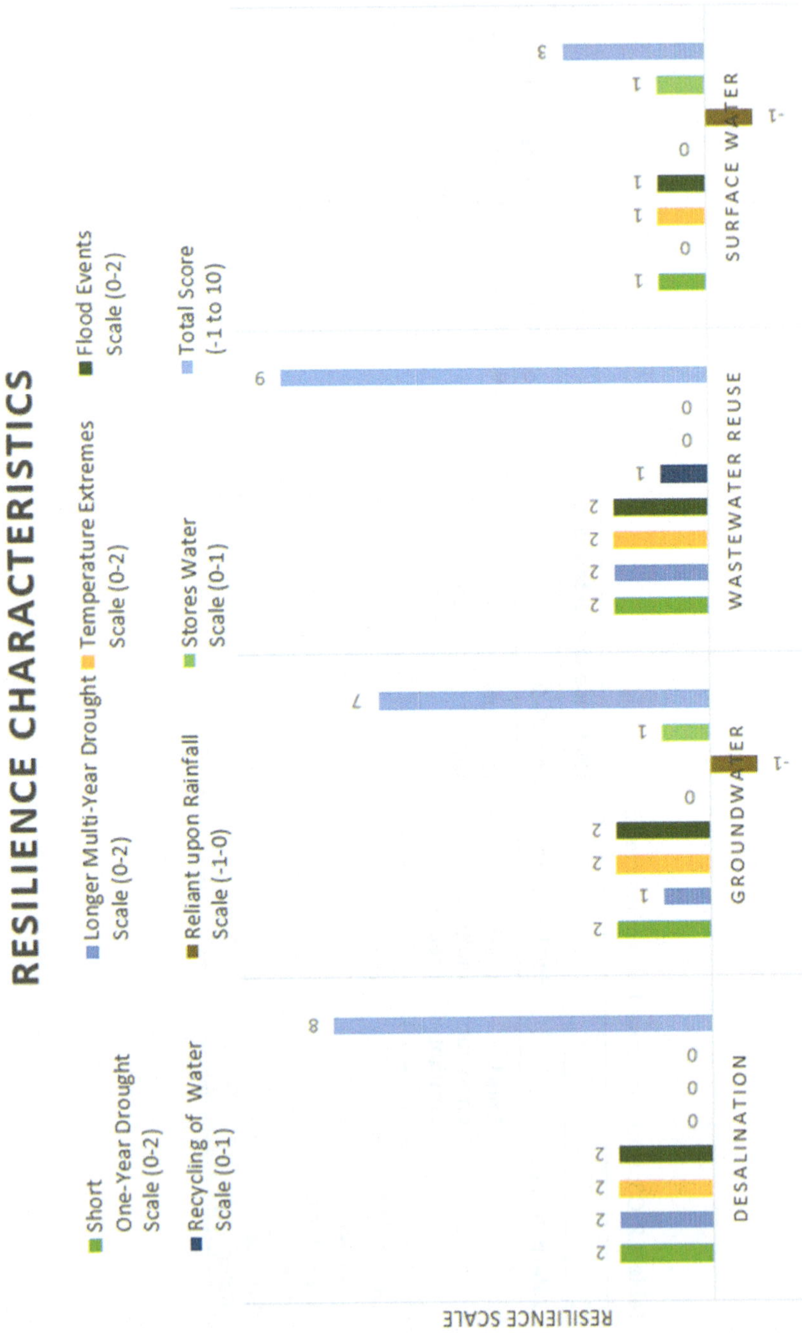

Figure 8.6 Distribution of resilience characteristics.

abstractions are perceived to also have a high resilience to flood events (Defra, 2015).

8.3.3.3 Option 3: wastewater reuse treatment plant

Wastewater reuse provides a high degree of resilience to droughts and other extreme events due to the nature of the closed-loop system and should be considered as a baseline supply option as opposed to an additional source (Defra, 2015). However, similar to desalination, as the plant capacity is technically locked-in, it is unable to produce more water if demand increases beyond its capacity.

8.3.3.4 Option 4: surface water transfer scheme

Surface water transfer schemes from river to reservoir are reliant upon rainfall and as the water resource is exposed, they have medium resilience to temperature extremes (Defra, 2015). However, reservoirs are able to store excess water when supply is plentiful, especially during high rainfall or flooding events, which can be utilised during low rainfall periods. This scheme has medium resilience to short term droughts but has low resilience for longer, multi-term droughts.

8.3.3.5 Summary of resilience characteristics analysis

Table 8.2 summarises the results of the resilience assessment of the four options. The sum of the scores of the seven characteristics for each option show that wastewater reuse scores the highest, and hence can be deemed as the most resilient option. Figure 8.6 illustrates these results schematically.

8.3.4 Criteria 4 (C4): environmental impacts

The environmental impacts of each of the options is assessed qualitatively in the following sections and scored subjectively based on low (1), moderate (2) and severe (3) negative impacts on the environment, relative to each other, and are summarised in Figure 8.7.

8.3.4.1 Option 1: desalination plant

Plants between 100 000–200 000 m^3/day consume 3.5–4 kWh/m^3 of energy (Zarzo & Prats, 2018), equating to 1.4–1.8 kg CO_2 per cubic meter of produced water which makes the carbon footprint of large-scale desalination plants substantial (Elimelech & Phillip, 2011). This is particularly concerning since about 80% of South Africa's primary energy needs are provided by coal, which is unlikely to change significantly in the next two decades (Energy RSA, 2018).

Studies have shown that a major concern with SWRO desalination, as is being considered in Cape Town, is the effect that seawater intake will have on marine organisms – entrainment can kill a large number of fish and small planktonic organisms if open surface intakes are not implemented safely (Elimelech &

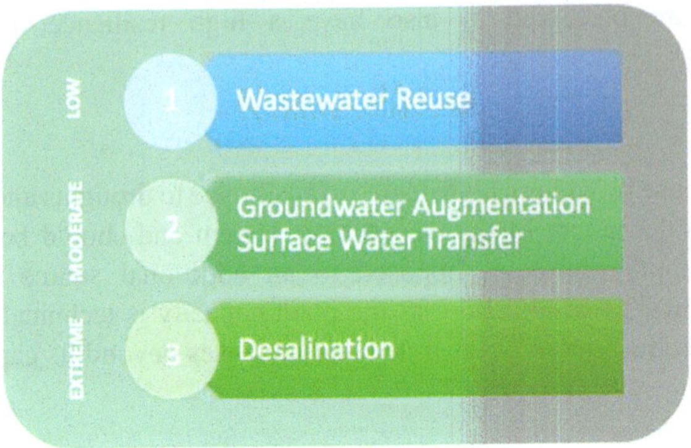

Figure 8.7 Summary of environmental impact analysis.

Phillip, 2011). Furthermore, the increased salinity of SWRO brines, which is about twice that of seawater, and the chemicals used in the desalination process, also pose environmental risks to the marine ecosystem (Elimelech & Phillip, 2011).

8.3.4.2 Option 2: groundwater augmentation scheme

There are several environmental concerns with groundwater augmentation, particularly the impacts it may have on terrestrial ecosystems, biodiversity and water table levels. Currently it appears that there is a lack of widespread knowledge on groundwater data and monitoring in Cape Town, therefore it is unknown what effect drawdown will have on the water table (Parsons, 2018), especially if it is being abstracted faster than it is being recharged (EEA, 2018). Furthermore, there are fears that drilling boreholes in environmentally sensitive areas such as the Table Mountain Group Aquifer will threaten critically endangered species with extinction, degrade ecosystems and wetlands and drastically alter the hydrology of catchments, all while violating environmental management regulations of the National Environmental Management Act (Slingsby, 2018). There is also some concern that large-scale abstraction can affect river flows (DWAF, 2007a).

8.3.4.3 Option 3: wastewater reuse treatment plant

There are several benefits of reusing effluent water, such as reducing the volumes of treated sewage effluent and industrial discharges into the environment, and reducing the dependency on surface water thereby enhancing flows through the system (Toze, 2006). However, there are a number of potential risk factors that need to be considered. The physical characteristics of the recycled water, such as the pH, salinity, dissolved oxygen and suspended solid content, have an impact on the soil environment in which it is used, particularly when the water is used for irrigation (Toze, 2006). Furthermore, the presence of enteric pathogens, including viruses, bacteria, protozoa and helminths, in recycled water can contaminate the water bodies that are in contact with it and impact the ecosystems dependent on it (Toze, 2006). However, these risks can be mitigated by sufficiently treating the effluent to suit the requirements of its purpose, be it for potable consumption or non-potable activities such as irrigation.

8.3.4.4 Option 4: surface water transfer scheme

Transferring water from the Berg River to the Voëlvlei Dam has the potential to provide augmentation of low river flows to mitigate losses in the dam during the drought (Defra, 2015). However, there will be negative changes in the water balance between the donor and receiving catchments and the potential transfer of alien species or diseases (Environment Agency, 2006). This will impact the water quality and ecology of the receiving watercourse and the ecosystems that depend on it.

8.3.5 Criteria 5 (C5): social considerations

Consumer perception regarding the price and quality of water reaching households and places of business can affect the ultimate decision on the type of water supply. The perception of the source and quality of water it produced was only significantly different for wastewater reuse when compared to the other options. Wastewater reuse can be used directly for potable supplies, however there is general consensus globally that there is presumption against it, and in order to use it for this purpose there would need to be considerable change in public perception and acceptance of it, together with regulatory changes, which would take time to enact (Defra, 2015). Using treated wastewater for non-potable use such as irrigation is less controversial, although there are not as many examples globally of current practise and, as noted earlier, it would also have to be treated sufficiently to meet environmental regulations (Defra, 2015). With that being said, the quality of water has been disregarded as a sub-criterion for consideration in this research.

The issue of pricing water, particularly in the South African context, is also complex. A balance has to be found between how much of the water required needs to be treated and delivered, and how much the consumers can afford or are willing to pay for it. The price of water, which is directly correlated to the level of tariff applied and more deeply connected to the issue of social equity and access to water, varied between the options and hence is a critical social consideration for policymakers. Average tariffs are determined by dividing the total cost of providing the service by the volume of water sold (DWS, 2018), therefore the higher the cost per unit of water produced, the higher the tariff will be.

Cape Town's tariff structure is dependent on large volumes of water being sold at higher levels in order to subsidise water at lower levels (DWS, 2018). If the price of water increases, it is likely to push wealthier households to invest in decentralised

Table 8.3 Summary of option characterisation result

Criteria	**C1: yield (m^3/day)**	**C2: cost (\$/m^3)**	**C3: resilience (scale of –1 to 10)**	**C4: environmental impacts scale of low (1), moderate (2), extreme (3)**	**C5: social considerations price of water/tariff scale of low (1), medium (2), high (3)**	**C6: ease of implementation time in years to implement the project**
Alternatives	Quantitative	Quantitative	Qualitative	Qualitative	Qualitative	Quantitative
Desalination	150 000	0.87	8	3	1	5
Groundwater	125 000	0.34	7	2	2	3
Wastewater reuse	70 000	0.07	9	1	3	4
Surface water	60 000	0.08	3	2	3	2

water solutions to avoid paying higher tariffs. This will impact the ability of the city to subsidise water at lower levels and hence have knock-on social equity effects on households who depend on it and potentially restrict the ability of the city to improve levels of access to water.

For the purpose of analysis in this chapter, higher tariffs are given a score of 1, medium tariffs a score of 2 and lower tariffs a score of 3, where 1 indicates lower social equity and access to water and 3 indicates higher social equity and access to water, as shown in Table 8.3.

8.3.6 Criteria 6 (C6): ease of implementation

Project implementation is a complex, context-specific process. It is dependent on a number of factors including the source of water and technology applied, procurement and contract regulations and type, project financing options and local or national regulatory requirements (Defra, 2015). In South Africa, the regulatory requirements surrounding public procurement is stringent, thereby resulting in a complex and lengthy process (DWS, 2018).

For the purpose of analysis here, it is assumed that the varying factor amongst the options is the time taken to implement the project from the time a decision is made to proceed. As shown in Table 8.3, the time scales estimated are indicative based on previous local and international experience (Defra, 2015; DWS, 2018).

8.3.7 Summary of results

The preceding sections analysed each of the options with respect to the six key evaluation criteria, the results of which are summarised in Table 8.3. The following sections will use the results from the option characterisation analysis to feed into the sensitivity analysis using the Analytic Hierarchy Process.

8.4 ANALYSIS AND RESULTS

As part of the MCDA, the results of the sensitivity analysis that was conducted under the lenses of three perspectives, using the options evaluation framework outlined in the introduction and the findings from the option characterisation analysis, are detailed in the following sections.

8.4.1 Prioritising criteria under the three perspectives

Prioritising criteria to define each of the perspectives in lieu of real stakeholder preference data is an important part of the process outlined in the options evaluation framework. The results of this analysis are summarised in Figure 8.8 and detailed in the ensuing sub-sections.

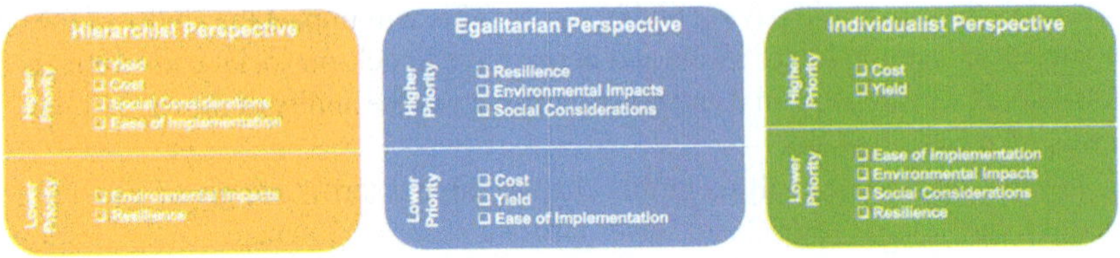

Figure 8.8 Summary of hierarchist, egalitarian and individualist priorities.

8.4.1.1 *The hierarchist perspective*

Hoekstra (2000) states that hierarchists regard water scarcity as a problem of supply rather than demand, which is considered to be a given need. While demand can be managed to enhance efficiency, they are cognisant that this could be met with economic and social constraints. Their water management strategy is therefore related to resource management and looking at ways to increase supply by introducing high technology on a large scale in order to meet demand (Hoekstra, 2000). However, hierarchists do not believe that consumers should have to repay all investment costs and rather aim for water charges to cover operational and maintenance costs only – they do not believe in market pricing (Hoekstra, 2000). Furthermore, a rapid increase in water prices would also disturb socio-economic stability. It can therefore be deduced that yield, cost, social considerations, and the related criteria of the ease of implementation of a project to a lesser extent, are leading principles from a hierarchists perspective.

Whilst hierarchists recognise the environmental impacts of certain supply options, such as groundwater extraction, and believe that they should be mitigated as much as possible, they do not reject the idea of implementing them all together (Hoekstra, 2000). Furthermore, they perceive nature as tolerant and robust within limits, meaning that disturbances can be absorbed as long as they do not reach critical levels and can be managed by defining adequate standards (Middelkoop *et al.*, 2004).

8.4.1.2 *The egalitarian perspective*

Egalitarians perceive nature as fragile and climate as sensitive, considering temporal and spatial variability when assessing water resources (Hoekstra, 2000). Their water management strategy is therefore driven by natural processes whereby water guides landscape planning (Middelkoop *et al.*, 2004). They also favour pro-active, preventative and adaptive measures instead of mitigation strategies (Middelkoop *et al.*, 2004). It can be deduced that environmental impacts and resilience are leading principles from an egalitarian perspective.

Water scarcity is perceived to be caused by demand growth and pollution and therefore can be managed by policy incentives and changes in social preferences

(Hoekstra, 2000). Egalitarians advocate that the price of water should include a tax to cover this and the environmental impacts of excessive water usage. However, because social equity is an important facet of egalitarianism and access to water and sanitation is a principal policy goal, water should be free to people who cannot afford it (Hoekstra, 2000). Water can be priced in a stepped-tariff structure whereby the price of water increases the more volume is consumed in a bid to subsidise water in the lower steps (DWS, 2018). Regardless of whether large scale measures are foreseen at great costs, egalitarians are strongly opposed to projects where the social and ecological costs outweigh the potential benefits (Hoekstra, 2000).

8.4.1.3 The individualist perspective

Individualists perceive water as an economic good and believe that it should be managed as such (Hoekstra, 2000). Provided they are cost-effective, all options to improve water supply and efficiency to reduce demand are considered to be realistic (Hoekstra, 2000). Therefore, projects such as groundwater extraction and desalination are considered if they are profitable, regardless of their environmental impacts. Two overriding principles emerge from this; economy is more important than environment and the environment can be exploited for economic use.

Hoekstra (2000) states that water prices are established by market mechanisms and subsidies are strongly discouraged, which is current common practice worldwide. Furthermore, an active policy in public water and sanitation is not pursued because it is believed that economic development will improve this inherently (Hoekstra, 2000). Climate change risks are accepted as an economic trade-off – any disturbances of the hydrological cycle are of minor importance. Resilience is therefore not a priority from an individualists' perspective.

8.4.2 Results of analytic hierarchy process

Utilising the priorities of the criteria to define each of the perspectives, the complex decision problem of selecting an option that will enhance the resilience of the WCWSS was evaluated via pairwise comparison of criteria and options with respect to criteria where decision-makers can express preference for one option over another. The results of the sensitivity analysis conducted under the lenses of the hierarchist, egalitarian and individualist perspectives using the Analytic Hierarchy Process are summarised in Figures 8.9 and 8.10. Figure 8.10 demonstrates that even as the global priority vectors change according to the priority of the criteria under each perspective, the cumulative sum of the criteria across each of the four options results in wastewater reuse as the preferred option in all three perspectives.

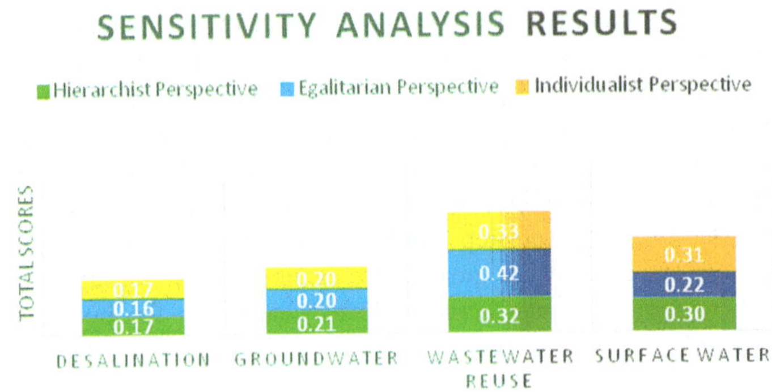

Figure 8.9 Summary of final results of the sensitivity analysis.

Hierarchist Perspective: Global Priority Matrix							
	C1: Yield	C2: Cost	C3: Resilience	C4: Environmental Impacts	C5: Social Considerations	C6: Ease of Implementation	Total
	(m^3/day)	($/$m^3$)	(Scale of -1 to 10)	Scale of Low (1), Moderate (2), Extreme (3)	Scale of Low (1), Medium (2), High (3)	Time in Years	
Alternatives	Quantitative	Quantitative	Qualitative	Qualitative	Qualitative	Quantitative	
Desalination	0.13	0.01	0.02	0.003	0.01	0.01	0.17
Groundwater	0.10	0.03	0.01	0.01	0.04	0.03	0.21
Wastewater Reuse	0.02	0.12	0.02	0.03	0.11	0.01	0.32
Surface water	0.02	0.10	0.002	0.01	0.11	0.06	0.30
Total	0.26	0.26	0.05	0.05	0.26	0.11	1.00

Egalitarian Perspective: Global Priority Matrix							
	C1: Yield	C2: Cost	C3: Resilience	C4: Environmental Impacts	C5: Social Considerations	C6: Ease of Implementation	Total
	(m^3/day)	($/$m^3$)	(Scale of -1 to 10)	Scale of Low (1), Moderate (2), Extreme (3)	Scale of Low (1), Medium (2), High (3)	Time in Years	
Alternatives	Quantitative	Quantitative	Qualitative	Qualitative	Qualitative	Quantitative	
Desalination	0.05	0.00	0.08	0.02	0.01	0.00	0.16
Groundwater	0.04	0.01	0.06	0.04	0.04	0.01	0.20
Wastewater Reuse	0.01	0.04	0.10	0.16	0.10	0.01	0.42
Surface water	0.01	0.04	0.01	0.04	0.10	0.03	0.22
Total	0.10	0.10	0.26	0.26	0.26	0.04	1.00

Individualist Perspective: Global Priority Matrix							
	C1: Yield	C2: Cost	C3: Resilience	C4: Environmental Impacts	C5: Social Considerations	C6: Ease of Implementation	Total
	(m^3/day)	($/$m^3$)	(Scale of -1 to 10)	Scale of Low (1), Moderate (2), Extreme (3)	Scale of Low (1), Medium (2), High (3)	Time in Years	
Alternatives	Quantitative	Quantitative	Qualitative	Qualitative	Qualitative	Quantitative	
Desalination	0.12	0.02	0.02	0.003	0.002	0.01	0.17
Groundwater	0.09	0.04	0.01	0.01	0.01	0.03	0.20
Wastewater Reuse	0.02	0.22	0.02	0.03	0.02	0.02	0.33
Surface water	0.02	0.18	0.002	0.01	0.02	0.07	0.31
Total	0.25	0.46	0.05	0.05	0.05	0.13	1.00

Figure 8.10 Summary of final results of the sensitivity analysis performed using the AHP.

8.5 DISCUSSION AND RECOMMENDATIONS

8.5.1 Decision making in water supply infrastructure planning

As Cape Town is looking at long-term solutions to enhance and diversify its water supply to reduce its reliance on surface water, all four options assessed satisfy these objectives. The results of the sensitivity analysis found that wastewater reuse is the preferred option in the hierarchist, egalitarian and individualist perspectives, emphasising that regardless of the water management ethos, reuse should be the foundation for resilient water systems. The results indicate that this is true for cost, resilience, environmental and social criteria under all three scenarios.

The benefit of the options evaluation framework is that it provides a holistic outlook of the solutions by assessing reliability, resilience and sustainability-related criteria. Incorporating resilience as its own key criterion indicates a progressive step in shifting the paradigm towards more integrated, adaptive mindsets in light of climate change. This emphasises the importance of accounting for this characteristic in the process of planning for an uncertain future.

However, by only considering the physical infrastructure aspect of the resilience conceptualisation, the framework does not account for the management and financing arrangements involved in service provision, and the behaviour of the water consumers, especially as resources become limited. The success of the combination of demand and supply-side interventions in Australian cities following the Millennium drought are proof that both strategic directions are integral in enhancing the long-term sustainability and resilience of a water management system. It is therefore recommended that the approach is modified in future to incorporate these aspects into the framework to further enhance resilience.

Furthermore, although the sensitivity analysis was conducted to explore different water management perspectives in lieu of direct stakeholder involvement, it by no means replaces its necessity. The uniqueness, and hence the importance of the local context in Cape Town, highlighted that in developing water management frameworks it is crucial to identify and utilise the forces which drive policy-making decisions in order to determine the most appropriate strategy. This is best achieved through stakeholder engagement. However, the benefit of the AHP method is that it allowed for the weights of the criteria to be altered based on preferences. Therefore, in the event that stakeholder engagement occurred in the future, the AHP could be adapted to represent the findings of this process.

It is important to acknowledge the real data gaps in the options characterisation analysis. Each criterion was evaluated against a simplified set of either objective quantifiable performance data or subjective qualitative preference data. In a more detailed study, it is recommended that each criterion be evaluated with more rigor, perhaps by incorporating sub-criteria such as public perception under social considerations. Minimising subjectivity where appropriate, and using advice from academics or subject matter industry experts and/or stakeholder engagement, particularly with regards to social considerations, is also recommended.

8.5.2 Water access and social equity

The water crisis has highlighted two key issues, namely that Cape Town is water scarce, and that dealing with the legacy of Apartheid has had vast implications on the socio-economic development of the city. The issue of water availability has received much attention recently, driven by the stresses exacerbated by the drought and the vast international coverage this has garnered as a result of its enormity. Droughts are sporadic, climate-driven events occurring over unknown periods of time. This research has investigated the problem extensively by looking at ways to enhance supply and build resilience into the existing water system in order to be better prepared in the event of likely future droughts.

The underlying, ever-present issue of water accessibility is driven by social inequality. This is a continuous, human-caused stress which is a deep-rooted problem linked to South Africa's complicated, inter-generational history. Although approximately half of Cape Town's population lives in informal settlements, they consume only about 5% of the city's total water supply (Dawson, 2018). There is clearly a gap in equitable access to water. The broader questions are whether there is a reinforcing link between availability and accessibility, and whether both these issues can be combated simultaneously (Figure 8.11).

The stepped-tariff structure is reliant on affluent households consuming large volumes of water to subsidise households in informal settlements. Enhancing and diversifying the supply therefore increases confidence in the water system during times of drought which will disincentive affluent households from going off-grid and opting for decentralised, private water sources, that is, boreholes. Although demand management strategies aim to improve efficiency in the system and reduce individuals demand for water over time, urbanisation and related

Figure 8.11 Expanding the water systems boundary to enhance availability and accessibility.

demand-growth factors, such as increased food production and energy consumption, will simultaneously increase demand in the long-term. Therefore, it can be assumed that the stepped-tariff structure will be able to sustain itself as long as it is made more drought resilient.

The issue with the current water management strategy is that enhancing the systems long-term resilience and availability will only benefit those who are being serviced by it within the existing systems boundary. Managing the issue of availability in this instance therefore does not manage the issue of accessibility, particularly in a society in which the issue of accessibility is complicated and multi-dimensional. To combat both issues concurrently, it is thus recommended that further research be undertaken to investigate an integrated, systems-based approach with multiple objectives, as it has the potential to address the unique development challenges that Cape Town is facing to facilitate change in informal urban areas by expanding the existing systems boundary.

Armitage *et al.* (2014) investigated a broad approach which includes a range of settlement types by using five principles selected from the National Water Act (RSA, 1998), the South African Constitution (RSA, 1996) and the Dublin Principles (United Nations, 1992) to define water sensitivity as the management of the country's urban water resources based on the premise that:

(1) Water is a finite and valuable resource, essential to sustaining all life.
(2) The constitution of South Africa states that access to adequate potable water is a basic human right that is, it is a 'social good'.
(3) South Africa is a water scarce country.
(4) Water has economic value, and should also be recognised as an 'economic good'.
(5) Management of water should be based on a participatory approach involving all stakeholders, that is, users, planners, policy-makers.

Using these principles as the underlying basis, the concept of Water Sensitive Urban Design (WSUD) can be adapted to suit the South African context to include developmental and equity considerations to improve both availability and accessibility (Armitage *et al.*, 2014). Stormwater management activities such as the Sustainable Drainage Systems (SuDS) approach, which enhances storage, amenity and biodiversity and mitigates flooding by infiltrating stormwater at the source, are popular types of infrastructure-related activities that can be implemented as part of WSUD (Armitage *et al.*, 2014). Other activities include those assessed in this research such as wastewater reuse and Managed Artificial Recharge (MAR) of groundwater supplies, as well as water conservation and demand management, and rainwater and stormwater harvesting (Armitage *et al.*, 2014).

There are several developmental challenges with implementing WSUD projects in informal settlements related to equity, dignity, ownership and respect, which will need to be addressed. Delivering these projects will be particularly challenging in areas where basic services do not exist, which adds a layer of complexity to the problem. This is linked to the larger problem of the provision of housing and the availability of suitable land for development, which emphasises the need for a systems approach based on the assumption that development issues are connected. Associated with this are the spread of potential health risks due to the creation of alternative pathways for waterborne diseases which will need to be controlled and mitigated.

The multiple benefits of implementing WSUD projects outweigh the challenges. Aside from mitigating the negative effects of scarcity and increasing the sustainability and resilience of the water system, the ability to develop social and intergenerational equity can bring about significant change for the people of Cape Town. Improving accessibility to water in informal settlements has the potential to produce sustainable and equitable economic growth, improve health and sanitation, gender equality, education and safety in these areas. In essence, water can be used as the medium to connect the extremely divided and disparate settlements that are inherent in the fabric of the country to bring about equitable transformation for all.

In terms of future infrastructure planning, policymakers in Cape Town are aware that climatic projections show a shift towards a drier, more drought-prone climate which means that the possibility of extreme drought events are also increasing (Wolski *et al.*, 2017). Urban populations are also expected to grow, as will demand. By investing in wastewater reuse, water-efficiency, conservation and WSUD initiatives, Cape Town has an opportunity to build resilience into the water system from a physical infrastructure perspective by increasing and diversifying the availability of the supply. However, the concept of resilience-building is not simply about sustainable water management but also about water provision and accessibility. To enable this transformation, a paradigm shift in both cultural-cognitive and normative dimensions of the water system is critical (Ferguson *et al.*, 2013). Therefore, aside from this being an opportunity for the city to invest in physical infrastructure, it is also an opportunity to invest in equity and reduce both resource and social vulnerabilities.

8.6 CONCLUSION

Water scarcity is a growing systemic risk and it is becoming increasingly important to prepare for the impacts that climatic and socio-economic changes may have on future water resource availability. The primary objective of this work was to develop an options evaluation framework which can aid policymakers in the early stages of the decision-making process. The aim of this framework was to critically appraise water supply options to provide adaptive, resilient solutions to enhance water security in water stressed cities of the Global South.

To develop a holistic framework, it was crucial to understand which driving forces contribute to the problem of water scarcity to be able to create boundaries for the outcomes of the framework. The recent drought in Cape Town, South Africa was used as a case study to understand these forces. Through this process, it was found that unprecedented climate variability, population growth and urbanisation, and demand management strategies that effectively delayed the need to develop more dam storage capacity, were to the city's detriment in the face of a 1-in-300-year drought.

The key evaluation criteria of the framework developed in this research were closely linked to the concepts of resilience and sustainability. The results of the sensitivity analysis conducted to evaluate the options using the Analytic Hierarchy Process found that wastewater reuse was the preferred option under all three lenses. This emphasised that regardless of the water management ethos, reuse should be considered as a baseline supply option for resilient water systems.

The focus of this chapter was on one aspect of the conceptualisation of building resilience, namely the provision of physical infrastructure, by evaluating seven specific resilience characteristics using the Resilience Scale Scoring Mechanism tool developed in this research. However, the broader concept of resilience can be interpreted as also involving the interaction of the resource, the management and financing arrangements involved in service provision, and the behaviour of the water consumers, especially as resources become limited. The impacts that these aspects will collectively have on building resilience in water systems to enhance overall robustness in the face of future uncertainties should be investigated further in the future.

Finally, this chapter emphasised the importance of establishing context in developing water management strategies. Inequity plays out in water access very obviously, and highlights that water and social justice are intrinsically linked. The underlying, ever-present stress of water accessibility in Cape Town is driven by social inequality linked to South Africa's complicated, inter-generational history. By adopting an integrated, systems-based approach such as the Water Sensitive Urban Design concept, adapted to address the unique development challenges that Cape Town is facing, the potential to facilitate change in informal urban areas by expanding the existing systems boundary is immense. In essence, water can be used as the medium to connect the extremely divided and disparate settlements that are inherent in the fabric of the country to bring about equitable transformation for all.

REFERENCES

Ainger C. and Fenner R. (2014). Sustainable Infrastructure: Principles into Practice, Sustainable Infrastructure. ICE Publishing, London. https://doi.org/10.1680/sipp.57548

Araos M., Ford J., Berrang-Ford L., Biesbroek R. and Moser S. (2017). Climate change adaptation planning for Global South megacities: the case of Dhaka. *Journal of*

Environmental Policy & Planning, **19**, 682–696. https://doi.org/10.1080/1523908X.2016.1264873

Armitage N., Fisher-Jeffes L., Carden K., Winter K., Naidoo V., Spiegel A., Mauck B. and Coulson D. (2014). Water Sensitive Urban Design (WSUD) for South Africa: Framework and Guidelines. Water Research Commission, Pretoria, p. 234.

Basholo Z. (2016). Overview of Water Demand Management Initiatives: A City of Cape Town Approach. Available from: www.greencape.co.za/assets/Water-Sector-Desk-Content/CoCT-WCWDM-presentation-Z-Basholo-Western-Cape-Water-Forum-160204-2016.pdf, (accessed 13 November 2020).

Bichai F., Ryan H., Fitzgerald C., Williams K., Abdelmoteleb A., Brotchie R. and Komatsu R. (2015). Understanding the role of alternative water supply in an urban water security strategy: an analytical framework for decision-making. *Urban Water Journal*, **12**, 175–189. https://doi.org/10.1080/1573062X.2014.895844

Birkland T. and Waterman S. (2008). Is Federalism the Reason for Policy Failure in Hurricane Katrina? *Journal of Federalism*, **38**(4), 692–714.

Bohatch T. (2017). What's Causing Cape Town's Water Crisis? GroundUp News, Cape Town.

Bosman D. (2017). Desalination: status quo, future options and cost. Available from: www.greenagri.org.za/assets/documents-/Water-Indaba/Water-Indaba-2017-Presentation-17-Dawid_Bosman-0-2.pdf, (accessed 8 January 2018).

Butler D., Farmani R., Fu G., Ward S., Diao K. and Astaraie-Imani M. (2014). A new approach to urban water management: safe and sure. *Procedia Engineering*, **89**, 347–354. doi: 10.1016/j.proeng.2014.11.198

Caldera U. and Breyer C. (2017). Learning curve for seawater reverse osmosis desalination plants: capital cost trend of the past, present and future. *Water Resources Research*, **53**, 10,523–10,538. https://doi.org/10.1002/2017WR021402

Cheslow D. (2018). Cape Town's Water Crisis Marks Divide between Rich and Poor. NPR, Washington, DC.

City of Cape Town. (2017a). Socio-economic profile. Available from: www.westerncape.gov.za/assets/departments/treasury/Documents/Socio-economic-profiles/2017/city_of_cape_town_2017_socioeconomic_profile_sep-lg_-_26_january_2018.pdf, (accessed 23 July 2018).

City of Cape Town. (2017b). Water Services Development Plan – IDP Water Sector Input Report. Available from: https://resource.capetown.gov.za/documentcentre/ocuments/City%20strategies,%20plans%20and%20frameworks/Water%20Services%20Development%20Plan.pdf, (accessed 13 November 2020).

City of Cape Town. (2018). City of Cape Town Link. City Cape Town. Available from: www.capetown.gov.za/Family%20and%20home/residentialutility-services/residential-water-and-sanitation-services/this-weeksdam-levels, (accessed 23 July 2018).

CMI. (2017). Desalination Technologies and Economics: CAPEX, OPEX & Technological Game Changers to Come. Available from: www.cmimarseille.org/knowledgelibrary/desalination-technologies-and-economics-capex-opextechnological-game-changers-0, (accessed 8 January 2018).

Cotterill J. (2018). South Africa: How Cape Town beat the drought. Financial Times.

Currie P. K., Musango J. K. and May N. D. (2017). Urban metabolism: a review with reference to Cape Town. *Cities*, **70**, 91–110. https://doi.org/10.1016/j.cities.2017.06.005

Dawson A. (2018). Cape Town has a new apartheid. Washington Post.

Defra. (2015). Strategic Water Infrastructure and Resilience – Final Report. Available from: http://randd.defra.gov.uk/Document.aspx?Document=13784_WT1535StrategicWaterInfrastructureandResilienceFinalreport.pdf, (accessed 13 November 2020).

Department of Water Affairs and Forestry (DWAF). (2007a). Western Cape Water Supply System: Reconciliation Strategy Study – Scenario Planning for Reconciliation of Water Supply and Requirement. Department of Water Affairs and Forestry, Pretoria, South Africa.

Department of Water Affairs and Forestry (DWAF). (2007b). Western Cape Water Supply System: Reconciliation Strategy Study – Treatment of Effluent to Potable Standards for Supply from the Faure Water Treatment Plant. Department of Water Affairs and Forestry, Pretoria, South Africa.

Department of Water Affairs and Forestry (DWAF). (2009). Water for Growth & Development Framework (No. Version 7). Department of Water Affairs and Forestry, Pretoria, South Africa.

Department of Water Affairs and Forestry (DWAF). (2012). Volume 1: First Phase Augmentation of Voelvlei Dam (No. 3). Department of Water Affairs, Pretoria, South Africa.

DWS D. (2018). City of Cape Town Water Outlook 2018 Report. Available from: https://resource.capetown.gov.za/documentcentre/Documents/City%20research%20reports%20and%20review/Water%20Outlook%202018%20-%20Summary.pdf, (accessed 23 July 2018).

EEA. (2018). Impacts due to over-abstraction. European Environment Agency. Available from: www.eea.europa.eu/themes/water/waterresources/impacts-due-to-over-abstraction, (accessed 8 January 2018).

Elimelech M. and Phillip W. A. (2011). The future of seawater desalination: Energy, technology, and the environment. *Science*, **333**, 712–717. https://doi.org/10.1126/science.1200488

eNCA. (2018). Cape Town gets 10bn litres of water. Available from: www.enca.com/south-africa/cape-town-gets-10bn-litres-ofwater, (accessed 24 July 2018).

Energy RSA. (2018). Energy Sources: Coal | Department: Energy | Republic of South Africa. Available from: www.energy.gov.za/files/coal_frame.html, (accessed 8 January 2018).

Environment Agency. (2006). Do We Need Large-scale Water Transfers for Southeast England? Environment Agency, Bristol, UK.

Ferguson B. C., Brown R. R., Frantzeskaki N., de Haan F. J. and Deletic A. (2013). The enabling institutional context for integrated water management: Lessons from Melbourne. *Water Research*, **47**, 7300–7314. https://doi.org/10.1016/j.watres.2013.09.045

Fong P. S. W. and Choi S. K. Y. (2000). Final contractor selection using the analytical hierarchy process. *Construction Management and Economics*, **18**, 547–557. https://doi.org/10.1080/014461900407356

Hajkowicz S. and Collins K. (2007). A review of multiple criteria analysis for water resource planning and management. *Water Resource Management*, **21**, 1553–1566. https://doi.org/10.1007/s11269-006-9112-5

Hoekstra A. (2000). Appreciation of water: four perspectives. *Water Policy*, **1**, 605–622. https://doi.org/10.1016/S1366-7017(99)00013-6

Holling C. S. (1996). Engineering resilience versus ecological resilience. In: Engineering Within Ecological Constraints, P. C. Shultze (ed.), National Academy Press, Washington DC, USA.

Lai E., Lundie S. and Ashbolt N. J. (2008). Review of multi-criteria decision aid for integrated sustainability assessment of urban water systems. *Urban Water Journal*, **5**, 315–327. https://doi.org/10.1080/15730620802041038

Luker E. (2017). Transitioning Towards Water Supply Diversification: Possibilities for Groundwater in Cape Town, South Africa. University of British Columbia, Vancouver, Canada. https://doi.org/10.14288/1.0354496

MacAlister C. and Subramanyam N. (2018). Climate change and adaptive water management: innovative solutions from the global South. *Water International*, **43**, 133–144. https://doi.org/10.1080/02508060.2018.1444307

Mauck B. A. (2017). The Capacity of the Cape Flats Aquifer and its Role in Water Sensitive Urban Design in Cape Town. Thesis. University of Cape Town, South Africa.

McKenzie D. and Swails B. (2018). Day Zero deferred, but Cape Town's water crisis is far from over. CNN. Available from: www.cnn.com/2018/03/09/africa/cape-town-day-zero-crisisintl/index.html (accessed 23 July 2018).

Mekonnen M. M. and Hoekstra A. Y. (2016). Four billion people facing severe water scarcity. *Science Advances*, **2**, e1500323. https://doi.org/10.1126/sciadv.1500323

Middelkoop H., Asselt M. B. A. V., Klooster S. A. V. T., Deursen W. P. A. V., Kwadijk J. C. J. and Buiteveld H. (2004). Perspectives on flood management in the Rhine and Meuse rivers. *River Research and Applications*, **20**, 327–342. https://doi.org/10.1002/rra.782

Mukheibir P. and Ziervogel G. (2007). Developing a Municipal Adaptation Plan (MAP) for climate change: the city of Cape Town. *Environment and Urbanization*, **19**(1), 143–158.

Olivier D. W. (2017). Cape Town's water crisis: driven by politics more than drought. The Conversation.

Parsons R. (2018). Cape drought: What's happening to our groundwater? Daily Maverick. Available from: www.dailymaverick.co.za/article/2018-07-30-cape-drought-whats-happening-to-our-groundwater/, (accessed 14 November 2020).

Postel S. L. (2000). Entering an era of water scarcity: the challenges ahead. *Ecological Applications*, **10**, 941–948. https://doi.org/10.1890/1051-0761(2000)010[0941:EAEOWS]2.0.CO;2

Rijsberman F. R. (2006). Water scarcity: Fact or fiction? Agricultural Water Management, Special Issue on Water Scarcity: Challenges and Opportunities for Crop Science, **80**, 5–22. https://doi.org/10.1016/j.agwat.2005.07.001

RSA. (1996). The Constitution of the Republic of South Africa. Available from: www.justice.gov.za/legislation/constitution/SAConstitution-webeng.pdf (accessed 23 July 2018).

RSA. (1998). National Water Act. Act No. 36 of 1998. Available from: www.waternet.co.za/policy/le_nwa.html#:~:text=No.,36%20of%201998&text=The%20aim%20of%20water%20resource,benefit%20of%20all%20water%20users, (accessed 14 November 2020).

Saaty R. W. (1987). The analytic hierarchy process – what it is and how it is used. *Mathematical Modelling*, **9**, 161–176. https://doi.org/10.1016/0270-0255(87)90473-8

Sieff K. (2018). Divided by Drought. Washington Post.

Singels E., Cousins S. and Kraaij T. (2018). Invasive alien plants in South Africa pose huge risks, but they can be stopped. The Conversation.

Slingsby J. (2018). Rush to Drill for Water Threatens our Future Water Supply. GroundUp News, Cape Town.

Slingsby J. and Botha M. (2018). Aliens are Greatest Threat to Cape Town's Water Security. GroundUp News, Cape Town.

Smith L. (2004). The murky waters of the second wave of neoliberalism: corporatization as a service delivery model in Cape Town. *Geoforum*, **35**, 375–393. https://doi.org/10.1016/j.geoforum.2003.05.003

Sousa-Alves M. D. (2015). Update and additions to existing long-term water conservation and water demand management strategy. City of Cape Town. Available from: https://resource.capetown.gov.za/documentcentre/Documents/City%20strategies,%20plans%20and%20frameworks/WCWDM_Strategy_doc.pdf, (accessed 14 November 2020).

Szabó A. (2015). The value of free water: analyzing South Africa's free basic water policy. *Econometrica*, **83**, 1913–1961. https://doi.org/10.3982/ECTA11917

Tadross M. and Johnston P. (2012). Sub-Saharan African Cities: A five-City Network to Pioneer Climate Adaptation through Participatory Research and Local Action Local Governments for Sustainability – Africa, Cape Town, South Africa.

Thompson M. (1988). Socially viable ideas of nature: a cultural hypothesis. In: Man, Nature and Technology: Essays on the Role of Ideological Perceptions, E. Baark and U. Svedin (eds), Palgrave Macmillan, London, pp. 57–79. https://doi.org/10.1007/978-1-349-09087-7_4

Toze S. (2006). Reuse of effluent water-benefits and risks. *Agricultural Water Management*, **80**(1–3)147–159.

United Nations. (2017). Goal 6: Sustainable Development Knowledge Platform. Available from: https://sustainabledevelopment.un.org/sdg6, (accessed 23 August 2018).

United Nations. (1992). The Dublin Statement on Water and Sustainable Development – UN Documents: Gathering a body of global agreements. Presented at the International Conference on Water and the Environment, Dublin, Ireland.

Western Cape Government. (2018). Your guide to use 50 litres of water per day. Available from: www.westerncape.gov.za/general-publication/your-guide-use-50-litres-water-day, (accessed 26 August 2018).

Wolski P. (2018). Facts are few, opinions plenty on drought severity again. CSAG. Available from: www.csag.uct.ac.za/2018/01/22/facts-are-few-opinions-plentyon-drought-severity-again/, (accessed 23 July 2018).

Wolski P., Hewitson B. and Jack C. (2017). Why Cape Town's drought was so hard to forecast. Available from: https://theconversation.com/why-cape-towns-drought-was-so-hard-to-forecast-84735, (accessed 14 November 2020).

World Economic Forum. (2018). The Global Risks Report 2018. World Economic Forum, Geneva.

Younos T. (2005). The economics of desalination. *Journal of Contemporary Water Research & Education*, **132**, 39–45.

Zarzo D. and Prats D. (2018). Desalination and energy consumption. What can we expect in the near future? *Desalination*, **427**, 1–9. https://doi.org/10.1016/j.desal.2017.10.046

Zhou Y. and Tol R. S. J. (2005). Evaluating the costs of desalination and water transport. *Water Resource Research*, **41**(3), W03003. https://doi.org/10.1029/2004WR003

Chapter 9

Advances in experimental modelling of urban flooding

M. Rubinato[1,2,3], C. Lashford[1,2] and M. Goerke[3]

[1]*Faculty of Engineering, Environment and Computing; School of Energy, Construction and Environment; Coventry University, Coventry, UK*
[2]*CAWR; Centre for Agroecology, Water and Resilience; Coventry University, Ryton Gardens, Wolston Lane, Ryton-on-Dunsmore, Coventry, UK*
[3]*IKT-Institute for Underground Infrastructure, Exterbruch 1; 45886 Gelsenkirchen, Germany*

9.1 INTRODUCTON: URBAN FLOODING

Flooding impacts more people annually than any other natural hazards, and the 21st century has seen a rise in the number of extreme meteorological events (CRED, 2019; Depietri & McPhearson, 2018). Flooding, including coastal, fluvial and pluvial, costs the global economy US$19.7 billion in 2018, with densely populated urban areas typically exposed to 'flash flooding' as a result of intense rainfalls that exceed the anthropogenic and natural drainage capacity of a city (CRED, 2019) (Figure 9.1). Recent years have seen a number of significant floods impacting cities globally; between June and July 2016, flooding is estimated to have impacted over 32 million people in China alone (Tang *et al.*, 2017). While these events, and many others globally, cannot be entirely mitigated, there is an intrinsic vulnerability to flooding associated with living in urban areas which is further exacerbated by increasing urbanisation, and a changing climate (Miller & Hutchins, 2017; Rubinato *et al.*, 2020).

Editors: Xiaochang C. Wang and Guangtao Fu
doi: 10.2166/9781789060768_0235

Figure 9.1 Example of recent (August 2019) pluvial flooding events in urban areas: on the left, flow over streets in Cavarzere, (VE), Italy; on the right, manhole cover removed by the high pressure caused by overflow of the drainage system in Calalzo di Cadore, (BL), Italy. ©LeonardoRubinato and ©MariaGraziaSattin.

With the IPCC (2014) stating that anthropogenic factors have 'unequivocally' altered future climate, the natural global hydrological balance is shifting. Global water scarcity will be exacerbated in the future, particularly in areas that currently face significant droughts, while global flood risk is also likely to increase as a result of a changing water balance (Liu and Jensen, 2018; McMichael *et al.*, 2006; Pozzi *et al.*, 2013; Spinoni *et al.*, 2014; Trenberth, 2011).

The UK is likely to see an increase in extreme rainfall events, with the UK Met Office suggesting an average annual temperature rise for Central England of up to 5.8°C, with winter rainfall increasing up to 33% and summer rainfall decreasing by 57%, by 2070, for a high emission scenario at the 90th Percentile (Met Office, 2018). Further highlighting the likely increase in the risk of flooding as a result of a changing climate, Kundzewicz *et al.* (2014) noted that the one in 20 year return interval rainfall event is likely to become more frequent globally (aside from the Sahara) for all emission scenarios, with the event occurring every four years in South East Asia by the end of the 21st Century.

To coincide with an increasing risk of high intensity rainfall events, the rise in urban population results in aggravated threat of flooding, meaning that historical approaches to flood management are likely to be unsuitable in the future (Reynard *et al.*, 2017). It is anticipated that without adaption, there is likely to be up to US$24 trillion increase in global annual costs due to flooding by the end of the century (Jevrejeva *et al.*, 2018).

Urban flooding is increasingly a risk; statistics demonstrate that 55% of the total global population lives in urban areas, and this figure is expected to increase to 60% by 2030, and the number of global 'megacities' will increase from 33 to 43 during the same time period (UN, 2018). While urban redevelopment can manage some of the growth, the increasing population will drive cities to sprawl into the peri-urban environment (Haaland & van den Bosch, 2015; Miller *et al.*, 2014; Pili *et al.*, 2017). Urbanisation causes a change in the natural hydrological balance by constructing impermeable surfaces, the compaction of soil and the removal of vegetation (Brilly *et al.* 2006; Haaland & van den Bosch, 2015; Lashford *et al.*, 2019; Li *et al.*, 2018; Rubinato *et al.*, 2019). Additionally, the installation of a traditional, pipe-based drainage system, which can efficiently remove water from urban areas to nearby watercourses, increases flood risk by altering the hydrological response to rainfall (Lundy & Wade, 2011; Miller *et al.* 2014).

Since the installation of the hydraulically efficient London Sewerage System in the 1850s, resulting from the need for suitable sanitation post Industrial Revolution, pipe based drainage has been adopted globally (Hughes, 2013; Walsh *et al.*, 2005; Xie *et al.*, 2017). Pre industrial-age approaches to drainage still exist, for example the Cloaca Maxima, Rome, Italy, however they have been largely subsumed by post-industrial drainage, further ensuring water is 'out of sight, out of mind', aiming to rapidly remove water from urban areas (Perales-Momparler *et al.*, 2015). Many cities consequently now rely on ageing drainage systems built in the 19th and 20th Century that are now, with an increasing urban landscape and a changing climate, incapable of managing the current urban pluvial flood risk (Djordjevic *et al.*, 2011; Egger & Maurer, 2015; Guo, 2006). Additionally, the 'clogging' of conventional systems with debris inhibits their potential to effectively remove water, causing a back log through the system and an increased flood risk. The 2007 UK summer floods is a typical example of such scenarios (Fenner, 2000).

Most urban drainage systems were designed to manage smaller volumes of runoff than faced today, and therefore are vulnerable to failure during high-intensity rainfall events (Butler & Parkinson, 1997; Butler *et al.*, 2018; Mark *et al.*, 2004). The integration of pipe based drainage system at new build sites is still part of typical design culture, with runoff to the sewer system typically flowing underground via gully pots and pipes before reaching the watercourse (Woods-Ballard *et al.*, 2015). This poses an increased flood risk for the outfall as a result of a reduced lag time and increased peak flow at the receiving water course (Qin *et al.*, 2013).

Table 9.1 outlines the design flood frequency for pipe based systems according to the British Standards (British Standards Institution, 2008) and the European Standard EN 752. All drainage systems in a city centre should manage all storms up to and including the one in 30 year storm event, while in comparison, designs are up to the one in 10 year return period for the USA (Guo, 2006). Many cities are consequently at risk of flooding due to insufficient capacity of drainage, which is an even bigger problem in less developed countries due to lower drainage standards in urban areas (Fratini *et al.*, 2012; Guo, 2006; Mark *et al.*, 2004).

Table 9.1 Conventional drainage design storm frequency scenario for different locations (adapted from British Standards Institution 2008 and the European Standard EN 752)

Location	Design Storm Frequency
Rural areas	One in 10 years
Residential areas	One in 20 years
City centres	One in 30 years

In addition to having the primary concern of increased flood risk at the source and the outfall, conventional drainage has also created a water quality issue. Improving runoff quality prior to being released into the watercourse is a neglected aspect of conventional drainage (Hoang & Fenner, 2015). Consequently, runoff transports a variety of urban pollutants without treatment into the watercourse which has an impact on the biodiversity of urban streams (Zhang *et al.*, 2013).

This chapter aims at summarising previous research conducted to improve the accuracy of urban flood modelling, experimentally and numerically, and highlights the need of full-scale and field case studies that could provide hidden insights not achievable via scaled models for a better application and better benefits for multiple urban developments.

9.2 EXPERIMENTAL AND NUMERICAL URBAN FLOOD MODELLING

9.2.1 Input parameters and boundary conditions

Recent advances in technology, and consequently in the performance of physical models, have generated an abundance of data which are crucial to calibrate and validate numerical models (Mignot *et al.*, 2019). The essential data requirements of flood inundation models can be summarised as follows into multiple categories.

Topographic data of the channels and floodplains to act as model bathymetry (Peña & Nardi, 2018) are essential when setting up a numerical model. A high quality Digital Terrain Model (DTM) representing the ground surface with surface objects removed is the basic topographic data requirement. As provided in literature (Ramsbottom & Wicks, 2003), vertical accuracy of about 0.5 m and a spatial resolution of at least 10 m for DTMs are required for rural floodplain modelling, while a vertical accuracy of 5 cm with a spatial resolution of 0.5 m are needed to resolve gaps between buildings (Ozdemir *et al.*, 2013; Smith *et al.*, 2006; Van Ootegem *et al.*, 2016) for numerical modelling over urban floodplains, considering that the knowledge of the micro-topography over large areas may become much more significant.

Time series of bulk flow rates and rainfall are crucial for characterising numerical boundary conditions (Morales-Hernández *et al.*, 2013). Ideally, flow rates should be accurate to 5%, however, this figure may differ consistently and errors may be much higher during flood events. Recent studies have also investigated more complex features such as the rain falling directly into streets (Paquier & Bazin, 2014). Rainfall has been considered for many years as a major uncertainty source, in the accurate prediction of urban flooding events (Niemczynowicz, 1999), but recent progress associated with the use of radar and microwave networks have reduced the uncertainties by facilitating the gathering of sufficient information on temporal and spatial variation of rainfall processes in urban catchments (Fletcher *et al.*, 2013; Schellart *et al.*, 2012).

Roughness coefficients for channels and floodplains, which may be spatially distributed (Kirstetter *et al.*, 2016), are another category of essential parameters to use when accurately setting up a numerical model, as well as bottom roughness coefficients in the sewers and floodplain (Bellos *et al.*, 2017). By introducing these parameters, energy losses not represented explicitly in the model equations can be parameterised. Considering that in practice they are usually estimated by calibration, it can be difficult to separate the contribution associated to friction from that attributable to compensation for model structural and input flow errors. It is necessary to calibrate numerical models, adopting two separate roughness coefficients for sewers and floodplains.

Despite the progress in producing essential datasets for calibration and validation of numerical models, there is still a paucity of datasets that are essential to accurately represent detailed features associated with the high complexity of the urban environment (for example multiple flow paths through crossroads, sewers-surface interactions, parks, Sustainable urban Drainage Systems (SuDS)). These features are crucial to assess the hydraulic performance of drainage systems in urban areas and to verify the probability of flooding. The uncertainty associated with the nature of flood events, which are typically characterised with high intensity and a short duration, results in challenges to identify possible locations of flooding in advance. This creates complications when attempting to record the data required for the calibration and validation of numerical models. Furthermore, even if municipalities have the availability of equipment to consistently record flow, water levels and pressure in sewers and streets, it is expensive to optimise the maintenance procedure by checking all these sensors repeatedly over several years, causing eventual gaps or non-reliable datasets.

To cope with this lack of data, physical full and scaled models are important tools to replicate urban flooding conditions as well as field case studies where feasible. Over the last decade, multiple studies have been performed and the next section summarises them based on multiple variables such as flow patterns, hydraulic and geometrical conditions, scale factors tested, and SuDS techniques adopted.

9.2.2 Flow patterns. hydraulic conditions and geometrical setups

As previously mentioned in Section 9.1, by 2030 60% of the world's population will be living in urban areas, increasing to 68% by 2050. Cities across the world continue to develop, targeting different social needs and following various planning criteria established by governments and municipalities, generating different urban scenarios to incorporate flood resilience into urban planning (Bertilsson *et al.*, 2019). As a consequence, flow patterns and their interactions in urban environments are very complex.

To date, experimental studies to estimate these complex flows and/or energy losses in urban drainage systems have been conducted to take into account subcritical (Nania *et al.*, 2011; Riviere *et al.*, 2011; Schindfessel *et al.*, 2015) and supercritical flow regimes (Creëlle *et al.*, 2017; Kemper and Schlenkhoff, 2019; Riviere *et al.*, 2014), open-channel and pressurised flow (Martins *et al.*, 2017; Rubinato *et al.*, 2018a), interactions between the minor and major systems (Beg *et al.*, 2018; Fraga *et al.*, 2017; Gomez & Russo, 2005, 2009; Gomez *et al.*, 2019; JinNoh *et al.*, 2016; Lopes *et al.*, 2014, 2015; Martins *et al.*, 2014, 2018; Rubinato, 2015; Rubinato *et al.*, 2011; Rubinato *et al.*, 2013, 2014, 2017a, b, 2018b; Vasconcelos *et al.*, 2006) and with the presence of obstacles or streets (Arrault *et al.*, 2016; Finaud-Guyot *et al.*, 2018) and building blocks (Güney *et al.*, 2014; Smith *et al.*, 2016). Examples of experimental facilities are shown in Figure 9.2.

Mignot *et al.* (2019) have pointed out that these multiple studies address very few identical flow patterns and these studies, despite providing crucial specific insights to improve the experimental and numerical modelling of urban flood flows, are

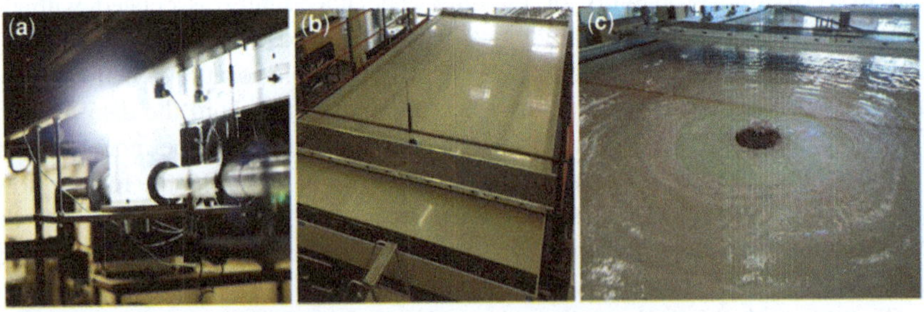

Figure 9.2 Example of experimental urban flood models with sub-surface interaction (Rubinato 2015). (a) View of the scaled drainage system of the sub/surface experimental facility at the University of Sheffield; (b) Birdseye view of the urban floodplain of the sub/surface experimental facility at the University of Sheffield and (c) experimental simulation of sewer overflow. ©MatteoRubinato.

limited to specific hydraulic conditions or have been conducted to optimise a specific case scenario, and hence may be limited to be up-scaled or applied within a variety of urban environments.

Furthermore, experimental models have been developed to date to provide datasets to represent local energy (head) losses (friction and local losses) (Butler & Davies, 2011; Butler *et al.*, 2018) at junctions and urban drainage features for improving the performance of existing numerical models and better predicting future scenarios.

Multiple parameters have been found to affect head losses during the last two decades, including the channel water depths between the upstream branches and the downstream channel (Hsu & Lee, 1998); upstream and downstream subcritical or supercritical hydraulic conditions (Gargano & Hager, 2002; Hager & Gisonni, 2005; Zhao *et al.*, 2008); the joining angle between any lateral pipes and the main pipe (Pfister & Gisonni, 2014); and the ratio between pipe diameter and manhole diameter (Ramamurthy & Zhu, 1997). Bearing in mind that these outputs were mainly obtained replicating physical scale models of singular structures (e.g. single manholes), the impacts of these parameters are likely to be different when considering an entire drainage system including several structures, with a hypothetical 'domino' effect towards the downstream section of the area under investigation.

Additionally, previous research has provided an insight into consequences of floodwater flowing on streets in urban areas (Beg *et al.*, 2020). Based on the amount of water and the geometrical conditions of the roads (e.g. slope), vehicles parked on the sides of streets can be swept away, causing various hazards to people and properties (Xia *et al.*, 2014).

Stability criteria for vehicles in floodwaters were determined by Gordon and Stone (1973), Keller and Mitsch (1993), Shu *et al.* (2011) and Xia *et al.* (2011), and experimental data were used to validate the derived formulae based on the principles of similarity and scale ratios. Limitations due to scaling effects have also made it not possible to apply the results to corresponding prototype vehicles in the study conducted by Teo *et al.* (2012).

Runoff on streets is also a hazard for pedestrians attempting to cross the roads under a variety of water depths and velocities (Martinez-Gomariz *et al.*, 2016). Experimental studies for human body stability in laboratory setups by using real human bodies started with Foster and Cox (1973), and have been expanded by Abt *et al.* (1989) who adopted different ground surfaces and Jonkman and Penning-Rowsell (2008) who tested high velocities and psychological factors. In the UK, Defra and the Environment Agency provided a method to quantify the hazards for pedestrians based on velocity, depth and presence of debris within the flows (Defra, 2006). Russo (2009) focused on the impact of pedestrians of different ages and weights, with diverse visibility conditions and the use of hands, wearing dissimilar footwear.

9.2.3 Sustainable drainage systems

Section 9.1 identified the challenges regarding traditional drainage methods and flood risk reduction, consequently the UK has moved to produce non-statutory guidance for SuDS, advocating their role at new build sites, to minimise an increase in flood risk post-development (Defra, 2015). This was developed as a result of the 2007 UK floods, which cost the UK economy approximately £3 billion, and paved the way for the UK Flood and Water Management Act (HM Government, 2010), which, as part of Schedule 3 of the act, further identifies SuDS as being a critical approach to sustainable flood management for the future (Pitt, 2008).

Due to the need to make cities more sustainable, a number of experimental studies have also been completed to examine the impact that SuDS can have at reducing runoff and ultimately flooding, in an urban setting (Fach & Dierkes, 2011; Nnadi *et al.*, 2012; Sanudo-Fontaneda *et al.*, 2018; van Woert *et al.*, 2005). It is understood that by integrating 'green infrastructure' into urban design, or by simply reducing the total coverage of impermeable surfaces, it is possible to reduce localised, pluvial flooding (see Figure 9.3) (Eckart *et al.*, 2017).

At the laboratory scale, experimental control tests of SuDS have been undertaken over the last 20 years (Sanudo-Fontaneda *et al.*, 2018; van Woert *et al.*, 2005). Models to determine the effectiveness of both extensive and intensive green roofs have demonstrated their ability to reduce runoff received from traditional tiled roofs (Bouzouidja *et al.*, 2018; Lee *et al.*, 2013; van Woert *et al.*, 2005). Stovin *et al.* (2015a) monitored a number of rainfall scenarios over a four year period to analyse the water retention capabilities of a green roof, taking into account the effects of the vegetation type, density and structure of the green roof. Subsequently, it was possible to determine roughness coefficients in a numerical model to determine energy losses (Stovin *et al.*, 2015b) (see Section 9.2.1).

Figure 9.3 Example of a SuDS system, comprising of rock-lined swales and vegetated detention ponds, in Leicester, England. ©CraigLashford.

Similar experimental laboratory tests have been conducted on permeable paving (Nnadi *et al.*, 2012; Sanudo-Fontaneda *et al.*, 2014). By altering the topographic variables, runoff surface length and surface slope of a laboratory scale permeable pavement design, it is possible to maximise runoff reduction (Sanudo-Fontaneda *et al.*, 2014). To further optimise a design, Nnadi *et al.* (2014) analysed the possible influence of membrane layers, such as more traditional geotextile or OASIS®, not only on total outflow, but also water quality. However, experimental models are still required to better understand the scale of reduction, and to maximise the benefits that can be achieved by integrating SuDS into the urban landscape.

9.2.4 Scale factors: the need for full-scale and field models

Urban areas are complex, and most experimental setups previously described either reproduce a simplified urban city or a synthetic urban area (simple streets, 45 and 90° junctions, rectangular buildings) by adopting scale factors based on specific similitudes, hence these are called physical scale models.

Despite providing important insights and better understanding of urban flooding scenarios, physical scale models include assumptions and 'artefacts' that typically do not completely resemble real-world prototype observations. This is due to governing non-dimensional parameters (i.e. force ratios) which are not completely identical between the model and its prototype (Heller, 2011). Consequently, the application of these features may include a slight alteration of the flow regime (laminar, turbulent and transition phase), or of the relative importance of frictional resistance, generating datasets that could be dissimilar from those linked with real environments (Heller, 2011).

Considering, for example, the ageing conditions of existing sewer systems in the UK and worldwide networks, an extended number of variables (e.g. ground water level, traffic, number of house connections, materials, age, and diameters) are affecting the identification of techniques and solutions for the rehabilitation of existing infrastructure. Experimental models tend to view systems separately, focusing on a specific problem without taking into account the interaction and possible synergies between subsystems (Tscheikner-Gratl *et al.*, 2015). For example, companies building or repairing sewers on-site are required to follow steps regarding excavation, backfill and compaction. Existing studies (Del Borghi *et al.*, 2008; Uche *et al.*, 2013) focused on the use of concrete and aggregates for pipes bending, but did not include the paving of the roads using bitumen, a product typically included, hence materials and their properties may differ from scale models to real sites. Furthermore, CCTV (closed-circuit television) inspection may be expensive, and the quality of the analysis conducted on existing sewer systems may be dependent on the quality of the pictures taken as well as the skills and the experience of the technician analysing them (Wirahadikusumah *et al.*, 1998).

Figure 9.4 Upstream and downstream view of the full experimental facility constructed at ICAIR, plus a detailed view of a full-scale manhole. ©The University of Sheffield and ©SimonTait.

To address this gap, full-scale facilities have been developed, or are under development, such as the Integrated Civil and Infrastructure Research Centre (ICAIR) at the University of Sheffield (Figure 9.4) and the full sub-surface model at the IKT – Institute for Underground Infrastructure in Germany (Figure 9.5).

Their common aim focuses on producing results that could identify inaccurate assumptions or specifications, and supporting numerical models for the simulation of precipitation and runoff events in urban areas. By obtaining this outcome, numerical models could be better calibrated and validated, and could

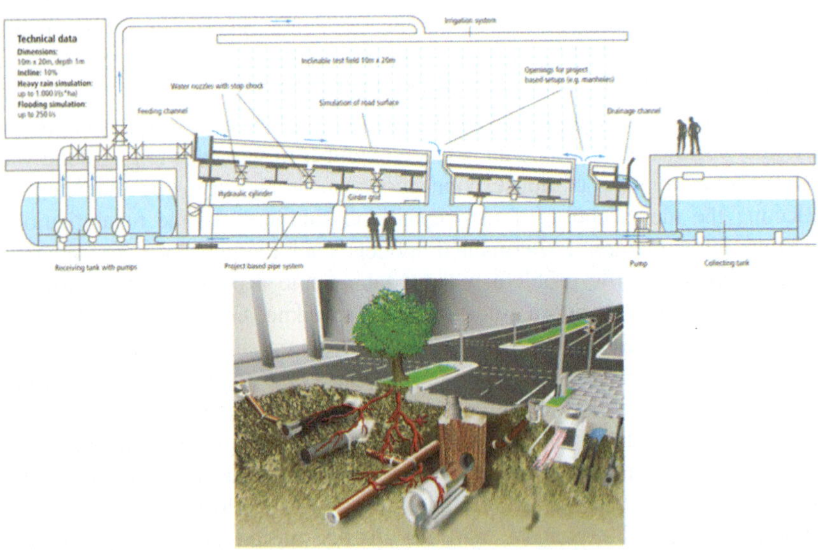

Figure 9.5 Scheme of the full-scale experimental facility designed and under development at IKT-Institute for Underground Infrastructure. ©IKT-Institute for Underground Infrastructure (Goerke, 2019).

increase their ability to assess the hydraulic performance of existing and new drainage systems, making a stronger impact in municipalities and reducing the potential damages associated with urban flooding. By increasing accuracy, full-scale experimental facilities (Figures 9.4 and 9.5) can take into consideration typical hydraulic conditions (backwater effects, overflow situations, rain-runoff interaction) on the road surface in connection with underneath sewer systems, that if replicated within a scale model, may not be precisely replicated with all associated features included. Furthermore, multiple existing materials and geometries adopted in different countries can be tested for efficiency without the issue of having to scale properties that could be fundamental in the replication of the phenomenon to investigate.

Outside of the laboratory, there has been a recent shift to better understand sustainable design through the use of 'living labs', i.e. experimental field study space constructed into the built environment (Hutley *et al.*, 2011; Lima & Ribeiro, 2016). While there has been an acknowledgement through much of the 21st Century of the role that SuDS can play at limiting urban runoff, their integration with both new and existing urban drainage schemes has been limited, particularly in England (Melville-Shreeve *et al.*, 2018). It is imperative therefore that SuDS design is optimised through experimental field modelling, as well as that at the laboratory scale, to ensure standards are retained, and the most effective design is integrated (Tedoldi *et al.*, 2016). There are now a number of best practice installation, which have also been used for experimental field studies (Fach & Dierkes, 2011; Garcia-Serrana *et al.*, 2017). By integrating a two-phased calibration and validation approach, it is possible to model field data with an acceptable level of accuracy, accounting for epistemic and aleatory uncertainty (Beven, 2016; Lamera *et al.*, 2014; Mei *et al.*, 2018; Versini *et al.*, 2015).

Using experimental data previously described is crucial to validate numerical models and hence develop confidence on the predictions for future events of similar magnitude. To date, only a few experimental datasets have been available for numerical model validation (bulk flow measurements taken at a small number of points in the model domain, often including the catchment outlet, water levels obtained by CCTW cameras). However, the 2D nature of modern distributed models requires spatially distributed observational data (e.g. water depth or flow velocity) at a scale adequate to model predictions for successful validation. Full-scale facilities already constructed and those under development can expand the possibility to secure these records over the course of a simulated flood event. Successfully validated experimental field scale studies allow for a better understanding of external parameters, which often cannot be replicated in a laboratory setting, for example the ability to understand the influence of maintenance of sites, the availability of a long term, ideally continuous, dataset and the ability to utilise dynamic catchment-scale rainfall and ultimately, flow.

Sites such as the 12.5 km^2 Pontbren catchment in Powys, Wales, which was maintained between 2004 and 2009, provide an experimental field site to better understand the influence of different land management methods (Marshall *et al.*, 2009; Marshall *et al.*, 2014; Wheater & Evans, 2009). The Pontbren project aimed to provide a long-term, continuous record of how different farming techniques and catchment management approaches altered the hydrology of the area, by monitoring soil structure change, infiltration rates, peak flow and rainfall (Bulygina *et al.*, 2013; McIntyre & Marshall, 2010). A further example of catchment-scale experimental research is by Heal *et al.* (2006), at the Dunfermeline Eastern Expanse, a 5 km^2 site consisting of three ponds (the largest being 15,495 m^3) and an 18,633 m^2 wetland area. The research enabled a better understanding of how each pond and the wetland system managed stormwater quantity and quality over a four-year period, informing future maintenance to ensure standards for water quality improvements, and quantity reduction are retained.

However, there are issues to be faced when using active urban drainage systems as test models that can be associated with the likelihood of equipment being damaged or stolen, depending on the site chosen (Gomani *et al.*, 2010; Rivett *et al.*, 2008; Ruhl *et al.*, 2001). There is a need for continuous data to understand long-term trends of SuDS systems, as runoff is inherently dynamic, but also for example, to understand how natural seasonal vegetation growth cycles influence runoff, however this is not always possible due to extraneous circumstances, such as data-drop-out, battery-life span and equipment failure (Roinas *et al.*, 2014). It is therefore common for urban modelling to utilise less expensive monitoring equipment, which can be of a reduced resolution, to negate the cost-impact of tampering of equipment (Chetpattananondh *et al.* 2014; Rivett *et al.*, 2008).

9.3 CONCLUSIONS

Climate change, urbanisation and ageing conditions make future urban flood events very difficult to predict. However, despite this huge challenge that engineers, urban planners and policymakers will have to deal with, it is crucial to be able to develop a better understanding of this uncertain phenomenon, to facilitate decision-making and the identification of tools and strategies to adopt, and to achieve the desirable goal of reducing urban flooding and its negative impacts.

This chapter has grouped research studies conducted to date to enhance the quality of experimental and numerical urban flood modelling for a better prediction of future flood events. It is clear that there is a great emphasis in the literature on improving the accuracy of models, and continuous progress is being made on understanding areas for improvement to reduce impacts on existing and new infrastructure. This development is directly associated with the availability of data, which is limited or may not exist for specific hydraulic conditions. The understanding of precise links between each variable involved during flooding conditions still requires more assessments and more research is needed and should be prioritised to assess and reduce flood impacts. To achieve more consistent and less incomplete flood impact assessments, the collection of more data would be highly valuable to build upon.

Continuous development of technology aids the improvement of new tools and measurement techniques (Nichols *et al.*, 2020; Rojas Arques *et al.*, 2018) to design and inspect more accurate scaled and full-scale models. Multiple aspects need to be advanced in the future, and stakeholders should all combine resources and strengths to:

- Provide a better understanding of urban flooding and its causes, studying features not fully investigated to date such as manhole covers removal due to high pressure in overflowing manholes; sediment transport in pipes and in streets and the interaction of the two systems during flooding scenarios; micro plastics settled in the road-deposited dust and urban surface soil. Data obtained should be made open access so that numerical modellers across the world can calibrate and validate their models. There is a need for practically orientated technological solutions to provide protection for the environment and property and to maintain the functionality of the entire infrastructure. Concepts for both public and private wastewater systems which result in the most damage-free removal of the precipitation water and minimise the risk of flooding events (e.g. temporary retention of large volumes of water, delayed passing-on of precipitation and combined wastewater, throttling of influxes and outflows) are, in particular, thus gaining in importance.
- Make cities more liveable, combining both 'soft' and 'hard' engineering solutions to achieve sustainability of urban systems. Sustainable Drainage Systems and reuse and recycling technologies are in continuous progress and efforts should be made to facilitate the transition to these new systems.
- Identify procedures for better monitoring, repairing and rehabilitating existing infrastructure. Methods are required to reduce groundwater infiltration into sewers, to reduce the risk of sewer blockages to improve their serviceability. Rehabilitation techniques such as cured-in place pipes are continuously increasing their confidence, however to be able to generate benefits in many countries around the world, they need to guarantee not only the structural repair of the old pipe, but also ecological compatibility with different materials, under varying chemicals and environmental conditions from one country to another. New materials, such as glass-fibre-reinforced plastics GRP, needle-felt, plastic coatings and reactive resins, for example, are coming into use everywhere for the rehabilitation of existing systems. The behaviour of these materials raises new questions, particularly with respect to such aspects as durability,

high-cycle fatigue performance, ecological efficiency and sustainability – and also in respect of anticipated new requirements. Changing environmental conditions (e.g. longer dry periods, increasingly frequent heavy rainfall events), and also greater volumes of road traffic (e.g. increases in heavy-goods traffic) definitively influence material performance.

Nowadays, new technologies continue to be introduced at a rapid rate, therefore new accurate and sophisticated equipment as well as faster computers can be used to enhance the quality of physical modelling and its outcomes. To further improve existing ideas, communities should be promoting awareness and improving environmental education. The public should be invited to respond to plans and proposals designed by authorities and private companies, providing invaluable local experience to refine planned physical models because community engagement is an essential component of sustainable flood risk management. Public participation can thus directly inform decisions and support the execution of actions rather than only raise awareness.

REFERENCES

Abt S. R., Wittler R. J., Taylor A. and Love D. J. (1989). Human stability in a high flood hazard. *Water Resources Bulletin*, **25**(4), 881–890.

Arrault A., Finaud-Guyot P., Archambeau P., Bruwier M., Erpicum S., Pirotton M. and Dewals B. (2016). Hydrodynamics of long-duration urban floods: experiments and numerical modelling. *Natural Hazards and Earth System Sciences*, **16**(6), 1413–1429.

Beg Md N. A., Carvalho R. F., Tait S., Brevis W., Rubinato M., Schellart A. and Leandro J. (2018). A comparative study of manhole hydraulics using stereoscopic PIV and different RANS models. *Water Science and Technology*, **2017**(1), 87–98.

Beg Md N. A., Rubinato M., Carvalho R. F. and Shucksmith J. (2020). CFD Modelling of the transport of soluble pollutants from sewer networks to surface flows during urban flood events. *Water*, **12**(9), 2514; https://doi.org/10.3390/w12092514

Bellos V., Kourtis I., Moreno-Rodenas A. and Tsihrintzis V. (2017). Quantifying roughness coefficient uncertainty in urban flooding simulations through a simplified methodology. *Water*, **9** (12), 944. https://doi.org/10.3390/w9120944

Bertilsson L., Wiklund K., Tebaldi I. M., Rezende O. M., Pires Veról A. and Gomes Miguez M. (2019). Urban flood resilience – A multi-criteria index to integrate flood resilience into urban planning. *Journal of Hydrology*, **573**, 970–982.

Beven K. (2016). Facets of uncertainty: Epistemic uncertainty, non-stationarity, likelihood, hypothesis testing and communication. *Hydrological Sciences Journal*, **61**(9), 1652–1665. http://dx.doi.org/10.1080/02626667.2015.1031761

Bouzouidja R., Séré G., Claverie R., Ouvrard S., Nuttens L. and Lacroix D. (2018). Green roof aging: quantifying the impact of substrate evolution on hydraulic performances at the lab-scale. *Journal of Hydrology*, **564**, 416–423.

Brilly M., Rusjan S. and Vidmar A. (2006). Monitoring the impact of urbanisation on the Glinscica stream. *Physics and Chemistry of the Earth*, **31**(17), 1089–1096.

British Standards Institution. (2008). Drain and Sewer Systems Outside Buildings. Hydraulic Design and Environmental Considerations, BS EN 752: 2008. British Standards Institution, London.

Bulygina N., McIntyre N. and Wheater H. (2013). A comparison of rainfall-runoff modelling approaches for estimating impacts of rural land management on flood flows. *Hydrology Research*, **44**(3), 467–483.

Butler D. and Davies J. (2011). Urban Drainage. CRC Press, London. https://doi.org/10.1201/9781315272535

Butler D. and Parkinson J. (1997). Towards sustainable urban drainage. *Water Science and Technology*, **35**(9), 53–63.

Butler D., James Digman C., Makropoulos C. and Davies J. (2018). Urban Drainage. CRC Press, Boca Raton. https://doi.org/10.1201/9781351174305

Centre for Research on the Epidemiology of Disasters. (2019) Natural Disasters: 2018. Available from: www.cred.be/publications

Chetpattananondh K., Tapoanoi T., Phukpattaranont P. and Jindapetch N. (2014). A self-calibration water level measurement using an interdigital capacitive sensor. Sensors and Actuators. A, Physical, **209**, 175–182. http://dx.doi.org/10.1016/j.sna.2014.01.040

Creëlle S., Engelen L., Schindfessel L., Cunha Ramos P. and De Mulder T. (2017). Experimental investigation of free surface gradients in a 90° angled asymmetrical open channel confluence. In: Advances in Hydroinformatics, P. Gourbesville, J. Cunge and G. Caignaert (eds.), Springer, Singapore, pp. 803–819.

Defra. (2015). Non-statutory Technical Standards for Sustainable Drainage Systems. March 2015, Report PB14308, Department of Environment & Rural Affairs (Defra), London. UK.

Defra and Environment Agency (EA). (2006). Flood and coastal defence R&D programme, R&D outputs: Flood risks to people (Phase 2), Defra Report, London.

Del Borghi A., Gaggero P. L., Gallo M. and Strazza C. (2008). Development of PCR for WWTP based on a case study. *International Journal of Life Cycle Assessment*, **13**, 512–521.

Depietri Y. and McPhearson T. (2018). Changing urban risk: 140 years of climatic hazards in New York City. *Climatic Change*, **148**(1–2), 95–108.

Djordjević S., Butler D., Gourbesville P., Mark O. and Pasche E. (2011). New policies to deal with climate change and other drivers impacting on resilience to flooding in urban areas: The CORFU approach. *Environmental Science & Policy*, **14**(7), 864–873. http://dx.doi.org/10.1016/j.envsci.2011.05.008

Eckart K., McPhee Z. and Bolisetti T. (2017). Performance and implementation of low impact development – A review. *Science of the Total Environment*, **607–608**, 413–432.

Egger C. and Maurer M. (2015). Importance of anthropogenic climate impact, sampling error and urban development in sewer system design. *Water Research*, **73**, 78–97.

Fach S. and Dierkes C. (2011). On-site infiltration of road runoff using pervious pavements with subjacent infiltration trenches as source control strategy. *Water Science and Technology*, **64**(7), 1388–1397.

Fenner R. (2000). Approaches to sewer maintenance: A review. *Urban Water Journal*, **2**(4), 343–356.

Finaud-Guyot P., Garambois P. A., Araud Q., Lawniczak F., François P., Vazquez J. and Mosé R. (2018). Experimental insight for flood flow repartition in urban areas. *Urban Water Journal*, **15**(3), 1–9.

Fletcher T. D., Andrieu H. and Hamel P. (2013). Understanding, management and modelling of urban hydrology and its consequences for receiving waters: A state of the art. *Advances in Water Resources*, **51**, 261–279. https://doi.org/10.1016/j.advwatres.2012.09.001

Foster D. N. and Cox R. J. (1973). Stability of children on roads used as floodways, Technical Report No.73/13, Water Research Laboratory of the University of New South Wales, Manly Vale, Australia.

Fraga I., Cea L. and Puertas J. (2017). Validation of a 1D–2D dual drainage model under unsteady part-full and surcharged sewer conditions. *Urban Water Journal*, **14**(1), 74–84.

Fratini C., Geldof G., Kluck J. and Mikkelsen P. (2012). Three Points Approach (3PA) for urban flood risk management: a tool to support climate change adaptation through transdisciplinarity and multifunctionality. *Urban Water Journal*, **9**(5), 317–331.

García-Serrana M., Gulliver J. and Nieber J. (2017). Non-uniform overland flow-infiltration model for roadside swales. *Journal of Hydrology*, **552**, 586–599.

Gargano R. and Hager W. (2002). Supercritical flow across sewer manholes. *Journal of Hydraulic Engineering – ASCE*, **11**, 1014–1017.

Goerke M. (2019). Planning and conception of a heavy rain laboratory and its practical use. Proceedings of 4th Stormwater and Thaw Waters Management Conference, 4–6th September, Zakopane, Poland.

Gomani M., Dietrich O., Lischeid G., Mahoo H., Mahay F., Mbilinyi B. and Sarmett J. (2010). Establishment of a hydrological monitoring network in a tropical African catchment: An integrated participatory approach. *Physics and Chemistry of the Earth*, **35**(13–14), 648–656.

Gómez M. and Russo B. (2005). Comparative study among different methodologies to determine storm sewer inlet efficiency from test data: HEC22 methodology vs. UPC method. In: Water Resour. Manag. III, M. de Conceicao Cunha and C. A. Brebbia (eds). WIT Press, Algarve, Portugal, pp. 623–632.

Gómez M. and Russo B. (2009). Methodology to estimate hydraulic efficiency of drain inlets. *Proceedings of the Institute of Civil Engineers – WaterManagement*, **164**(2), 85–91.

Gómez M., Russo B. and Tellez-Alvarez J. (2019). Experimental investigation to estimate the discharge coefficient of a grate inlet under surcharge conditions. *Urban Water Journal*, **16**(2), 85–91.

Gordon A. D. and Stone P. B. (1973). Car stability on road floodways, Technical Report 73/12, University of New South Wales, Manly Vale, Australia.

Güney M. S., Tayfur G., Bombar G. and Elci S. (2014). Distorted physical model to study sudden partial dam break flows in an urban area. *Journal of Hydraulic Engineering – ASCE*, **140**(11), 05014006.

Guo Y. (2006). Updating rainfall IDF relationships to maintain urban drainage design standards. *Journal of Hydraulic Engineering – ASCE*, **11**(5), 506–509.

Haaland C. and van den Bosch C. (2015). Challenges and strategies for urban green-space planning in cities undergoing densification: A review. *Urban Forestry & Urban Greening*, **14**(4), 760–771.

Hager W. H. and Gisonni C. (2005). Supercritical flow in sewer manholes. *Journal of Hydraulic Research*, **43**, 660–667. https://doi.org/10.1080/00221680509500385

Heal K. V., Hepburn D. A. and Lunn R. J. (2006). Sediment management in sustainable urban drainage ponds. *Water Science and Technology*, **53**(10), 219–227.

Heller V. (2011). Scale effects in physical hydraulic engineering models. *Journal of Hydraulic Research*, **49**, 293–306. https://doi.org/10.1080/00221686.2011.578914

HM Government (2010). Flood and Water Management Act 2010. 2010 c.29. Available from: www.legislation.gov.uk/ukpga/2010/29/contents/enacted

Hoang L. and Fenner R. (2015). System interactions of stormwater management using sustainable urban drainage systems and green infrastructure. *Urban Water Journal*, **13** (7), 1–20.

Hsu C. C. and Lee W. J. (1998). Flow at 90° equal-width open-channel junction. *Journal of Hydraulic Engineering – ASCE*, **124**, 186–191.

Hughes M. (2013). The Victorian London sanitation projects and the sanitation of projects. *International Journal of Project Management*, **31**(5), 682–691.

Hutley L. B., Beringer J., Isaac P. R., Hacker J. M. and Cernusak L. A. (2011). A sub-continental scale living laboratory: Spatial patterns of savanna vegetation over a rainfall gradient in northern Australia. *Agricultural and Forest Meteorology*, **151**(11), 1417–1428. http://dx.doi.org/10.1016/j.agrformet.2011.03.002

Intergovernmental Panel on Climate Change, and (2014). Climate Change 2014: Synthesis Report. In: Contribution of Working Groups I, II and III to the Fifth Assessment Report of the Intergovernmental Panel on Climate Change, Core Wriing Team, R. K. Pachauri, L. A. Meyer (eds.), IPCC, Geneva, Switzerland, p. 151.

Jevrejeva S., Packson L. P., Grinsted A., Lincke D. and Marzeion B. (2018). Flood damage costs under the sea level rise with warming of 1.5°C and 2°C. *Environmental Research Letters*, **13**, 074014. http://dx.doi.org/10.1088/1748-9326/aacc76

JinNoh S., Lee S., An H., Kawaike K. and Nakagawa H. (2016). Ensemble urban flood simulation in comparison with laboratory-scale experiments: impact of interaction models for manhole, sewer pipe and surface flow. *Advances in Water Resources*, **97**, 25–37.

Jonkman S. N. and Penning-Rowsell E. (2008). Human instability in flood flows. *Journal of the American Water Resources Association*, **44**(4), 1208–1218.

Keller R. J. and Mitsch B. (1993). Safety aspects of design roadways as floodways, Research Report No. 69, Urban Water Research Association of Australia, Australia.

Kemper S. and Schlenkhoff A. (2019). Experimental study on the hydraulic capacity of grate inlets with supercritical surface flow conditions. *Water Science and Technology*, **79**, 1717–1726.

Kirstetter G., Hu J., Delestre F., Darboux F., Lagree P. Y., Popinet S., Fullana J. M. and Josserand C. (2016). Modeling rain-driven overland flow: empirical versus analytical friction terms in the shallow water approximation. *Journal of Hydrology*, **536**, 1–9.

Kundzewicz Z. W., Kanae S., Seneviratne S. I., Handmer J., Nicholls N., Peduzzi P., Mechler R., Bouwer L. M., Arnell N., Mach K., Muir-Wood R., Brakenridge G. R., Kron W., Benito G., Honda Y., Takahashi K. and Sherstyukov B. (2014). Flood risk and climate change: global and regional perspectives. *Hydrological Sciences Journal*, **59**(1), 1–28. http://doi.org/10.1080/02626667.2013.857411

Lamera C., Becciu G., Rulli M. and Rosso R. (2014). Green roofs effects on the urban water cycle components. *Procedia Engineering*, **70**, 988–997.

Lashford C., Rubinato M., Cai Y., Hou J., Albofathi S., Coupe S., Charleworth S. and Tait S. (2019). SuDS & Sponge Cities: A Comparative Analysis of the Implementation of Pluvial Flood Management in the UK and China. *Sustainability*, **11**(1), 213.

Lee J., Moon H., Kim T., Kim H. and Han M. (2013). Quantitative analysis on the urban flood mitigation effect by the extensive green roof system. *Environmental Pollution*, **181**, 257–261.

Li N., Qin C. and Du P. (2018). Optimization of China sponge city design: the case of Lincang Technology Innovation Park. *Water*, **10**(9), 1189. http://dx.doi.org/10.3390/ w10091189

Lima E. and Ribeiro S. K. (2016). Monitoring sustainability at Rio de Janeiro Federal University. *Proceedings of the Institute of Civil Engineers – Municipal Engineer*, **169** (4), 189–198. http://dx.doi.org/10.1680/jmuen.15.00012

Liu L. and Jensen M. (2018). Green infrastructure for sustainable urban water management: practices of five forerunner cities. *Cities*, **74**, 126–133.

Lopes P., Shucksmith J., Leandro J., Fernandes de Carvalho R. and Rubinato M. (2014). Velocities profiles and air-entrainment characterization in a scaled circular manhole. Proceedings of 13th International Conference on Urban Drainage, 7–12 September 2014, 13th ICUD, Sawarak, Malaysia.

Lopes P., Leandro J., Carvalho R. F., Páscoa P. and Martins R. (2015). Numerical and experimental investigation of a gully under surcharge conditions. *Urban Water Journal*, **12**, 468–476. http://dx.doi.org/10.1080/1573062X.2013.831916

Lundy L. and Wade R. (2011). Integrating sciences to sustain urban ecosystem services. *Progress in Physical Geography*, **35**(5), 653–669

Mark O., Weesakul S., Apirumanekul C., Aroonnet S. and Djordjević S. (2004). Potential and limitations of 1D modelling of urban flooding. *Journal of Hydrology*, **299**(3–4), 284–299.

Marshall M., Francis O., Frogbrook Z., Jackson B., McIntyre N., Reynolds B., Solloway I., Wheater H. and Chell J. (2009). The impact of upland land management on flooding: results from an improved pasture hillslope. *Hydrological Processes*, **23**, 464–475.

Marshall M., Ballard C., Frogbrook Z., Solloway I., Mcintyre N., Reynolds B. and Wheater H. (2014). The impact of rural land management changes on soil hydraulic properties and runoff processes: results from experimental plots in upland UK. *Hydrological Processes*, **28**(4), 2617–2629.

Martínez-Gomariz E., Gómez M. and Russo B. (2016). Experimental study of the stability of pedestrians exposed to urban pluvial flooding. *Natural Hazards*, **82**(2), 1259–1278.

Martins R., Leandro J. and de Carvalho R. F. (2014). Characterization of the hydraulic performance of a gully under drainage conditions. *Water Science and Technology*, **69**, 2423–2430, http://dx.doi.org/10.2166/wst.2014.168

Martins R., Rubinato M., Kesserwani G., Leandro J., Djordjevic S. and Shucksmith J. (2017). Validation of 2D shock capturing flood models around a surcharging manhole. *Urban Water Journal*, **14**(9), 892–899.

Martins R., Rubinato M., Kesserwani G., Leandro J., Djordjevic S. and Shucksmith J. (2018). On the characteristics of velocity fields on the vicinity of manhole inlet grates during flood events. *Water Resoures Research*, **54**(9), 6408–6422.

McIntyre N. and Marshall M. (2010). Identification of rural land management signals in runoff response. *Hydrological Processes*, **24**(24), 3521–3534.

McMichael A., Woodruff R. and Hales S. (2006). Climate change and human health: present and future risks. *Lancet*, **367**(9513), 859–869.

Mei C., Liu J., Wang H., Yang Z., Ding X. and Shao W. (2018). Integrated assessments of green infrastructure for flood mitigation to support robust decision-making for sponge city construction in an urbanized watershed. *Science of the Total Environment*, **639**, 1394–1407.

Melville-Shreeve P., Cotterill S., Grant L., Arahuetes A., Stovin V., Farmani R. and Butler D. (2018). State of SuDS delivery in the United Kingdom. *Water and Environment Journal*, **32**(1), 9–16.

Met Office (2018) UKCP18 Science Overview Report. Available from: www.metoffice.gov.uk/pub/data/weather/uk/ukcp18/science-reports/UKCP18-Overview-report.pdf

Mignot E., Li X. and Dewals B. (2019). Experimental modelling of urban flooding: a review. *Journal of Hydrology*, **568**, 334–342.

Miller J. and Hutchins M. (2017). The impacts of urbanisation and climate change on urban flooding and urban water quality: a review of the evidence concerning the United Kingdom. *Journal of Hydrology: Regional Studies*, **12**, 345–362.

Miller J., Kim H., Kjeldsen T., Packman J., Grebby S. and Dearden R. (2014). Assessing the impact of urbanization on storm runoff in a peri-urban catchment using historical change in impervious cover. *Journal of Hydrology*, **515**, 59–70.

Morales-Hernández M., Murillo J. and García-Navarro P. (2013). The formulation of internal boundary conditions in unsteady 2-D shallow water flows: application to flood regulation. *Water Resources Research*, **49**(1), 471–487.

Nanía L. S., Gómez M., Dolz J., Comas P. and Pomares J. (2011). Experimental study of subcritical dividing flow in an equal-width, four-branch junction. *Journal of Hydraulic Engineering – ASCE*, **137**(10), 1298–1305.

Nichols A., Rubinato M., Cho Y. H. and Wu J. (2020) Optimal use of titanium dioxide colourant to enable water surfaces to be measured by Kinect Sensors. *Sensors*, **20** (12), 3507, https://doi.org/10.3390/s20123507Y

Niemczynowicz J. (1999). Urban hydrology and water management – present and future challenges. *Urban Water Journal*, **1**. 1–14. https://doi.org/10.1016/S1462-0758(99)00009-6

Nnadi E. O., Newman A. P., Duckers L., Coupe S. J. and Charlesworth S. (2012). Design and validation of a test rig to simulate high rainfall events for infiltration studies of permeable pavement systems. *Journal of Irrigation and Drainage Engineering – ASCE*, **138**(6), 553–557. http://dx.doi.org/10.1061/(ASCE)IR.1943-4774.

Nnadi E. O., Coupe S. J., Sañudo-Fontaneda L. A. and Rodriguez-Hernandez J. (2014). An evaluation of enhanced geotextile layer in permeable pavement to improve stormwater infiltration and attenuation. *International Journal of Pavement Engineering*, **15**(10), 925–932. http://dx.doi.org/10.1080/10298436.2014.893325

Ozdemir H., Sampson C. C., de Almeida G. A. M. and Bates P. D. (2013). Evaluating scale and roughness effects in urban flood modelling using terrestrial LIDAR data. *Hydrology and Earth System Sciences*, **17**(10), 4015–4030.

Paquier A. and Bazin P. H. (2014). Estimating uncertainties for urban floods modelling. *La Houille Blanche* **6**, 13–18.

Peña F. and Nardi F. (2018). Floodplain terrain analysis for coarse resolution 2D flood modelling. *Hydrology*, **5**(4), 52.

Perales-Momparler S., Andrés-Doménech I., Andreu J. and Escuder-Bueno I. (2015). A regenerative urban stormwater management methodology: the journey of a Mediterranean city. *Journal of Cleaner Production*, **109**, 174–189.

Pfister M. and Gisonni C. (2014). Head Losses in Junction Manholes for Free Surface Flows in Circular Conduits. *Journal of Hydraulic Engineering*, **140**(9): 06014015.

Pili S., Grigoriadis E., Carlucci M., Clemente M. and Salvati L. (2017). Towards sustainable growth? A multi-criteria assessment of (changing) urban forms. *Ecological Indicators*, **76**, 71–80.

Pitt M. (2008). Learning Lessons from the 2007 Floods. The Pitt Review. Cabinet Office, London.

Pozzi W., Sheffield J., Stefanski R., Cripe D., Pulwarty R., Vogt J., Heim R., Jr, Brewer M., Svoboda M., Westerhoff R., van Dijk A., Lloyd-Hughes B., Pappenberger F., Werner M., Dutra E., Wetterhall F., Wagner W., Schubert S., Mo K., Nicholson M., Bettio L., Nunez L., van Beek R., Bierkens M., Goncalves de Goncalves L. G., Gerd Zell de Mattos J. and Lawford R. (2013). Toward global drought early warning capability: expanding international cooperation for the development of a framework for monitoring and forecasting. *Bulletin of the American Meteorological Society*, **94**(6), 776–785.

Qin H. P., Li Z. X. and Fu G. (2013). The effects of low impact development on urban flooding under different rainfall characteristics. *Journal of Environmental Management*, **129**, 577–585.

Ramamurthy A. S. and Zhu W. (1997). Combining flows in 90 junctions of rectangular closed conduits. *Journal of Hydraulic Engineering – ASCE*, **123**, 1012–1019.

Ramsbottom D. and Wicks J. (2003). Catchment Flood Management Plans: Guidance on Selection of Appropriate Hydraulic Modelling Methods. Environment Agency, Bristol, UK.

Reynard N., Kay A., Anderson M., Donovan B. and Duckworth C. (2017). The evolution of climate change guidance for fluvial flood risk management in England. *Progress in Physical Geography*, **41**(2), 222–237.

Rivett M., Ellis P., Greswell R., Ward R., Roche R., Cleverly M., Walker C., Conran D., Fitgerals P., Wilcox T. and Dowle J. (2008). Cost-effective mini drive-point piezometers and multilevel samplers for monitoring the hyporheic zone. *Quarterly Journal of Engineering Geology and Hydrogeology*, **41**(1), 49–60.

Rivière N., Travin G. and Perkins R. J. (2011). Subcritical open channel flows in four branch intersections. *Water Resources Research*, **47**(10), W10517.

Rivière N., Travin G. and Perkins R. J. (2014). Transcritical flows in three and four branch open-channel intersections. *Journal of Hydraulic Engineering – ASCE*, **140**(4), 04014003.

Roinas G., Mant C. and Williams J. (2014). Fate of hydrocarbon pollutants in source and non-source control sustainable drainage systems. *Water Science and Technology*, **69** (4), 703–709.

Rojas A. S., Rubinato M., Nichols A. and Shucksmith J. (2018). Cost effective measuring technique to simultaneously quantify 2D velocity fields and depth-averaged solute concentrations in shallow water flows. *Journal of Flow Measurement and Instrumentation*, **64**, 213–223. https://doi.org/10.1016/j.flowmeasinst.2018.10.022

Rubinato M. (2015). Physical scale modelling of urban flood systems. Ph.D. Thesis, The University of Sheffield, URL http://etheses.whiterose.ac.uk/9270/

Rubinato M., Shucksmith J. and Saul A. J. (2011). Hydraulic performance of a scale model facility and optimization through the use of real time sensors. Proceedings of 12th International Conference on Urban Drainage Modelling, 11–17 Sep 2011, Porto Alegre, Brazil.

Rubinato M., Shucksmith J., Saul A. J. and Shepherd W. (2013). Comparison between Infoworks results and a physical model of an urban drainage system, *Water Science and Technology*, **68**(2), 372–379.

Rubinato M., Shucksmith J. and Saul A. J. (2014). Experimental investigation of between above and below ground drainage systems through a manhole. Proceedings of 11th International Conference on Hydroinformatics, New York, USA, 17–21 August 2014.

Rubinato M., Martins R., Kesserwani G., Leandro J., Djordjevic S. and Shucksmith J. (2017a). Experimental calibration and validation of sewer/surface flow exchange equations in steady and unsteady flow conditions. *Journal of Hydrology*, **552**, 421–432.

Rubinato M., Martins R., Kesserwani G., Leandro J., Djordjevic S. and Shucksmith J. (2017b). Experimental investigation of the influence of manhole grates on drainage flows in urban flooding conditions. In: Proceedings of 14th IWA/IAHR International Conference on Urban Drainage, 10–15 September, Prague, Czech Republic.

Rubinato M., Martins R. and Shucksmith J. (2018a). Quantification of energy losses at a surcharging manhole. *Urban Water Journal*, **15**(3), 234–241.

Rubinato M., Seungsoo L., Martins R. and Shucksmith J. (2018b). Surface to sewer flow exchange through circular inlets during urban flood conditions. *Journal of Hydroinformatics*, **20**(3), 564–576.

Rubinato M., Nichols A., Peng Y., Zhang J., Lashford C., Cai Y., Lin P. and Tait S. (2019). Urban and river flooding: Comparison of flood risk management approaches in the UK and China and an assessment of future knowledge needs. Water Science and Engineering, **12**(4), 274–283.

Rubinato M., Luo M., Zheng X., Pu J. H. and Shao S. (2020). Advances in modelling and prediction on the impact of human activities and extreme events on environments. *Water*, **12**(6), 1768. https://doi.org/10.3390/w12061768

Ruhl C., Schoellhamer D., Stumpf R. and Lindsay C. (2001). Combined use of remote sensing and continuous monitoring to analyse the variability of suspended-sediment concentrations in San Francisco Bay, California. Estuarine, Coastal and Shelf Science, **53**(6), 801–812.

Russo B. (2009). Design of surface drainage systems according to hazard criteria related to flooding of 396 urban areas. PhD Thesis. Technical University of Catalonia, Barcelona (Spain).

Sañudo-Fontaneda L., Rodriguez-Hernandez J., Calzada-Pérez M. and Castro-Fresno D. (2014). Infiltration behaviour of polymer-modified porous concrete and porous asphalt surfaces used in SuDS techniques. *CLEAN – Soil, Air, Water*, **42**(2), 139–145.

Sañudo-Fontaneda L. A., Jato-Espino D., Lashford C. and Coupe S. J. (2018). Simulation of the hydraulic performance of highway filter drains through laboratory models and stormwater management tools. *Environmental Science and Pollution Research International*, **25**(20), 19228–19237. http://dx.doi.org/10.1007/s11356-017-9170-7

Schellart A. N. A., Shepherd W. J. and Saul A. J. (2012). Influence of rainfall estimation error and spatial variability on sewer flow prediction at a small urban scale. *Advances in Water Resources*, **45**, 65–75, https://doi.org/10.1016/j.advwatres.2011.10.012

Schindfessel L., Creëlle S. and De Mulder T. (2015). Flow patterns in an open channel confluence with increasingly dominant tributary inflow. *Water*, **7**(9), 4724–4751.

Shu C. W., Xia J. Q., Falconer R. A. and Lin B. L. (2011). Incipient velocity for partially submerged vehicles in floodwaters. *Journal of Hydraulic Research*, **49**(6), 709–717.

Smith M. J., Edwards E. P., Priestnall G. and Bates P. D. (2006). Exploitation of new data types to create Digital Surface Models for flood inundation modelling, FRMRC Research report UR3, June 2006. University of Nottingham, UK, 78pp. http://dx.doi.org/10.13140/RG.2.2.29963.08487

Smith G. P., Rahman P. F. and Wasko C. (2016). A comprehensive urban floodplain dataset for model benchmarking. *International Journal of River Basin Management*, **14**(3), 345–356.

Spinoni J., Naumann G., Carrao H., Barbosa P. and Vogt J. (2014). World drought frequency, duration, and severity for 1951–2010. *International Journal of Climatology*, **34**(8), 2792–2804.

Stovin V., Vesuviano G. and De-Ville S. (2015a) Defining green roof detention performance. *Urban Water Journal*, **14**(6), 574–588.

Stovin V., Poë S., De-Ville S. and Berretta C. (2015b). The influence of substrate and vegetation configuration on green roof hydrological performance. *Ecological Engineering*, **85**, 159–172.

Tang G., Zeng Z., Ma M., Liu R., Wen Y. and Hong Y. (2017). Can near-real-time satellite precipitation products capture rainstorms and guide flood warning for the 2016 summer in South China? *IEEE Geoscience and Remote Sensing Letters*, **14**(8), 1208–1212.

Teo Y., Falconer R., Lin B. and Xia J. (2012). Investigations of hazard risks relating to vehicles moving in flood. *Water Resources Management*, **1**, 52–66.

Tedoldi D., Chebbo G., Pierlot D., Kovacs Y. and Gromaire M. (2016). Impact of runoff infiltration on contaminant accumulation and transport in the soil/filter media of Sustainable Urban Drainage Systems: A literature review. *Science of the Total Environment*, **569–570**, 904–926.

Trenberth K. (2011). Changes in precipitation with climate change. *Climate Research*, **47**(1–2), 123–138.

Tscheinkner-Gratl F., Sitzenfrei R., Rauch W. and Kleidorfer M. (2015). Integrated rehabilitation planning of urban infrastructure systems using a street section priority model. *Urban Water*, **13**(1), 28–40.

Uche J., Martínez A., Castellano C. and Subiela V. (2013). Life cycle analysis of urban water cycle in two Spanish areas: Inland city and island area. *Desalination and WaterTreatment*, **51**, 280–291

United Nations (2018). World Urbanization Prospects: The 2018 Revision. Available from: https://population.un.org/wup/Publications/Files/WUP2018-KeyFacts.pdf, accessed 18 August 2018.

Van Ootegem L., Van Herck K., Creten T., Verhofstadt E., Foresti L., Goudenhoofdt E., Reyniers M., Delobbe L., MurlaTuyls D. and Willems P. (2016). Exploring the potential of multivariate depth-damage and rainfall-damage models. *Journal of Flood Risk Management*, **11**(S2), S916–S929.

van Woert N. D., Rowe D. B., Andresen J. A, Rugh C. L., Fernandez R. T. and Xiao L. (2005). Green roof stormwater retention: effects of roof surface, slope, and media depth. *Journal of Environmental Quality*, **34**(3), 1036–1044. http://dx.doi.org/10.2134/jeq2004.0364

Vasconcelos J. G., Wright S. J. and Roe P. L. (2006). Improved simulation of flow regime transition in sewers: two-component pressure approach. *Journal of Hydraulic Engineering – ASCE*, **132**, 553–562.

Versini P., Ramier D., Berthier E. and de Gouvello B. (2015). Assessment of the hydrological impacts of green roof: From building scale to basin scale. *Journal of Hydrology*, **524**, 562–575.

Walsh C., Fletcher T. and Ladson A. (2005). Stream restoration in urban catchments through redesigning stormwater systems: looking to the catchment to save the stream. *Journal of the North American Benthological Society*, **24**(3), 690–705.

Wheater H. and Evans E. (2009). Land use, water management and future flood risk. *Land Use Policy*, **26**, S251–S264.

Wirahadikusumah R., Abraham D. M., Iseley T. and Prasanth R. (1998). Assessment technologies for sewer system rehabilitation. *Automation in Construction*, **7**(4), 259–270.

Woods-Ballard B., Wilson S., Udale-Clarke H., Illman S., Scott T., Ashley R. and Kellagher R. (2015). The SuDS Manual. CIRIA, London, UK.

Xia J. Q., Teo F. Y., Lin B. L. and Falconer R. A. (2011). Formula of incipient velocity for flooded vehicles. *Natural Hazards*, **58**(1), 1–14.

Xia J., Falconer R. A., Xiao X. and Wang Y. (2014). Criterion of vehicle stability in floodwaters based on theoretical and experimental studies. *Natural Hazards*, **70** (2), 1619–1630.

Xie J., Chen H., Liao Z., Gu X., Zhu D. and Zhang J. (2017). An integrated assessment of urban flooding mitigation strategies for robust decision making. *Environmental Modelling & Software*, **95**, 143–155. http://dx.doi.org/10.1016/j.envsoft.2017.06.027

Zhang W., Zhang X. and Liu Y. (2013). Analysis and simulation of drainage capacity of urban pipe network. Research Journal of Applied Sciences, Engineering and Technology, **6**(3), 387–392.

Zhao C. H., Zhu D. Z. and Rajaratnam N. (2008). Computational and Experimental Study of Surcharged Flow at a 90 Combining Sewer Junction. *Journal of Hydraulic Engineering – ASCE*, **134**, 688–700.

Chapter 10

Integrated modelling and control of urban wastewater systems

Fanlin Meng[1] *and Ramesh Saagi*[2]

[1]*Centre for Water Systems, College of Engineering, Mathematics and Physical Sciences, University of Exeter, Exeter, UK*

[2]*Division of Industrial Electrical Engineering and Automation (IEA), Lund University, Lund, Sweden*

10.1 INTRODUCTION

With the protection of the aquatic environment being increasingly valued and understood, environmental water quality standards have become more comprehensive and stringent in the last decades. For example, the EU Water Framework Directive (WFD) (European Parliament and Council of the European Union, 2000) establishes a holistic legislative framework consolidating relevant environmental water quality standards and sets an overarching aim of good (chemical and ecological) status required for all water bodies by 2027 at the latest. Urban wastewater systems (UWWSs) are a key pollution source to the surface water quality, mainly by the effluent discharges from wastewater treatment plants (WWTPs) and combined sewer overflows (CSOs). In the traditional management regime, WWTP discharges have been the focus of regulation and cost is expected to escalate to meet the greater regulatory expectations. For example, £27 billion ($46 billion) was estimated to be invested to install additional treatment capacity (e.g., biological, adsorption or ultrafiltration processes for the removal of metals, pharmaceuticals, nutrients and

Editors: Xiaochang C. Wang and Guangtao Fu
doi: 10.2166/9781789060768_0259

ammonia etc.) in the UK between 2010 and 2030 (Severn Trent Water Limited, 2013) to meet the 'good status' requirement of the WFD (Georges *et al.*, 2009). In addition to the financial burden, enhanced treatment (e.g., increased aeration or carbon source addition, and treatment process extension) can increase Greenhouse Gas (GHG) emissions (Flores-Alsina *et al.*, 2011; Georges *et al.*, 2009; Sweetapple *et al.*, 2014a, 2014b), thus contributing to climate change. The increased wastewater treatment under the WFD is estimated to increase GHG emissions by over 110 million kilograms per year in the UK (Georges *et al.*, 2009). As such, it is difficult to comply with a stricter effluent permit without raising GHG emissions (and cost) by the conventional strategy of enlarging the capacity of existing treatment processes.

In contrast to the strict regulation of effluent discharges from WWTPs, spills of untreated wastewater from CSOs are separately controlled by simple measures such as spill frequency (Environment Agency, 2011; U.S. Environmental Protection Agency, 1995), even though the highly concentrated wastewater spills have an acute toxic effect and can be lethal to the aquatic community (Kay *et al.*, 2008; Phillips *et al.*, 2012; Weyrauch *et al.*, 2010). Indeed, research has clearly shown the poor correlation between reducing CSO spill frequency or volume and improving receiving water quality (Lau *et al.*, 2002). It was estimated that some 8000 of approximately 25 000 CSOs in England and Wales were causing water problems at the beginning of the 1990s (Clifforde *et al.*, 2006) and many remain underperforming even today (Nardell, 2012). The investment needed to improve CSOs is considerable, for example, £2.9 billion ($4.9 billion) was estimated for the UK (Clifforde *et al.*, 2006) and £26.5 billion ($45 billion) for the USA (U.S. Environmental Protection Agency, 1999).

To address urban water pollution in a more sustainable manner, nonconventional engineering solutions have been investigated and practiced. These include source pollution control such as green/grey infrastructure and optimal operation of the wastewater systems to minimise overflows to sensitive sites and/or to maximise treatment efficiency against dynamic wastewater inflows. However, sewer and WWTP are often operated separately and poorly coordinated. For example, operational strategies of the sewer system are usually developed to minimise the volume of wastewater spill and retain for treatment with limited account of the capacity of the WWTP (U.S. Environmental Protection Agency, 1995). Likewise, technological measures targeted at the WWTP, such as resource recovery and recycling schemes (Guest *et al.*, 2009; Jin *et al.*, 2015; Mccarty *et al.*, 2011), innovative wastewater treatment technologies (Castro-Barros *et al.*, 2015; Strous *et al.*, 1997; U.S. Environmental Protection Agency, 2013) and efficient operation and control techniques (Sweetapple *et al.*, 2014a; Thornton *et al.*, 2010), are developed with little consideration of the interactions between the WWTP and the sewer. This may lead to under-performing solutions as the overall impact of the urban wastewater system on the receiving water is not fully appraised (Lau *et al.*, 2002).

Integrated modelling of the sewer system, WWTP and receiving water body is a valuable tool in providing a holistic view of system performance (Bach *et al.*, 2014; Butler & Schütze, 2005; Meirlaen, 2002; Vanrolleghem *et al.*, 2005). Traditionally, models for the individual components of the UWWS were developed in a separate way with limited consideration of the impact from/to other components. However, the interactions are non-negligible to the system performance. For example, the surface runoff directly determines the wastewater load transported in the combined sewer systems, which in turn affects the amount of CSOs to the receiving water and inflow to the WWTP; sewage septicity and sulphide generated in the sewer system are associated with sludge bulking in the WWTP; and the operation in the primary clarifier affects the treatment performance in the activated sludge reactor, which in turn influences the solid settling property in the secondary clarifier (Schütze *et al.*, 2002). Integrated modelling can represent the interactions between different components and has already been used to demonstrate the potential for significant improvements in river water quality by coordinated optimisation of the operational or control strategy within an UWWS without the need for upgrade or redesign of the treatment system (Fu *et al.*, 2008; Meng *et al.*, 2016; Saagi *et al.*, 2017; Schütze *et al.*, 2002). Apart from surface water quality analysis, multiple features of system performance (e.g., GHG emissions, cost, sulfur compounds and micropollutants) can also be evaluated using mathematical modelling (Fu *et al.*, 2008; Guo *et al.*, 2012; Snip *et al.*, 2014; Sweetapple *et al.*, 2014b; Vezzaro *et al.*, 2014a) and be considered simultaneously in optimising system operation by multi-objective optimisation tools (Deb *et al.*, 2002). Research findings indicate a trade-off between river water quality and GHG emissions. That is, better river water quality is likely to be achieved with more GHG emissions. To address this, integrated real-time control (RTC) can be applied to achieve responsive operation within the UWWS according to the assimilation capacity of the environment. In this chapter, a brief review is provided on the state-of-the-art of integrated modelling and control of UWWSs.

10.2 INTEGRATED MODELLING OF UWWS

10.2.1 A brief history of integrated modelling

The idea of integrated modelling was first proposed by Beck (1976) and one of the early models was developed by Beck and Finney (1987) for the study of operational strategies in UWWSs to reduce stress on downstream rivers. Studies revealed the interactions between different subsystems in an UWWS and called for the development of holistic approaches for improving the quality of receiving waters (Harremoës *et al.*, 1993; House *et al.*, 1993). The INTERURBA (Interactions between sewers, treatment plants and receiving waters in urban areas) workshop in 1992 reviewed the state-of-the-art in wastewater system simulation and identified requirements and challenges in UWWS modelling

(Lijklema *et al.*, 1993). The workshop stressed the need to design and operate sewers and WWTPs based on the impacts on the receiving water. Meanwhile, an important development in WWTP modelling had taken place. The efforts to standardise activated sludge models by the IWA Task Group on Mathematical Modelling for Design and Operation of Biological Wastewater Treatment led to the generation of the Activated Sludge Model No. 1 (ASM1) (Henze *et al.*, 1987). This was a milestone in WWTP modelling, which also paved the way towards improved simulation of the sewers and rivers.

There was a renewed interest in integrated modelling of UWWSs during the late 1990s, fuelled by the availability of technical know-how and increased computational power. The enforcement of the EU WFD necessitated a shift from the traditional emission-based strategies to receiving water quality-based approaches for the management of UWWS (Vanrolleghem *et al.*, 2005), hence generating new interests in integrated modelling and control. The INTERURBA II conference in 2001 was an indication of the continued, strong interest (Rauch *et al.*, 2002). Parallel to standardised WWTP modelling, efforts were also made in developing a standard and consistent framework for river water quality modelling, for example, RWQM1 was developed which was compatible with the ASMs (Reichert *et al.* 2001b; Shanahan *et al.*, 2001; Vanrolleghem *et al.*, 2001). With the availability of standard models of subsystems, efforts towards integration strengthened further. Software platforms were developed by academic groups and commercial software developers for integrated modelling (e.g., Achleitner *et al.*, 2007; DHI, 2020; Ifak, 2016; Mannina *et al.*, 2006; Saagi *et al.*, 2017; Schütze *et al.*, 1999; Vezzaro *et al.*, 2014a). Integrated models were reported to be developed for different urban catchments, for example, Odenthal, Germany (Erbe *et al.*, 2002), Copenhagen, Denmark (Harremoës *et al.*, 2002; Vezzaro *et al.*, 2014b), Lambro river, Italy (Benedetti *et al.*, 2007), and Dommel, the Netherlands (Weijers *et al.*, 2012). Integrated models have been used for holistic analysis of the performance of UWWSs and the development of mitigation and adaptation strategies for tackling challenges such as climate change, population growth and urbanisation (Astaraie-Imani *et al.*, 2012; Doglioni *et al.*, 2009; Fu *et al.*, 2009).

10.2.2 Overview of integrated models

In order to describe the entire UWWS, models to describe the underlying sub-sections are developed and integrated. As illustrated by an example of the integrated urban wastewater system (Figure 10.1), the major sub-sections generally considered for modelling are:

- urban catchment;
- sewer network;
- wastewater treatment plant;
- receiving water system.

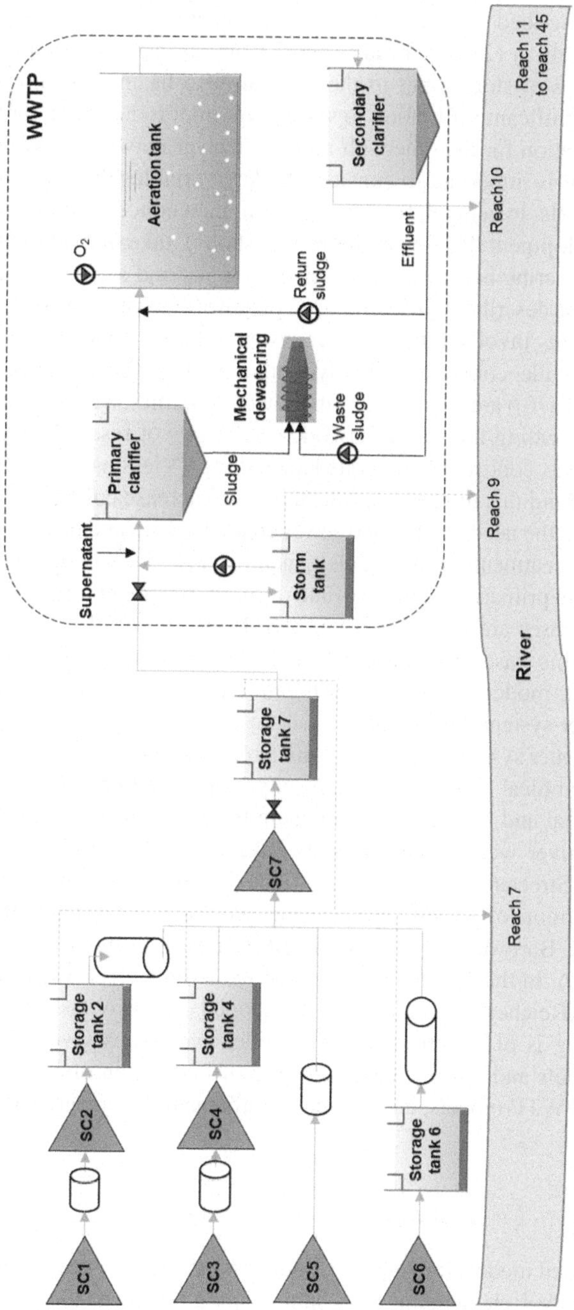

Figure 10.1 Layout of an example integrated urban wastewater system (adapted from Meng *et al.*, 2016).

Urban catchment is the source of all the wastewater generated in a city. Major components described are: (1) wastewater generation from households and industries; as well as (2) stormwater runoff from surfaces during rain events. In addition, seepage of stormwater to the soil can also be important, especially for systems with significant infiltration to sewers. In order to be able to provide all the required information for the other sub-sections (sewer network, WWTP, receiving water system), it is important to consider the key variables that must be described using these models. In general, flow rate and pollutant concentrations are described.

Model development for sewer networks should take into consideration the possibility to describe both combined sewer systems and separate sewer systems. Model blocks for describing different types of sewer (gravity sewers, rising means) and other elements involved in the sewer system (storage tanks, pumping stations) are required. While conventionally hydraulic models (Saint-Venant's equation (Saint-Venant, 1870)) are used to describe sewer dynamics, a conceptual modelling approach (representing the sewer networks as a series of reservoirs) is commonly undertaken in the context of integrated modelling (Viessman *et al.*, 1989). The main reason behind this model simplification is to increase computational speed as well as to reduce the need for detailed sewer characteristics and data.

Wastewater treatment plant models generally describe various unit operations involved in the primary, secondary and tertiary treatment of wastewater. The biological processes are described using ASMs (Henze *et al.*, 1987). The choice of models in this case also depends on the ease of integration of wastewater treatment plant models with the upstream (sewer network) and downstream (receiving water system) sub-sections.

Receiving water systems can range from streams and rivers to oceans depending on the geographical location of the city. In literature, models describing physico-chemical and biological variations in the river water system are available. Traditionally, river water quality models have evolved from using the Streeter–Phelps model (Streeter & Phelps, 1925) for describing oxygen dynamics to a very detailed description of various biochemical processes. QUAL family of models (e.g., Brown & Barnwell, 1987) and WASP (Di Toro *et al.*, 1983) are examples of this approach. In the context of integrated modelling, river water quality model 1 (RWQM1) (Reichert *et al.*, 2001a), developed by IWA task group, on river water modelling is of significance. This model overcomes several limitations in the earlier models and also increases the ease of integration of river water quality models with WWTP models, an essential pre-requisite for integrated modelling.

10.2.3 Challenges in the development of integrated modelling

Although detailed models for all the sections exist, the idea of integrating them is far more complex and challenging than simply plugging them together. The major challenges and some approaches to handle them are described below.

- Adapting to the purpose – simplifying the model: The models for sewers, WWTPs and rivers are made for different purposes and cannot be integrated without modifications/simplifications to the model. For example, in the case of sewer models, 2D continuity and momentum equations (Saint-Venant equation) are replaced by conceptual models (Meirlaen *et al.*, 2002; Solvi, 2007). This not only allows for the integration of sewer models with other elements in the UWWS, but also significantly reduces the simulation times. Other simplifications in terms of spatial and time boundaries are also commonly employed (Vanrolleghem *et al.*, 2005).
- Adapting to the purpose – adding complexity to the model: While some aspects of the models are simplified, other areas need an improved description of the underlying processes. The difference in the level of detail and complexity between sewer models and WWTPs in terms of their ability to describe pollutants needs to be reconciled (Fronteau *et al.*, 1997). A major step towards reconciliation between WWTP and river quality models is RWQM1 (Reichert *et al.*, 2001a). The modelling framework for RWQM1 draws inspiration from ASM models for WWTPs and also facilitates interfacing with WWTP models.
- Model calibration and validation: Integrated models aggravate the identifiability and model calibration issues that already exist in the models for the individual sections (Reichert & Vanrolleghem, 2001; Sin *et al.*, 2005). Coupled with this is the need for extensive data collection campaigns needed to satisfactorily calibrate the UWWS models. Detailed frameworks are developed to address this issue and provide guidelines on the best practices for applying integrated models to real urban catchments (Breinholt *et al.*, 2008; Muschalla *et al.*, 2009). Nevertheless, one should recognize the fact that inherent identifiability issues and their effect on model calibration cannot be completely removed.

10.2.4 Modelling platforms for integrated UWWS

Some commonly available commercial/free software platforms for integrated modelling are briefly described below.

- The WEST modelling software offers WESTforIUWS (DHI, 2020), which can simulate the catchment, sewer, WWTP and river water system of an integrated UWWS. It offers the possibility to evaluate water quality-based objectives for both long- and short-term evaluation periods. Additionally, uncertainty and sensitivity analysis of the models can also be performed.
- SIMBA# water is developed by Institut fur Automation und Kommunikation (ifak), Germany (Ifak, 2016) and is used for simulation of the integrated UWWS. The software consists of a model library to simulate processes in sewers, WWTPs and rivers. Simplified hydrological models as well as

hydrodynamic models are available for the sewer network. Various modules for biochemical and physical processes in the WWTP and biochemical processes in the river are included. Additionally, it facilitates easy implementation of control studies. There is a possibility to program the controllers using industry standard languages such as structured text, petri nets etc.

- CITY DRAIN is a free open-source Matlab based toolbox for integrated UWWS evaluations (Achleitner *et al.*, 2007). It has hydrological models for sewers and river systems as well as a simplified WWTP model. It gives the users the possibility to create their own user defined blocks in addition to the existing model library. It allows for fast simulation of the UWWS owing to its simplified nature.
- BSM-UWS is a Matlab based platform primarily developed for benchmarking integrated control strategies (Saagi *et al.*, 2017). However, the toolbox includes a model library that can be used to model several other integrated UWWSs. It includes several possibilities to develop and evaluate control strategies using a standard set of evaluation criteria.

10.2.5 Guideline for integrated modelling – HSG guidelines

In order to encourage a widespread application of integrated modelling studies for UWWS, The Central European Simulation Research Group (Hochschulsimulationsgruppe-HSG, www.hsgsim.org) has developed a guideline document identifying the key steps for successful implementation of an integrated modelling project for UWWS (HSGsim, 2008). The document (in German) explains the key steps in detail and presents three successful case studies. An overview of the six-step procedure (Figure 10.2) summarised from Muschalla *et al.* (2009) is provided here.

(1) System analysis: This step includes understanding the present state (e.g., current regulatory compliance status, data availability) and also determining a future/target state that the modelling study is expected to identify. The future state can be an improvement in terms of compliance, economic costs, and process efficiency etc. Very often a combination of several criteria is aimed for. In most cases, such a system analysis study is undertaken well before the beginning of any modelling exercise.

(2) Processes and criteria: in this step, the outline provided from the previous step is further refined. The root cause for the problems in the current system is identified. Additionally, the underlying processes to achieve the defined target state are also described. Further, potential measures that can lead to reaching the target state are identified. These measures will be further modified/supplemented in later stages. A critical process in this step is to also identify the model boundaries and outline the interactions and model requirements for the different sub-sections (sewer

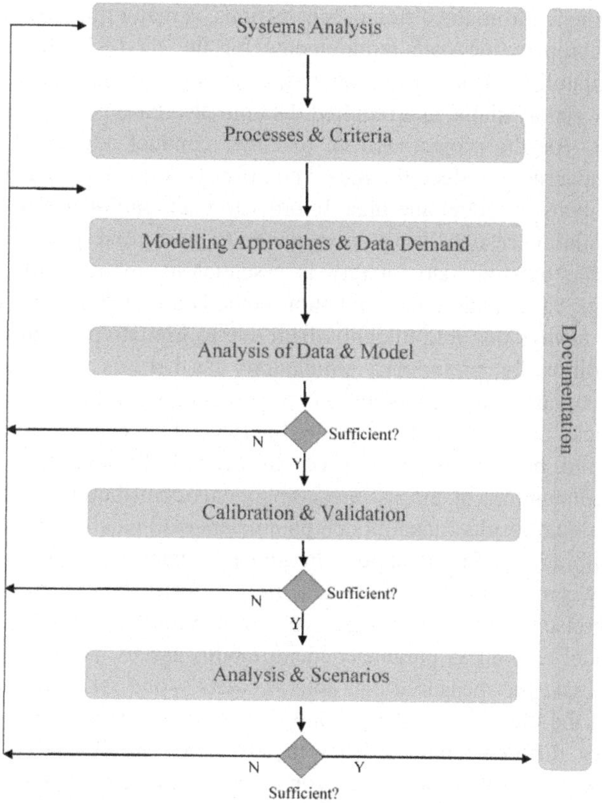

Figure 10.2 An overview of the key steps in the HSG procedure for integrated modelling (adapted from Muschalla *et al.*, 2009).

network, WWTP, receiving water system). An important question to answer will be – is an integrated model really required to achieve the target? While integrated modelling offers excellent value for several situations, there can be cases where modelling at the sub-section level can be sufficient.

(3) Modelling approaches and data demand: Based on the key processes and interactions identified in step 2, a thorough identification of the modelling requirements is conducted in this step. The model complexity should be sufficient enough to be able to describe the identified processes. Based on this analysis, it can be determined if the current models are sufficient or should be further modified/adapted to meet the requirements of the project. Another major aspect is the vast amount of data needed from different sub-sections for such a study. A major challenge will be to reconcile the differences in data quality and

frequency from the different sub-sections. A major milestone at this point is to compare the data requirements for the model with the actual data availability. A key decision based not only on the mismatch (if any) between available and required data but also based on time and economic costs for the project will be to either conduct additional measurement campaigns or reduce the model complexity to match the available data.

(4) Analysis of model and data: In order to gain confidence in the model and simulation results, it is a pre-requisite to have good quality data. Analysis of the existing data quality is essential to ascertain the data quality. While systematic data validation methods are available in literature, it is not always possible to apply them due to limitations in time and money. Additionally, preliminary simulations can help us understand the model behaviour and also point out any modelling errors using plausibility checks and process knowledge. Interfaces between the different sections should be carefully evaluated. In general, through this step, a good understanding of the various components contributing to the uncertainty in model results (data, model, parameters etc.) can be achieved.

(5) Calibration and validation: In general, there is always a discrepancy between the model results and the measured data. A model calibration exercise should be devised keeping in mind the simplifications in the model as well as parameter identifiability issues. A good approach is to separate the measured data into two sets – one set for model calibration and the other for model validation. Potentially, these two datasets should have different characteristics in terms of rainfall patterns. In order to identify the key parameters for model calibration, a sensitivity analysis study can be handy. The individual sub-sections should first be calibrated followed by further calibration after integrating them. Also, calibration of the hydraulic/hydrology model should be first carried out before calibrating the pollutant quality models.

(6) Scenario analysis: The measures identified in step 2 can be further refined and turned into more concrete scenarios by also considering the model inputs and the evaluation period. A baseline scenario should also be defined to compare with various future scenarios and evaluate improvements in the defined objectives. In addition to the scenarios already identified earlier, new scenarios can also be defined. However, the newly identified scenario should be validated by following the earlier steps again. In case a thorough calibration is performed, the scenarios can be analysed both using relative improvements as well as absolute numbers. However, with limited calibration efforts, only a relative comparison between the different scenarios can be made.

An additional process that should be carried out in all the above steps is the continuous documentation. The documentation should include the objectives, data needs, and model details etc. It will aid in providing a good understanding to different stakeholders and also to ensure reproducibility of the simulation results.

10.3 INTEGRATED CONTROL OF UWWS

The application of RTC in UWWSs has gradually matured in line with the development of automation control technology and the understanding of the treatment processes. Started from primary control of water levels, flow rates, pressures and temperatures, the application then developed further into concentration control, which requires at least a basic knowledge of the reactions and processes. Mathematical models are usually formulated based on the acquired knowledge and combined into the controller design. With the development of on-line nutrient sensors, the fixed set-point DO control in the reactor, which is a surrogate parameter for biological process control, evolved into variable set-point DO control with direct and more reasonable objectives such as ammonia removal rate (Olsson, 2012).

Besides the control of single treatment units (i.e., 'local control'), RTC can be applied to the whole WWTP to coordinate and optimise the control in the plant systematically (Duyy & Laboratorium, 1975; Serra *et al.*, 1993). This is termed 'global control' as defined in Schütze *et al.* (2002), referring to control where sensor information from within the same subsystem (i.e., sewer system, WWTP, receiving water) is used to determine the setting of a control device. Examples are ratio controlled return sludge pumping rate according to inflow rate to the WWTP (Bauwens *et al.*, 1996), and overflow threshold setting of CSOs based on volume and quality measurement in the sewer system (Petruck *et al.*, 1998). The 'integrated control' of the UWWS, as opposed to 'global control', is characterised by two aspects (Schütze *et al.*, 1999):

(1) Integration of objectives: Objectives of control within one part of the UWWS may be based on criteria measured in other subsystems;
(2) Integration of information: When taking a control decision within one part of the system, information about the present or predicted future state of another subsystem may be used, hence state information is transferred across subsystem boundaries.

Examples of integrated control are Aeration Tank Settling in the WWTP based on rainfall prediction or flow information in the sewer (Nielsen *et al.*, 1996); overflow threshold setting of detention basins by downstream river DO condition (Rauch & Harremoës, 1999); control of the inflow to the WWTP according to ammonia concentration in the downstream river (Meirlaen *et al.*, 2002); rule-based aeration in the WWTP based on the dilution capacity in the upstream river (Meng *et al.*, 2017); and a hierarchical control that overrides local controllers on CSO overflow threshold and inflow to the WWTP according to the loading condition in the sewer system, WWTP and the receiving water (Schütze *et al.*, 1999). Despite the advance in the research on RTC of the integrated UWWS, real-life implementation of the RTC technology is mostly limited to local (i.e., within the sewer system or WWTP) control (Escaler, 2004; Fuchs & Beeneken, 2006; Ito, 2004; Pleau *et al.*, 2005; Thornton *et al.*, 2010).

For predictive global or integrated UWWS control, a complex non-linear system model is usually used to determine time-varying set-points or control inputs according to process evolution. The RTC system is typically structured in three hierarchical levels, that is, field level, system level and supervisory level (Olsson, 2012; Schütze *et al.*, 2004). The sensor information from all units of the system is gathered and structured in the field level and transmitted to the system level. On this second level, reasoning modules, containing a heuristic knowledge of the process, would use the experience from previous similar and particular operating situations to provide suggested strategies. The strategies yielded on the system level would be sent upwards to the supervisory level, where a simulation model of the system would be employed to evaluate the strategies. The optimised strategy is then conveyed downwards for implementation (Olsson, 2012).

Predictive control can be optimised online or offline. In online optimal control, the simulation model is fed by real-time sensor information and provides estimation of the performance of control actions in a specified prediction horizon (e.g., 2 hours). The optimal control action(s), which performs best in achieving pre-defined goals, can be evaluated and implemented at every control time step (e.g., 5 minutes). As the computational time of detailed mechanistic models may be too great to be practical for online control optimisation, model simplification is often needed (Schütze *et al.*, 2004).

Despite the reasonable logic and successful application in some real-life cases (Pleau *et al.*, 2005; Scheer *et al.*, 2004), the use of online optimal control faces a number of problems related to practical applicability. In particular, when considering the system in its entirety, this can include the potential long-term effects associated with some water flow and quality changes (e.g., loss of nitrification in the treatment plant, sediment oxygen demand in the receiving water body) (Butler & Schütze, 2005). As an alternative, offline optimal control could be employed. As computational time is less of an issue in the offline approach, detailed modelling of the wastewater system can be used to analyse the long-term impacts of the control actions. The control algorithm could be pre-defined in the form of a set of 'if-then' rules or a decision matrix (Meng *et al.*, 2017). The quantification of the set-point values and parameters in the control algorithm can be optimised by different approaches, ranging from simple trial-and-error method to sophisticated stochastic optimisation tools (e.g., Genetic Algorithms) (Butler & Schütze, 2005).

Benchmarking tools consisting of predefined model inputs, a model library to describe the various sub-sections, a hypothetical system layout (which is described using the model library) and a standard set of evaluation criteria are available now. Such tools allow for easy adaptation as well as an objective means to evaluate various control strategies. Currently, such tools are available for the individual sub-sections – sewer system, WWTP as well as for the entire UWWS (Saagi *et al.*, 2017).

10.4 TOWARDS REAL-LIFE IMPLEMENTATION OF INTEGRATED OPERATION AND CONTROL

10.4.1 Regulatory implications

Research studies on integrated operation and control of UWWSs have been conducted with limited representation of real-life constraints from environmental policy. This is reflected by the simplified form of standard limits (e.g., maximum ammonia concentration) (Fu *et al.*, 2008; Schütze *et al.*, 2002) in describing river water quality, incomplete application of environmental standards (e.g., use wet weather-related standards only) (Lau *et al.*, 2002; Meirlaen, 2002), short evaluation period (e.g., 1 week) (Fu *et al.*, 2008; Schütze *et al.*, 2002), and no coverage of the impact of implementing an optimal integrated operational strategy on the compliance of wastewater discharge permits. As such, limited insights were provided on the regulatory risk of implementing integrated operational or control strategies.

Meng *et al.* (2016) were among the few attempts of assessing integrated operation and control of UWWSs under a realistic policy context. The integrated operational strategies of a semi-hypothetical UWWS were optimized against higher environmental water quality and lower operational cost (a surrogate indicator of GHG emissions). Compared to the baseline operational strategy, the optimal integrated operational strategies can improve the receiving water quality from ‘moderate’ to ‘good’ status defined by the EU WFD with lower operational cost, demonstrating the benefits of optimisation of system operation in an integrated manner. However, a new regulatory approach is necessary to achieve the desired environmental benefits. This is because tighter effluent discharge permit limits by the traditional regulatory approach can be achieved by various strategies, with some resulting in worse environmental water quality due to the increased CSOs which is weakly regulated. Operational-based permitting, which prescribes the optimal operational scheme expected to deliver the desirable outcomes, is a reliable regulatory approach.

10.4.2 Technical implications

Data availability: The importance of integrated modelling has been clearly established for a very long time (more than 40 years ago). Hence, there have been several attempts to model and evaluate integrated UWWSs. Various models specifically tailored for integrated studies are available in scientific literature as well as commercial and free simulation tools. However, a huge challenge is in terms of data availability and model calibration efforts. Given the large scale of such studies, dedicated efforts to identify available data, assess the requirement for additional data, execution of measurement campaigns and finally a systematic model calibration and validation are required. In all the steps, a balance between rigorous execution, model parsimony and economic costs should be made.

Uncertainty analysis: It is also equally important to understand the various sources of uncertainty and their propagation in order to be able to evaluate the model outcomes and arrive at practical solutions. Traditional uncertainty analysis tools are very extensive but given the high computational time and multiple sources of uncertainty arising in integrated modelling, it can be practically difficult to implement such methods. Research studies trying to address the gaps in identifying solutions for uncertainty assessment are in progress.

10.4.3 Institutional/organizational implications

The organizational structure of various entities managing the different sections of an urban wastewater system greatly influences the success of any integrated modelling project. In several places, the management of sewer network, WWTP and river water system is undertaken by different organizations with limited or no communication between them. In such cases, the first step towards implementing integrated modelling and control solutions is to bring the different stakeholders together on a common platform to facilitate exchange of ideas and knowledge. It can be an excellent opportunity for newly developing cities to implement such solutions as the institutional frameworks are not fully established and can be integrated from the start.

10.4.4 KALLISTO: A successful showcase

The river Dommel in the city of Eindhoven, which receives discharges from a WWTP (750 000 PE) and 200 CSOs, was failing the good status requirements by the EU WFD. To address the pollution challenges in a cost-effective manner, the Waterschap De Dommel launched the KALLISTO project to develop and implement impact-based integrated RTC systems for improving the water quality of the river Dommel. This system has already had RTC equipment in the interceptor sewer since early 1970s. A screening study was also performed according to the planning tool PASST, which showed that the UWWS is suited for control. As such, this system was an ideal case for the study of the state-of-the-art integrated control technology. The rainfall, flow rated data in the sewer and water quality-related data in the WWTP were monitored since 2006. The river was also monitored for the DO and ammonia level since 2006. These data were reviewed first to understand the UWWS and the need of the river and then used for model establishment and calibration.

WEST was used for building the integrated UWWS model. The sewer system was simulated in a simple manner with no water quality processes considered.

The CSOs were considered to have fixed concentrations of DO and ammonia. A simple calibration was conducted for the sewer model to check obvious errors in the underlying database rather than to get a perfect fit between the monitoring data and the simulation results. Pumping capacity was found to have a significant impact on the performance of the sewer and was a key parameter to calibrate. ASM2d was used to describe the biokinetics in the activated sludge treatment plant. The model was calibrated following the BIOMATH calibration protocol. Good modelling practice was adopted so that the focus was not on adjusting the parameters in the ASM model but rather the quality of data and information on the system characteristics and operation. DUFLOW was used to build a 1D hydrodynamic model for the simulation of the river. Discharges from the sewer and WWTP can be directly evaluated in the simulation. The integrated model was calibrated and run for one year starting from September 2009 with 30 minute intervals. Based on the integrated model, the following integrated operation and control strategies which maximised river water quality by utilising the storage capacity in the UWWS were evaluated. The optimal strategies will be implemented for this case study. Further details on the Kallisto project can be found in Weijers *et al.* (2012) and Langeveld *et al.* (2013).

10.5 CONCLUSIONS

Integrated modelling and control of UWWSs is a promising technology enabling our wastewater infrastructures to be flexible and intelligent. Different levels of integrated control can be implemented as described in this chapter. This chapter provides a brief review of the research development in integrated modelling and key aspects to be considered for establishing a reliable model under various sources of uncertainty. Despite that regulatory, organisational, technical and/or financial barriers still exist for widespread implementation of integrated control of UWWSs, there are already a few successful cases reported (e.g., the KALLISTO project) that provide valuable insights of this advanced technology, which will in turn boost its future development.

REFERENCES

Achleitner S., Möderl M. and Rauch W. (2007). CITY DRAIN© – An open source approach for simulation of integrated urban drainage systems. *Environmental Modelling and Software*, **22**(8), 1184–1195.

Astaraie-Imani M., Kapelan Z., Fu G. and Butler D. (2012). Assessing the combined effects of urbanisation and climate change on the river water quality in an integrated urban wastewater system in the UK. *Journal of Environmental Management*, **112**(15), 1–9.

Bach P. M., Rauch W., Mikkelsen P. S., Mccarthy D. T. and Deletic A. (2014). A critical review of integrated urban water modelling - urban drainage and beyond. *Environmental Modelling and Software*, **54**, 88–107.

Bauwens W., Vanrolleghem P. A. and Smeets M. (1996). An evaluation of the efficiency of the combined sewer – Wastewater treatment system under transient conditions. *Water Science and Technology*, **33**(2), 199–208.

Beck M. B. (1976). Dynamic modelling and control applications in water quality maintenance. *Water Research*, **10**(7), 575–595.

Beck M. B. and Finney B. A. (1987). Operational water quality management: problem context and evaluation of a model for river quality. *Water Resources Research*, **23** (11), 2030–2042.

Benedetti L., Meirlaen J., Sforzi F., Facchi A., Gandolfi C. and Vanrolleghem P. A. (2007). Dynamic integrated water quality modelling: a case study of the Lambro river, northern Italy. *Water SA*, **33**(5), 627–632.

Breinholt A., Santacoloma P. A., Mikkelsen P. S., Madsen H. and Grum M. (2008). Evaluation framework for control of integrated urban drainage systems. 11th International Conference on Urban Drainage, 31 August–5 September 2008, Edinburgh, Scotland, UK. IWA Publishing, London.

Brown L. C. and Barnwell T. O. (1987). The Enhanced Stream Water Quality Models QUAL2E and QUAL2E-UNCAS: Documentation and User Manual. United States Environmental Protection Agency, Environmental Research Laboratory, Athens, GA, USA.

Butler D. and Schütze M. (2005). Integrating simulation models with a view to optimal control of urban wastewater systems. *Environmental Modelling and Software*, **20**(4), 415–426.

Castro-Barros C. M., Daelman M. R. J., Mampaey K. E., van Loosdrecht M. C. M. and Volcke E. I. P. (2015). Effect of aeration regime on N_2O emission from partial nitritation-Anammox in a full-scale granular sludge reactor. *Water Research*, **68**, 793–803.

Clifforde I. T., Crabtree R. W. and Andrews H. O. (2006). 10 years experience of CSO management in the United Kingdom. WEFTEC 2006, 21–25 October 2006, Dallas, Texas, USA. Water Environment Federation, Alexandria, VA, USA.

Deb K., Member A., Pratap A., Agarwal S. and Meyarivan T. (2002). A fast and elitist multiobjective Genetic Algorithm: NSGA-II. *IEEE Transactions on Evolutionary Computation*, **6**(2), 182–197.

DHI. (2020). WWTP modelling that does it all. Available from: www.mikepoweredbydhi.com/products/west, (accessed 26 August 2020).

Di Toro D. M., Fitzpatrick J. J. and Thomann R. V. (1983). Documentation for Water Quality Analysis Simulation Program (WASP) and Model Verification Program (MVP). United States Environmental Protection Agency, Environmental Research Laboratory, Athens, GA, USA.

Doglioni A., Primativo F., Laucelli D., Monno V., Khu S. T. and Giustolisi O. (2009). An integrated modelling approach for the assessment of land use change effects on wastewater infrastructures. *Environmental Modelling and Software*, **24**(12), 1522–1528.

Duyy I. J. and Laboratorium T. (1975). Dynamic modelling and control simulation of a biological wastewater treatment process. *Water Research*, **10**(5), 461–467.

Environment Agency. (2011). How to Comply with Your Environmental Permit. Additional Guidance for: Water Discharge and Groundwater (from Point Source) Activity Permits (EPR 7.01). England and Wales. Environment Agency, Bristol, UK.

Erbe V., Risholt L. P., Schilling W. and Londong J. (2002). Integrated modelling for analysis and optimisation of wastewater systems – The Odenthal case. *Urban Water Journal*, **4** (1), 63–71.

Escaler I. (2004). Sensors and process control technology, RTCUDS Work Group Proceedings, NOVATECH 2004. 5th International Conference on Sustainable Techniques and Strategies in Urban Water Management, 6–10 June 2004, Lyon, France. IWA Publishing, London.

European Parliament and Council of the European Union. (2000). Directive 2000/60/EC Establishing a Framework for Community Action in the Field of Water Policy. Official Journal L 327/2. European Union, Brussels.

Flores-Alsina X., Corominas L., Snip L. and Vanrolleghem P. A. (2011). Including greenhouse gas emissions during benchmarking of wastewater treatment plant control strategies. *Water Research*, **45**(16), 4700–4710.

Fronteau C., Bauwens W. and Vanrolleghem P. A. (1997). Integrated modelling: Comparison of state variables, processes and parameters in sewer and wastewater treatment plant models. *Water Science and Technology*, **36**(5), 373–380.

Fu G., Butler D. and Khu S. T. (2008). Multiple objective optimal control of integrated urban wastewater systems. *Environmental Modelling and Software*, **23**(2), 225–234.

Fu G., Khu S. T. and Butler D. (2009). Optimal distribution and control of storage tank to mitigate the impact of new developments on receiving water quality. *Journal of Environmental Enigneering*, **136**(3), 335–342.

Fuchs L. and Beeneken T. (2006). Comparison of measured and simulated real-time control. 7th International Conference on Urban Drainage Modelling and the 4th International Conference on Water Sensitive Urban Design, 3–7 April 2006, Melbourne, Australia. Monash University, Clayton, Victoria, Australia.

Georges K., Thornton A. and Sadler R. (2009). Transforming Wastewater Treatment to Reduce Carbon Emissions. Environment Agency, Bristol, UK. Available from: https://assets.publishing.service.gov.uk/government/uploads/system/uploads/attachment_data/file/291633/scho1209brnz-e-e.pdf, (accessed 24 August 2020).

Guest J. S., Skerlos S. J., Barnard J. L., Beck M. B., Daigger G. T., Hilger H., Jackson S. J., Karvazy K., Kelly L., Macpherson L., Mihelcic J. R., Pramanik A., Raskin L., Van Loosdrecht M. C. M., Yeh D. and Love N. G. (2009). A new planning and design paradigm to achieve sustainable resource recovery from wastewater. *Environmental Science and Technology*, **43**(16), 6126–6130.

Guo L., Porro J., Sharma K. R., Amerlinck Y., Benedetti L., Nopens I., Shaw A., Van Hulle S. W. H., Yuan Z. and Vanrolleghem P. A. (2012). Towards a benchmarking tool for minimizing wastewater utility greenhouse gas footprints. *Water Science and Technology*, **66**(11), 2483–2495.

Harremoës P., Capodaglio A. G., Hellström B. G., Henze M., Jensen K. N., Lynggaard-Jensen A., Otterpohl R. and Søeberg H. (1993). Wastewater treatment plants under transient loading – Performance, modelling and control. *Water Science and Technology*, **27**(12), 71–115.

Harremoës P., Andersen H. S., Dupont R., Jacobsen P. and Rindel K. (2002). Analysis of scenarios for sewerage, wastewater treatment and prioritised load on environment from the Greater City of Copenhagen. *Water Science and Technology*, **45**(3), 95–100.

Henze M., Grady C. P. L., Jr, Gujer W., Marais G. v. R. and Matsuo T. (1987). Activated Sludge Model No. 1. Scientific and Technical Report Series. IWA Publishing, London, UK.

House M. A., Ellis J. B., Herricks E. E., Hvitved-Jacobsen T., Seager J., Lijklema L., Aalderink H. and Clifforde I. T. (1993). Urban drainage – Impacts on receiving water quality. *Water Science and Technology*, **27**(12), 117–158.

HSG Sim. (2008). Integrierte Modellierung von Kanalnetz, Kläranlageund Gewässer—HSG-Leitfaden der Arbeitsgruppe Integrierte Modellierung, 1st edn. Hochschulgruppe, Erfahrungsaustausch Dynamische Simulation in der Siedlungswasserwirtschaft (Integrated modeling of sewer networks, sewage treatment plants and bodies of water — HSG guidelines of the Integrated Modeling Working Group, 1st edn. University group, 'Exchange of experience dynamic simulation in urban water management). Available from: www.hsgsim.org, (accessed 07 July 2009).

Ifak. (2016). SIMBA#WATER. https://simba.ifak.eu/en/content/simba-sharp-water, (accessed 12 November 2016).

Ito K. (2004). Real time control project in Tokyo. RTCUDS Work Group Proceedings, NOVATECH 2004: 5th International Conference on Sustainable Techniques and Strategies in Urban Water Management, 6–10 June 2004, Lyon, France. IWA Publishing, London.

Jin Z., Gong H. and Wang K. (2015). Application of hybrid coagulation microfiltration with air backflushing to direct sewage concentration for organic matter recovery. *Journal of Hazardous Materials*, **283**(11), 824–831.

Kay D., Kershaw S., Lee R., Wyer M. D., Watkins J. and Francis C. (2008). Results of field investigations into the impact of intermittent sewage discharges on the microbiological quality of wild mussels (Mytilus Edulis) in a tidal estuary. *Water Research*, **42**(12), 3033–3046,

Langeveld J. G., Benedetti L., de Klein J. J. M., Nopens I., Amerlinck Y., van Nieuwenhuijzen A., Flameling T., van Zanten O. and Weijers S. (2013). Impact-based integrated real-time control for improvement of the Dommel river water quality. *Urban Water Journal*, **10**(5), 312–329.

Lau J., Butler D. and Schütze M. (2002). Is combined sewer overflow spill frequency/volume a good indicator of receiving water quality impact? *Urban Water Journal* **4**(2), 181–189.

Lijklema L., Tyson J. M. and Lesouef A. (1993). Interactions between sewers, treatment plants and receiving waters in urban areas: a summary of the INTERURBA'92 workshop conclusions. *Water Science and Technology*, **27**(12), 1–29.

Mannina G., Freni G., Viviani G., Sægrov S. and Hafskjold L. S. (2006). Integrated urban water modelling with uncertainty analysis. *Water Science and Technology*, **54**(6–7), 379–386.

Mccarty P. L., Bae J. and Kim J. (2011). Domestic wastewater treatment as a net eEnergy producer – Can this be achieved? *Environmental Science and Technology*, **45**(17), 7100–7106.

Meng F., Fu G. and Butler D. (2016). Water quality permitting: from end-of-pipe to operational strategies. *Water Research*, **101**(15), 114–126.

Meng F., Fu G. and Butler D. (2017). Cost-effective river water quality management using integrated real-time control technology. *Environmental Science and Technology*, **51** (17): 9876–9886.

Meirlaen J. (2002). Immission Based Real-Time Control of the Integrated Urban Wastewater System. PhD thesis, Ghent University, Belgium.

Meirlaen J., van Assel J. and Vanrolleghem P. A. (2002). Real time control of the integrated urban wastewater system using simultaneously simulating surrogate models. *Water Science and Technology*, **45**(3), 109–116.

Muschalla D., Schütze M., Schroeder K., Bach M., Blumensaat F., Gruber G., Klepiszewski K., Pabst M., Pressl A., Schindler N., Solvi A. M. and Wiese J. (2009). The HSG procedure for modelling integrated urban wastewater systems. *Water Science and Technology*, **60**(8), 2065–2075.

Nardell G. (2012). Combined Sewage Overflows: The UK in Deep Water at Luxembourg, Lexology. Available from: www.lexology.com/library/detail.aspx?g=3e23dd9c-1f4b-447e-97c4-1846101a2dd1, (accessed 27 August 2020).

Nielsen M., Carstensen J. and Harremoës P. (1996). Combined control of sewer and treatment plant during rain storm. *Water Science and Technology*, **34**(3–4), 181–187.

Olsson G. (2012). ICA and me – A subjective review. *Water Research*, **46**(6), 1585–1624.

Petruck A., Cassar A. and Dettmar J. (1998). Advanced real time control of a combined sewer system. *Water Science and Technology*, **37**(1), 319–326.

Phillips P. J., Chalmers A. T., Gray J. L., Kolpin D. W., Foreman W. T. and Wall G. R. (2012). Combined sewer overflows: an environmental source of hormones and wastewater micropollutants. *Environmental Science and Technology*, **46**(10), 5336–5343.

Pleau M., Colas H., Lavallee P., Pelletier G. and Bonin R. (2005). Global optimal real-time control of the Quebec urban drainage system. *Environmental Modelling and Software*, **20**(4), 401–413.

Rauch W. and Harremoës P. (1999). Genetic algorithms in real time control applied to minimize transient pollution from urban wastewater systems. *Water Research*, **33**(5), 1265–1277.

Rauch W., Bertrand-Krajewski J. L., Krebs P., Mark O., Schilling W., Schütze M. and Vanrolleghem P. A. (2002). Deterministic modelling of integrated urban drainage systems. *Water Science and Technology*, **45**(3), 81–94.

Reichert P. and Vanrolleghem P. A. (2001). Identifiability and uncertainty analysis of the River Water Quality Model No. 1 (RWQM1). *Water Science and Technology*, **43**(7), 329–338.

Reichert P., Borchardt D., Henze M., Rauch W., Shanahan P., Somlyódy L. and Vanrolleghem P. A. (2001a). River Water Quality Model No. 1. Scientific and Technical Report Series. IWA Publishing, London, UK.

Reichert P., Borchardt D., Henze M., Rauch W., Shanahan P., Somlyódy L. and Vanrolleghem P. A. (2001b). River Water Quality Model No. 1 (RWQM1): II. Biochemical process equations. *Water Science and Technology*, **43**(5), 11–30.

Saagi R., Flores-Alsina X., Kroll S., Gernaey K. V. and Jeppsson U. (2017). A model library for simulation and benchmarking of integrated urban wastewater systems. *Environmental Modelling and Software*, **93**, 282–295.

Saint-Venant A. J. C. B. (1870). Démonstration élémentaire de la formule de propagation dune onde ou dune intumescence dans un canal prismatique; et remarques sur les propagations du son et de la lumière, sur les ressauts, ainsi que sur la distinction des rivières et des torrents (Elementary proof of the propagation formula of a wave or an intumescence in a prismatic channel; and remarks on the propagation of sound and light, on the jumps, as well as on the distinction between rivers and torrents). *Comptes rendus des séances de l'Académie des Sciences*, **71**, 186–195 (in French).

Scheer M., Heppeler D., Krapp G., Nusch S. and Meßmer A. (2004). Real time control of an integrated system - sewer system and wastewater treatment plant. 6th International Conference on Urban Drainage Modelling, 15–17 September 2004, Dresden, Germany. IWA Publishing, London.

Serra P., Lafuente J., Moreno R., de Prada C. and Poch M. (1993). Development of a real-time expert system for wastewater treatment plants control. *Control Engineering Practice*, **1** (2), 329–335.

Severn Trent Water Limited. (2013). Changing Course through the Sustainable Implementation of the Water Framework Directive. Severn Trent Water Ltd., Coventry, UK.

Schütze M., Butler D. and Beck M. B. (1999). Optimisation of control strategies for the urban wastewater system – An integrated approach. *Water Science and Technology*, **39**(9), 209–216.

Schütze M., Butler D. and Beck B. (2002). Modelling, Simulation and Control of Urban Wastewater Systems. Springer, London, UK.

Schütze M., Campisano A., Colas H., Schilling W. and Vanrolleghem P. A. (2004). Real time control of urban wastewater systems – Where do we stand today? *Journal of Hydrology*, **299**(3–4), 335–348.

Shanahan P., Borchardt D., Henze M., Rauch W., Reichert P., Somlyódy L. and Vanrolleghem P. A. (2001). River Water Quality Model No. 1 (RWQM1): I. Modelling approach. *Water Science and Technology*, **43**(5), 1–9.

Sin G., van Hulle S. W. H., De Pauw D. J. W., van Griensven A. and Vanrolleghem P. A. (2005). A critical comparison of systematic calibration protocols for activated sludge models: a SWOT analysis. *Water Research*, **39**(12), 2459–2474.

Snip L. J. P., Flores-Alsina X., Plósz B. G., Jeppsson U. and Gernaey K. V. (2014). Modelling the occurrence, transport and fate of pharmaceuticals in wastewater systems. *Environmental Modelling and Software*, **62**, 112–127.

Solvi A. M. (2007). Modelling the Sewer-Treatment-Urban River System in View of the EU Water Framework Directive. PhD thesis, Ghent University, Belgium.

Streeter W. H. and Phelps E. B. (1925). Public Health Bulletin No. 146: A study of the pollution and natural purification of the Ohio river. US Public Health Service, Washington DC, USA.

Strous M., Van Gerven E., Zheng P., Kuenen J. G. and Jetten M. S. M. (1997). Ammonium removal from concentrated waste streams with the Anaerobic Ammonium Oxidation (Anammox) Process in different reactor configurations. *Water Research*, **31**(8), 1955–1962.

Sweetapple C., Fu G. and Butler D. (2014a). Multi-objective optimisation of wastewater treatment plant control to reduce greenhouse gas emissions. *Water Research*, **55**(15), 52–62.

Sweetapple C., Fu G. and Butler D. (2014b). Identifying sensitive sources and key control handles for the reduction of greenhouse gas emissions from wastewater treatment. *Water Research*, **62**(1), 249–259.

Thornton A., Sunner N. and Haeck M. (2010). Real time control for reduced aeration and chemical consumption: A full scale study. *Water Science and Technology*, **61**(9), 2169–2175.

US Environmental Protection Agency. (1995). EPA Combined Sewer Overflows Guidance for Long-Term Control Plan. US EPA, Washington, DC, USA.

U.S. Environmental Protection Agency. (1999). Combined Sewer Overflows Guidance for Monitoring and Modeling. US EPA, Washington, DC, USA.

U.S. Environmental Protection Agency. (2013). Emerging Technologies for Wastewater Treatment and In-Plant Wet Weather Management. US EPA, Washington, DC, USA.

Vanrolleghem P. A., Borchardt D., Henze M., Rauch W., Reichert P., Shanahan P. and Somlyódy L. (2001). River Water Quality Model No. 1 (RWQM1): III. Biochemical submodel selection. *Water Science and Technology*, **43**(5), 31–40.

Vanrolleghem P. A., Benedetti L. and Meirlaen J. (2005). Modelling and real-time control of the integrated urban wastewater system. *Environmental Modelling and Software*, **20**(4), 427–442.

Vezzaro L., Benedetti L., Gevaert V., De Keyser W., Verdonck F., De Baets B., Nopens I., Cloutier F., Vanrolleghem P. A. and Mikkelsen P. S. (2014a). A model library for dynamic transport and fate of micropollutants in integrated urban wastewater and stormwater systems. *Environmental Modelling and Software*, **53**, 98–111.

Vezzaro L., Christensen M. L., Thirsing C., Grum M. and Mikkelsen P. S. (2014b). Water quality-based real time control of integrated urban drainage systems: a preliminary study from Copenhagen, Denmark. *Procedia Engineering*, **70**, 1707–1716.

Viessman W., Lewis G. and Knapp J. W. (1989). Introduction to Hydrology, 3rd edn. Harper & Row, New York, NY, USA.

Weijers S. R., De Jonge J., van Zanten O., Benedetti L., Langeveld J. G., Menkveld H. W. and van Nieuwenhuijzen A. F. (2012). KALLISTO: Cost effective and integrated optimization of the urban wastewater system Eindhoven. *Water Practice and Technology*, **7**(2), 1–9.

Weyrauch P., Matzinger A, Pawlowsky-Reusing E., Plume S., von Seggern D., Heinzmann B., Schroeder K. and Rouault P. (2010). Contribution of combined sewer overflows to trace contaminant loads in urban streams. *Water Research*, **44**(15), 4451–4462.

Part III

Practices of Water-Wise Cities and Sustainable Water Systems

Chapter 11

Practices of 'Sponge City Construction' in China

Nanqi Ren[1], Xiuheng Wang[1], Jiazhuo Wang[2], Qionghua Zhang[3] and Xiaochang C. Wang[3]

[1]Harbin Institute of Technology, School of Environment, Harbin, China
[2]Chinese Academy of Urban Planning and Design, Beijing, China
[3]Xi'an University of Architecture and Technology, Xi'an, China

11.1 INTRODUCTION

As introduced in Chapter 3 of this book, Sponge City Construction (SCC) is a national strategy set by the Chinese government as a solution to urban water environmental problems within a systematic framework toward sustainable development. Similar to other countries and regions, the initial motivation of urban water management was mainly related to urban drainage. The applicable technology selection followed similar principles as those for Low Impact Development (LID) in North America (US EPA, 2000), Water-Sensitive Urban Design (WSUD) in Australia (Brown & Clarke, 2007), and Sustainable Urban Drainage Systems (SUDS) in Europe (Hoang, 2016). However, the connotation of Sponge City not only embraces urban green infrastructure to slow, spread, sink, and store surface runoff, infiltration-by-design to assist recharge of urban aquifers, mitigation of floods and waterlogging, and letting city surfaces breathe, but also stresses the systematic design of the broader urban area so that a city can

Editors: Xiaochang C. Wang and Guangtao Fu
doi: 10.2166/9781789060768_0283

be made sponge-like and have good ‘elasticity’ in adapting to environmental changes and responding to natural disasters and extreme events.

The formal proposal for the SCC as a national strategy for sustainable urban development was in December 2013 at the National Working Meeting on Urbanization. President Xi Jinping addressed the meeting and called for the promotion of Sponge City Construction through natural storage, natural infiltration, and natural purification. Following this proposal, the General Office of the State Council (2015) issued the central governmental Instruction on Promoting Sponge City Construction in 2015. This governmental document firstly sets the goal of SCC as to restore healthy urban aquatic ecology, conserve water resources, enhance the capability for urban flood control, expand effective investment in public services, improve urban life quality, and eventually support sustainable urban development under the harmony between man and nature. It was stressed that SCC should adhere to three basic principles, namely (1) to insist on the application of ecological technologies and realization of natural urban water circulation, (2) to insist on coordinative advancement following careful planning, and (3) to insist on governmental orientation and social participation.

This governmental document called for nationwide SCC actions in urban development by adopting comprehensive engineering measures, which could be denoted by six Chinese characters of ‘*Shen*’ (Infiltration), ‘*Zhi*’ (Stagnation), ‘*Xu*’ (Storage), ‘*Jing*’ (Purification), ‘*Yong*’ (Utilization), and ‘*Pai*’ (Discharge). The overall quantifiable target has been set as onsite absorption and/or utilization of 70% of the rainfall in the SCC project area. It was required that by 2020 the SCC target should be reached in more than 20% of the built-up urban area in the whole nation, and by 2030 the percentage should reach 80%.

In addition to these goals and targets, this governmental document stressed the fundamental policies to support the nationwide SCC actions, including (1) the leading role of city master plans and standards/regulations, (2) orderly advancement of SCC activities, (3) governmental social, and financial supporting policies, and (4) organization and implementation systems.

As SCC is a completely brand-new national action, it needs not only policy guidance but also reproducible models and successful experiences for all cities to follow. To meet such a need, the Ministry of Housing and Urban-Rural Development, the Ministry of Finance, and the Ministry of Water Resources selected 16 cities as the first batch in 2015 and then another 14 cities as the second batch in 2016, for the implementation of SCC pilot projects. Each of the pilot cities was subsidized by the central government with 400–600 million CHY (Chinese Yuan) per year (about 57–86 million USD per year) over three years. By the end of 2019, all these pilot cities made great progress following their SCC plans and played central roles in the advancement of SCC actions in China.

11.2 TECHNOLOGY STANDARDIZATION FOR SPONGE CITY CONSTRUCTION IN CHINA

11.2.1 National standard for assessment of sponge city construction

One important work to answer the central government's call for SCC actions is to provide a national standard. Led by the Ministry of Housing and Urban-Rural Development (MOHURD), a group of experts started to develop a standard for the assessment of SCC in 2016. After more than two years of extensive investigation and research, careful summarization of practical experiences, extensive consultation, and referring to relevant international regulations, a national 'Standard for Assessment of Sponge City Construction' (GB/T51345-2018) was put forward in December 2018 and became effective in August 2019 (MOHURD, 2018).

The most important feature of this Standard is its clear specifications on the assessment items and the related requirements, which provide basic guidance for selecting apt technological and engineering measures under the SCC goals.

11.2.1.1 Total annual runoff control

Runoff control is the first item of assessment on the SCC effect. It stresses the overall effect of surface runoff reduction through various engineering and ecological measures. In this regard, various targets have been set according to the annual rainfall conditions in different areas over the China mainland. For the Zone I area (the vast Northwest region and part of the North and Northeast region where annual rainfall is usually less than 500 mm), the total annual runoff control target is set to 85–90%, while for the Zone II to Zone V areas (other regions in an order of increasing annual rainfall) the lower limit of the total annual runoff control target is set to 80, 75, 70, and 60%, respectively, whereas the upper limits are all set to 85%.

It is strictly required that for SCC in newly developed urban areas, the total annual runoff control percentage should be no less than the lower limit. For SCC in reconstructed areas, it is preferable to follow this requirement if technically and economically feasible.

11.2.1.2 Source reduction

Source reduction for SCC projects includes the reductions of runoff source flow, pollutant sources, and runoff peak flow. Requirements are given to three typical urban areas, namely residential districts, roads and squares (including parking lots), and parks and green belts, respectively, in this standard.

For residential districts, the requirement for runoff flow reduction is the same as that described in section 11.2.1.1. Regarding pollutants in the runoff flow, it is required that the annual total SS (suspended solids) loading be reduced by no less than 70% for new development projects and no less than 40% for reconstruction and expansion projects. As for the reduction of runoff peak flow, the principle is that the peak runoff effluent from the project area should not be larger than that before the project construction. It is also required that for new development projects, the paved surface area should not exceed 40% of the whole project area, and for reconstruction and expansion projects, it should be no larger than the original percentage and not exceed 70% of the whole project area.

The above requirements are applicable for parking lots and squares, except for the last requirement on the percentage of paved surface area. For roads, it is advised that pollutant reduction and smooth drainage from the road surface are sufficiently considered in their design, especially for the roads that act as drainage channels in the rainy season. As parks and green belts are not hard surfaces, the requirement for runoff control should be no less than a reduction of 90% of the annual total runoff. In addition, parks and green belts should be capable of accommodating runoff from the neighbouring area.

11.2.1.3 Waterlogging prevention

Regarding the prevention of road surface flooding and waterlogging in urban areas, it is required that runoff peak flow be effectively reduced and/or staggered by the rational interconnection of gray and green facilities. No water stagnation is allowed to occur under the conditions corresponding to the design return period of the rainwater drainage pipelines, and no local waterlogging is allowed to occur under the conditions corresponding to the prescribed return period of rainfall intensity.

11.2.1.4 Urban water environmental quality

The protection of urban water quality is set as an important goal of SCC projects. It also reflects the overall effect of SCC from the viewpoint of water environmental improvement. The basic requirement is that the gray and green facilities employed can well perform their synergistic roles in runoff pollution control and water purification to efficiently prevent water pollution due to overflow from combined sewer systems. No wastewater is allowed to be discharged directly into any water body in dry weather. Water pollution due to the mixed discharge of rainwater and sewage or combined sewer overflow should be well controlled to protect the receiving water bodies from the occurrence of black odor. Alternative quantifiable criteria are set as a reduction of no less than 50% of the mixed discharge from the outlets of separate sewers, or overflow from the outlets of combined sewers. Also, the monthly average SS concentration of the effluent from treatment facilities should not exceed 50 mg/L.

Furthermore, no water body is allowed to be within the category of black and odorous water, namely, with typical water quality indicators as follows: water transparency >25 cm, DO >2.0 mg/L, ORP >50 mV, and NH_4-N <8.0 mg/L. The water quality in any water body after an SCC project should not become poorer than that before the project, and for rivers, the water quality downstream of the project area should not be poorer than that in the upstream in dry weather periods.

11.2.1.5 Urban ecological pattern management and aquatic shoreline protection

From the ecological viewpoint, SCC projects should well protect or improve urban ecological settings. It is thus required that the total area of natural waters should not decrease after the projects. The natural topography and landscape pattern should be well protected and/or recovered. All ecological sensitive areas, such as natural flood channels, flood plains, wetlands, forests, and grasslands, should not be encroached on for any purpose.

The protection of ecological shorelines is especially emphasized. For any newly constructed, restored, or expanded urban water body, the length of the ecological shoreline should be no less than 70% of the total shoreline length.

11.2.1.6 Variation of groundwater level

In urban areas, the groundwater level is much influenced by urban development due to impervious paving, which considerably reduces rainwater infiltration. It is expected that through SCC projects, the decline of groundwater levels can be well prevented. Therefore, the variation of the groundwater level in the project area should be an important aspect to reflect the overall effect of SCC projects. The basic requirement is that the downward trend of the average annual groundwater level should be under control.

11.2.1.7 Mitigation of urban heat island effect

The mitigation of the urban heat island effect is another important assessment item related to the overall effect of SCC projects. It is required that in summer, from June to September, the daily average air temperature difference between urban and suburban areas should show a downward tendency after the SCC projects in comparison with historical records.

11.2.2 Technical guidelines for sponge city construction

11.2.2.1 Technological strategy and basic framework

Following the abovementioned national standard for the assessment of sponge city construction, the technological routes can be considered from three aspects. The first aspect is the protection of the original ecological setting of the whole urban area, the second aspect is the restoration of the damaged ecological units, while the third aspect is the reduction of the impact of urban development on the ecological environment.

Natural water bodies such as rivers, lakes, and groundwater aquifers are the most important aquatic elements of the local ecological setting. These water bodies closely interrelate with other natural elements such as forests, grasslands, and wetlands, forming the basic physical framework of an ecosystem. As discussed in Chapter 3 of this book, any engineering constructions and facilities built in a city should not jeopardize the healthy circulation of water and materials in this ecosystem. To achieve such a goal, it is essential to prioritize the protection of the original ecological setting of the whole urban area. The model of Urban Water System 3.0 (Figure 3.4, Chapter 3) shows such a basic system setting. Therefore, strategically, a successful SCC project needs a scientifically reasonable plan at a higher level.

Currently, for some cities and their related watersheds in China, severe pollution of natural water bodies and deterioration of the eco-environment are important factors restricting sustainable urban development. The restoration of these damaged aquatic and ecological units should principally be put into the SCC objectives, especially the introduction of green technologies. However, as such problems are usually of large scale or cover broad areas, their solutions may also need to be within a master plan of a regional scale.

The third aspect, namely the reduction of the impact of urban development on the ecological environment, is usually related to specific objectives that may be achieved by implementing projects of limited scales. Adaptation of various technologies with the LID scheme to local conditions should be the principle in this regard.

Within the abovementioned strategic framework, many studies have been conducted to formulate applicable technical guidelines at national and local levels, such as the earlier document of 'Technical Guidelines (Tentative) for Sponge City Construction – Low-Impact Development Rainwater System Construction' by MOHURD (2014), and the recent document of 'Technical Specification for Construction of Sponge City' by Shanghai Municipal Housing and Urban-Rural Construction Management Committee (2019). Although the latter is a technical guideline applicable to the local situation in Shanghai, it can be a reference for other cities with adaption to the local conditions.

These documents gave technical guidance on SCC project planning, design, construction and acceptance, operation and maintenance, and monitoring and control. Below, we mainly introduce the principal technological requirements and their characteristics for SCC project planning and design.

11.2.2.2 Technical guidance for SCC project planning

In the project planning stage, attention should be paid mainly to the ultimate goal of SCC for the whole city and/or the project area as a district of the city. It is thus required that the SCC project plan be within the general scheme of the master plan of urban development, and coordinate closely with the related specific plans of urban construction. Project planning is usually conducted at three levels, namely general planning, unit planning, and detailed planning.

The general SCC planning is to formulate a comprehensive set of systematic solutions. It usually includes the following:

- investigation of the current situation based on the status quo survey and existing data, analysis of the major problems especially those related to urban drainage, and identification of the current and future needs;
- proposal of the overall strategy of SCC planning;
- determination of the urban development orientation and system layout based on the SCC concept, including the basic requirements for the urban functional group and ecological space layouts, urban space management, and control, and the general needs for building large 'Sponge' facilities in the urban area;
- formulation of a multitarget and multi-index framework with the annual total runoff control as the core objective;
- determination of the management and control objectives and associated indicators based on hydraulic zoning of the urban area with the overall SCC target of the whole area and the practical condition of each hydraulic zone as the constraints;
- proposal of systematic SCC solutions for the whole urban area;
- breakdown of SCC tasks and related subprojects, and proposal of an implementation strategy for each subproject coordinative to relevant existing urban construction plans; and
- formulation of a short-term SCC project implementation plan.

The abovementioned hydraulic zoning usually covers a natural watershed or sub-watershed, which is set as the primary control area in many cities and given priority in the general SCC planning. In contrast, there are so-called secondary control areas which often cover distinctive drainage districts. The SCC unit planning is, in many cases, targeting the secondary control areas. The principles discussed above for the general SCC planning are also applicable to the SCC unit planning. Conversely, at the unit planning level, the solutions worked out should be more problem-oriented, such as the protection of specific ecological spaces and proposal of green-gray technical strategies or solely gray technology approaches for solving specific problems.

Detailed planning is for deciding the pathway to reaching the SCC goals set in the general plan and/or unit plan, identifying the direction of project implementation in the specific area, and determining the reasonable facility scale and configuration. At the detailed planning level, clear indexes should be proposed for the SCC project. These indexes include the control indexes to be attained by project implementation and some recommended indexes that are not mandatory but can guide the ultimate goal. By clear delineation of control lines, the scope of important areas can be clarified, such as water regimes (river, lake, and so on), green spaces, and greening separation zones. Facility layouts in various plots of land are also included in the detailed planning.

The project implementation plan will further be formulated following the detailed planning. It should include the following:

- proposal of the main measures for achieving the SCC goals based on the practical condition and development needs in the project area;
- selection of the most appropriate types of facilities to be implemented in the project area;
- layout plan of major facilities and determination of their specifications associated with scenario analysis;
- assessment of the effects of runoff control in accordance with various scenarios by modelling simulation and using decision-support tools; and
- optimization of the project implementation plan by a comprehensive consideration of facility performance and effect, construction and operation/maintenance costs, and ecological and landscape benefits.

11.2.2.3 Technical guidance for SCC project design

SCC project design is to detail the engineering blueprints for the project implementation following the principles and basic requirements set at the SCC project planning stage. As described in the former sections, in China, the terminology of Sponge City principally covers a city as a whole or a district within a city. Therefore, what we say of an 'SCC project' at the planning stage of the project is for the whole city or city district, while the project design stage discusses specific issues of different scales from AREA (city area to district area) down to PLOT (part of the AREA with distinct features) and FACILITY (specific engineered and/or ecological structure or equipment). For the convenience of explanation, we call all the functional areas, structures, and equipment that perform the functions for achieving SCC goals 'Sponge Facilities' in this section.

The SCC project design at the area level usually covers a detailed configuration of Sponge Facilities either by engineering construction or utilization of its natural functions. This needs, first, an overall analysis of the geographical environment, including topography, elevation, soil, green areas, water regimes, and then an analysis of the attainable controlling indexes, such as green coverage, water surface coverage, and so on, along with an evaluation of the required restoration and construction requirements and their feasibility.

Residential areas, green spaces, roads, and squares are major plots within a city for SCC project design. Specific requirements are given to each of these plots in the project design.

For residential areas, smooth drainage of local rainfall is the basic principle of SCC project design. It is thus required that areas with continuously paved hard surfaces be minimized and sufficient permeable green area be provided at the downstream direction of the local drainage. In principle, the building floor elevation should be above the elevation of outdoor footpaths and roads within the residential area, and the elevation of these footpaths and roads should be above that of the main roads outside the residential area. The surface gradient should be well utilized so that all the footpaths, roads, open spaces, and outdoor parking lots within the residential area can be sloped towards the green areas. Drainage ditches and overflow outlets should be installed for large green areas. All the paved hard surfaces should be equipped with rainwater outlets at the lowest locations.

Green spaces are important buffer zones for a city due to their soft surface to uptake moisture and facilitate rainwater infiltration, the combined effects of water purification by soil media and vegetation, and the flexibility of areal layout and merging with other facilities to maximize their functions. For the reasonable design of green spaces, three aspects should be taken into account. The first is the target of surface runoff reduction, the second is pollution control, and the third is their ecological and landscape functions. It is thus required that in addition to soil and vegetation beds, rainwater discharge and storage facilities be coordinatively introduced into the green space design. Large green spaces (e.g., over 2 ha) should preferably include a certain area of water surface, while for smaller ones, biological retention facilities should be implemented to increase the storage capacity. Wet ponds and constructed wetlands are also preferable facilities for the reduction of pollutants.

Roads, especially motorways, are inevitable paved hard surfaces, and organized convergence and transfer of rainwater are the key points of their design. This usually needs combined gray-green measures, namely, a good drainage system and an associated green divider and a green buffer zone. For footpaths and roadside walkways, permeable pavements are recommended. Similar measures are also applicable to the design of squares.

Detailed guidance is given to the design of typical sponge facilities, such as green roofs, rain gardens, grass ditches, permeable pavements, ecological tree pools, storage facilities, initial rainwater disposal facilities, wet ponds, and vegetation buffer zones. The selection of these sponge measures is based on the technical and economic feasibility and adaptability to the local conditions.

11.2.2.4 Guideline for remediation of urban black and odorous waters

The so-called 'black and odorous water' is a problem drawing wide concern in China. This is a new category of heavily polluted waters, mostly in densely populated urban areas, with the characteristics of: (1) severe pollution with high concentrations of organic matter and nutrients (nitrogen and phosphorus) from domestic sewage and industrial wastewater, resulting in dissolved oxygen depletion, (2) probable emission of methane, hydrogen sulfide, and other insoluble gases from the bottom sediments under acidic and anaerobic conditions, making the water appear black with a smelly odor, and (3) poor water circulation and declined reoxygenation ability to stimulate blue and green algae bloom. According to the MOHURD data (unpublished), there were 2899 outbreaks of black and odorous waters (2548 as urban rivers/streams and 351 as urban lakes/ponds) in 295 cities of prefecture-level and above by. Therefore, the remediation of black and odorous waters has become an extremely urgent nationwide task as required by the national Action Plan for Prevention and Control of Water Pollution (State Council, 2015), as well as the program of sponge city construction.

Specifically, for the remediation of urban black and odorous waters, a technical guideline was put forward by the Ministry of Housing and Urban-Rural Development (MOHURD, 2015). As the black and odorous waters can no longer be evaluated using traditional methods, such as the classification of surface water quality following the Environmental Quality Standard for Surface Water (SEPA and GAQSIQ, 2002), new criteria for evaluating such heavily polluted waters were proposed as shown in Table 11.1 Four characteristic parameters, namely, water transparency, dissolved oxygen (DO), oxidation-reduction potential (ORP), and ammonia nitrogen (NH_3-N), were used for evaluating the severity of water pollution and categorizing the waters into two classes of 'mild' and 'severe' black and odorous.

The recommended technologies for the remediation of urban black and odorous waters are shown in Figure 11.1. This is a framework of 12 applicable measures of four categories for external source control, internal source removal, ecological restoration, among others.

11.2.3 Criteria for National Pilot Project Evaluation

The national standard for SCC assessment and the related technical guidelines provides the basic requirements for SCC project planning, design, construction, and evaluation in the whole country. However, as the national pilot projects implemented in the 30 selected cities (see the following sections for details) are to demonstrate replicable models for other cities to follow, higher requirements

Table 11.1 Criteria for evaluating black odorous waters (MOHURD, 2015).

Characteristic Parameter	Mild Black Odorous Water	Severe Black Odorous Water
Water transparency (cm)*	25–10	<10
DO (mg/L)	0.2–2.0	<0.2
ORP (mV)	−200–50	<-200
NH_3-N (mg/L)	8.0–15.0	>15

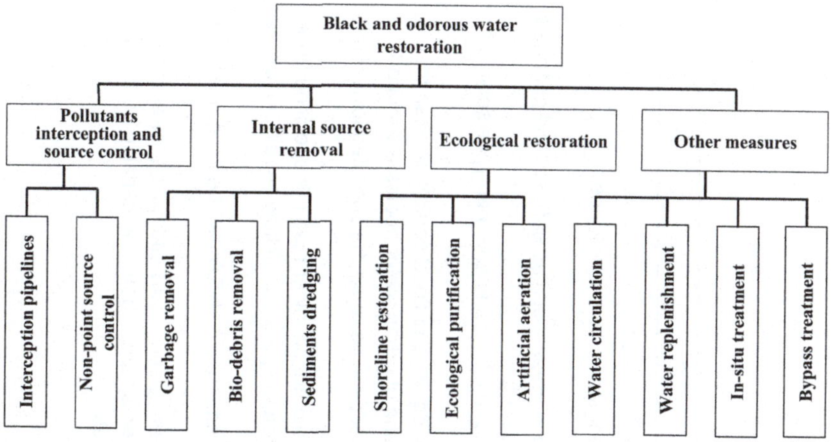

Figure 11.1 Recommended technologies for the remediation of urban black and odorous waters (adapted from MOHURD, 2015).

are given for these pilot SCC projects. Table 11.2 summarizes the criteria for evaluating the pilot SCC projects.

11.3 NATIONAL PILOT PROJECTS OF SPONGE CITY CONSTRUCTION

11.3.1 Cities selected for pilot projects implementation

Tables 11.3 and 11.4 outline the pilot cities of the first batch (selected in 2015) and second batch (selected in 2016), respectively, along with their geographical locations and features, administrative levels, and population sizes.

As shown in Figure 11.2, geographically, these pilot cities are distributed over the North, Central, South, East, Northeast, and Northwest regions of China, covering plain, mountainous, hilly, coastal, and island areas. As the topography of mainland China is generally featured by higher altitude in the west and lower altitude in the east, many large rivers, such as the Changjiang (Yangtze) River and Huanghe (Yellow) River, originate from the western plateau and flow eastward to the sea.

With the great differences among these pilot cities, in terms of their geographical locations and topographic features, the climatic and meteorological conditions also differ considerably from each other. The latitudes of these pilot cities lie between 46°N for Baicheng (No. 5 in Table 11.3) and 18°N for Sanya (No. 10 in Table 11.3). Therefore, the annual average temperature can differ by up to 20°C (5.5°C for Baicheng, and 25.8°C for Sanya). Similarly, the annual precipitation across mainland China varies greatly. In the southeast coastal area, it is more than 1600 mm per year and decreases toward the northwestern direction. In North

Table 11.2 Criteria for national SCC pilot project evaluation.

Parameter of Evaluation	Requirements*
Control of natural ecological setting	
• Natural water area increase (%)	• Protection and restoration of natural topography, landforms, and landscape patterns • No encroachment of ecologically sensitive areas, such as natural flood channels, floodplains, wetlands, forests, grasslands, etc. • Management and control requirements for the blue line and green line set in other relevant plans
• Expansion of ecological shoreline of urban water body (%)	• In addition to the shorelines required for docks and flood control, no less than 70% expansion of the ecological shoreline of urban water bodies is required
• Total reduction of annual runoff (%)	• For a newly built urban area, it should not be lower than the specified lower limit for the project region** • For a newly built residential area, it should not be lower than the specified lower limit, while for a renovated residential area the peak runoff flow should not exceed the original peak flow • For newly built roads, parking lots, and squares, it should not be lower than the specified lower limit • For newly built parks and protective green spaces, it should not be lower than the specified lower limit, and parks and green spaces should be capable of receiving rainfall and runoff from surrounding areas
Water resource use***	
• Rainwater harvesting and use (%)	
• Wastewater reuse (%)	

Water environmental control	
• Proportion of black and odorous waters in the built-up area (%)	• No sewage and wastewater discharge in dry weather • No black and odorous waters: water transparency >25 cm, DO >2.0 mg/L, ORP >50 mV, and NH_4-N <8.0 mg/L • Water quality should not be worse than that before the SCC project
• Frequency of overflow from combined sewer system (time/yr)	• Control of pollution from mixed discharges of separate sewers or combined sewer overflow to protect the receiving water body from developing black and odorous conditions during rainy days, or no less than 50% reduction of the annual overflow volume • Monthly average SS concentration of the treated discharge should not exceed 50 mg/L
Water safety guarantee	
• Number of waterlogged spots in the built-up area	• No occurrence of water accumulation under a rainfall corresponding to the design return period of the drainage system
• Compliance rate of waterlogging prevention and control (%)	• No occurrence of water logging under a rainstorm corresponding to the design return period of waterlogging control

*Refer to the Standard for Assessment of Sponge City Construction (GB/T51345-2018).
**As 85, 80, 75, 70, and 60%, respectively, for Zone I to Zone V according to the Standard for Assessment of Sponge City Construction (GB/T51345-2018).
***Non-mandatory requirements in the Standard for Assessment of Sponge City Construction (GB/T51345-2018).

Table 11.3 List of first batch SCC pilot cities (selected in 2015).

No.	City Name	Geographic Location and Feature	Administrative Level	Population Size
1	Chongqing	Southwest/Mountain area	Central municipality	>10 million
2	Shuining	Southwest/Hilly area	Prefecture-level city	3–5 million
3	Gui'an New District	Southwest/Hilly area	National new district	0.5–1 million
4	Xixian New District	Northwest/Semiarid plain area	National new district	1–3 million
5	Baicheng	Northeast/Plain area	Prefecture-level city	1–3 million
6	Zhenjiang	East China/Hilly area	Prefecture-level city	3–5 million
7	Pingxiang	East China/Hilly area	Prefecture-level city	1–3 million
8	Chizhou	East China/Hilly plain area	Prefecture-level city	1–3 million
9	Jinan	East China/Hilly plain	Provincial capital city	5–10 million
10	Xiamen	East China coast/Island area	Special municipality	5–10 million
11	Jiaxing	East China/Plain area	Prefecture-level city	3–5 million
12	Qian'an	North China/Plain area	Prefecture-level city	0.5–1 million
13	Hebi	North China/Plain area	Prefecture-level city	1–3 million
14	Wuhan	Central China/Low mountain area	Provincial capital city	>10 million
15	Changde	Central China/Low mountain area	Prefecture-level city	5–10 million
16	Nanning	South China/Hilly area	Provincial capital city	5–10 million

Table 11.4 List of second batch SCC pilot cities (selected in 2016).

No.	City Name	Geographic Location and Feature	Administrative Level	Population Size
1	Beijing	North China/Plain area	Central municipality	>10 million
2	Tianjin	North China/Plain area	Central municipality	>10 million
3	Dalian	Northeast coast/Hilly area	Special municipality	5–10 million
4	Shanghai	East China coast/Water-rich area	Central municipality	>10 million
5	Ningbo	East China coast/Water-rich area	Special municipality	5–10 million
6	Fuzhou	East China coast/Water-rich area	Provincial capital city	5–10 million
7	Qingdao	East China coast/Hilly area	Special municipality	5–10 million
8	Zhuhai	South China coast/Water-rich area	Prefecture-level city	1–3 million
9	Shenzhen	South China coast/Water-rich area	Special municipality	>10 million
10	Sanya	South China/Island area	Prefecture-level city	0.5–1 million
11	Yuxi	Southwest/High altitude area	Prefecture-level city	1–3 million
12	Qingyang	Northwest/Arid loess area	Prefecture-level city	1–3 million
13	Xining	Northwest/Arid loess area	Provincial capital city	1–3 million
14	Guyuan	Northwest/Arid loess area	Provincial capital city	1–3 million

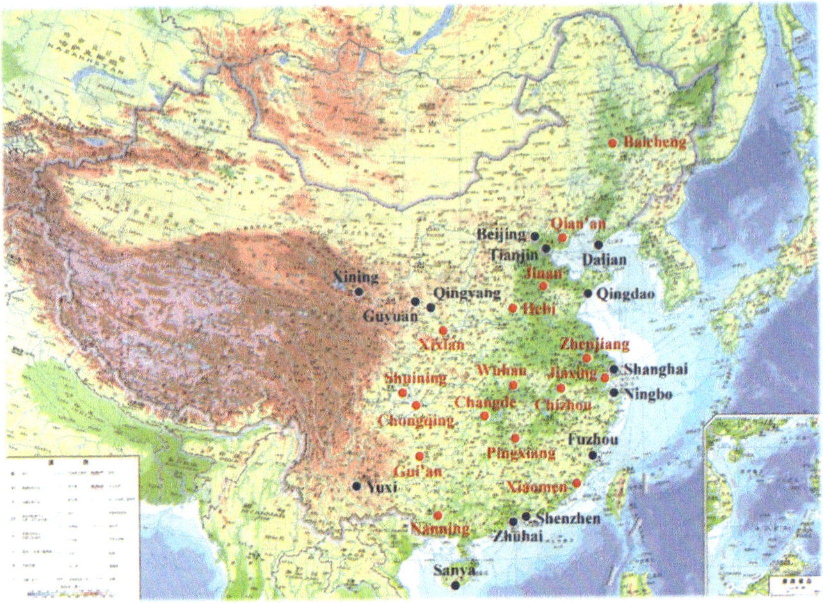

Figure 11.2 Distribution of SCC pilot cities of the first batch (red dots and city names) and second batch (blue dots and city names) in China (figure by authors).

China, the annual precipitation is usually about or less than 500 mm. There are also large areas in the Northwest region where the annual precipitation can fall below 100 mm. Across the 30 SCC pilot cities, the highest annual precipitation of as high as 1999.3 mm (long-term average) was recorded in Zhuhai (No. 8 in Table 11.4), located at the South China coast. Also, the lowest annual precipitation of 327 mm (long-term average) was recorded in Xining (No. 13 in Table 11.4), located at the Northwest arid loess area.

As a result of rapid urbanization in the past decades, many cities have developed into very populated megacities or large cities. As shown in Tables 11.3 and 11.4, categorized by population size, these pilot cities fall into five groups, namely, population >10 million (six cities), population of 5–10 million (eight cities), 3–5 million (three cities), 1–3 million (ten cities), and 0.5–1 million (three cities).

11.3.2 Classification of SCC strategies and demonstration technologies

The objective of implementing the SCC pilot projects in these cities was to explore reproducible success models and accumulate experiences for all cities in China to follow to realize the national goal of making cities sponge-like and adaptable to environmental changes and resilient to natural disasters. As discussed in section

11.3.1, the selection of these pilot cities considered the great variety of geographical and natural conditions over the whole country. Based on natural factors, the SCC strategies and associated demonstration technologies for these pilot cities are summarized in Table 11.5.

11.3.3 Progress of national pilot projects

After their selection in 2015 (the first batch) and 2016 (the second batch), the 30 pilot cities initiated the implementation of various SCC projects in accordance with the goals set in their proposals for the pilot projects, aided by the central government over three years. By the end of 2019, all these cities had completed their planned work and received inspection and evaluation organized by MOHURD.

11.3.3.1 City scale SCC planning

As a prerequisite, each of the pilot cities formulated a city-scale SCC plan in which the goal of urban construction and development toward a sponge-like city was set, the corresponding blueprint was plotted, and a general framework was proposed. Such a general plan is principally within the scope of the city's master plan of urban development and based on a comprehensive analysis of the current problems, especially those related to the water environment. As shown in Table 11.5, cities in different regions are facing a variety of problems closely related to climatic and geographic conditions. This leads to an identification of the bottleneck problems and the main SCC strategy.

For many pilot cities, the formulation of the city-scale SCC plan is a process to reconsider their policy of urban development. Specialists from various sectors have been involved in the work under the organization and coordination of the various municipal governments.

11.3.3.2 Selection of pilot project areas and formulation of implementation plans

The period for implementing the SCC pilot projects under the support of the central government was set for three years. As it would be impossible and unrealistic for a city to achieve the ultimate SCC goal within that limited time, most of the pilot cities have selected district areas to implement pilot projects to demonstrate the apt technologies and their effects on solving the major or bottleneck problems. Attention has been paid fully to the representativeness of the selected areas and their suitability for systematic solutions to the unique issues in a sub-watershed or a drainage zone. A coordinative relationship between pilot SCC projects and ongoing urban development projects is also an important factor considered.

In the selected areas of each pilot city, a detailed plan is formulated for implementing a series of pilot projects within a systematic framework. These pilot projects are designed for two major objectives. One is to demonstrate the

Table 11.5 Classification of SCC strategies and demonstration technologies based on natural factors.

Regional Characteristics	Major Common Problems	SCC Strategy and Technical Measures
Northeast cold area • *Weather*: Warm in summer, severely cold in winter, uneven precipitation, drought in spring, average annual precipitation 200–800 mm • *Geology*: Sand and gravel stratum, good permeability and shallow groundwater table • *Regions covered*: The provinces of Heilongjiang, Jilin, Liaoning, and north-eastern Inner Mongolia, with most cities located in plains and local hills • *Selected pilot cities*: Baicheng, Dalian	• Snowy and icy in winter, and adverse effect on roads, bridges, vegetation, and water bodies due to the use of snow-melting agents • Low temperature and cold in winter and low sewage treatment and water purification efficiencies due to low biological activity • Damage on permeable pavements due to repeated freezing in winter and thawing in spring • Urban streams mainly replenished in rainy season due to uneven rainfall and insufficient replenishment, resulting in fragile aquatic ecology and degraded water landscape	*Strategies* o Enhanced wastewater treatment in winter for efficient pollutants removal o Rational use of environment-friendly snow melting agents or mechanical snow removal to replace traditional snow-melting agents o Development and application of frost-resistant permeable pavement materials o Enhancement of pollutants interception and treatment, and reformation of urban drainage from combined systems to separated systems o Enlargement of wastewater treatment coverage and enhancement of rainwater and snowmelt water treatment *Main technical measures* o Recommend to use initial rainwater separators, regulation tanks, rainwater tanks, seepage wells, etc., and all green facilities but not green roofs due to weather and economic factors

		o Use of treatment facilities and/or green technologies suitable for coping with snowmelt water o Use of anti-freeze and thaw materials for permeable pavement
Northwest arid area • *Weather*: Hot in summer and cold in winter, dry climate and scarce precipitation, average annual precipitation <400 mm • *Geology*: Sandy soil, dry and permeable, collapsible loess in some areas, shallow groundwater table • *Regions covered*: The provinces of Shaanxi, Gansu, Ningqing, and Xinxin, with most cities distributed in the stepped plateau or low-level area between mountains • *Selected pilot cities*: Xixian New District, Qingyang, Xining, Guyuan	• Scarce rainfall, severe shortage of water resource • Fragile ecological environment and soil erosion as prominent problems • The adverse effects of collapsible loess to be considered in SCC projects • Relatively backward urban infrastructure • Requirements for water environmental quality improvement	*Strategies* o Storage supplemented with stagnation and infiltration, purification and reuse as the basic strategy o Maximized reuse of harvested rainwater and reclaimed wastewater for alleviating water shortage with attention paid to water quality control o Sufficient attention to the suitability of sponge facilities for collapsible loess areas o Onsite rainwater harvesting and storage to avoid soil erosion caused by surface runoff and centralized rainwater storage, and strengthening slope protection and plantation for soil and water conservation *Main technical measures* o Seepage ponds, wet ponds, seepage wells, and seepage pipes/ditches are less suitable to be used as sponge facilities in collapsible loess areas

(*Continued*)

Table 11.5 Classification of SCC strategies and demonstration technologies based on natural factors (*Continued*).

Regional Characteristics	Major Common Problems	SCC Strategy and Technical Measures
		o Selection of cold-tolerant, drought-tolerant, water-tolerant, salt-alkali-resistant, and pollution-resistant plants to adapt to the saline-alkali soil in this region
Central China warm area • *Weather*: Warm climate, distinct four seasons, uneven rainfall, flooding in summer and drought in spring, annual average precipitation 400–800 mm • *Geology*: Permeable soil, less groundwater recharge, reduced groundwater level, and increased groundwater depth • *Regions covered*: Region 1 covering Beijing, Tianjin, Hebei, Shandong, Henan, Anhui, Jiangsu, and other plain areas; Region 2 covering several foothill areas in Central China • *Selected pilot cities*: Beijing, Tianjin, Qian'an, Hebi in Region 1 and Chizhou, Jinan, Qingdao in Region 2	• Shortage of water sources, serious groundwater overexploitation, and appearance of groundwater funnels • Uneven rainfall, concentrated rainfall in summer, river flooding, and frequent waterlogging in the urban area • Insufficient surface runoff, large sewage discharges, serious water pollution, and deteriorated aquatic ecology • Inadequate urban drainage facilities and insufficient management and control • Siltation and deposits in drainage pipelines due to insufficient slopes, threat of flood, and waterlogging in piedmont areas	*Strategies* o Infiltration and purification as the main measures to facilitate groundwater recharge and rainwater reuse for mitigating water shortage o Enlargement of storage and stagnation space and improvement of rainwater pipe network for rainwater regulation and waterlogging control o Improving urban drainage facilities and upgrading wastewater treatment plants for water pollution control o Water reclamation by advanced treatment to promote water reuse and urban stream replenishment o Region 1 (plain area): Expansion of the coverage of sponge facilities for making the city more elastic

		○ Region 2 (piedmont area): Enlargement of storage space by reasonably utilizing natural terrain and channels for mitigating runoff scouring *Main technical measures* ○ Construction of any kind of sponge facilities as needed ○ Advanced treatment for effluent quality improvement and water reclamation by using constructed wetlands, denitrification filters, etc.
South China rainy area • *Weather*: Hot and rainy, heavy rainfall, annual average precipitation >800 mm • *Geology*: Sticky soil texture and low permeability, high groundwater level, shallow groundwater depth • *Regions covered*: Region 1 as river delta or plain river network areas; Region 2 as mountainous, piedmont, and hilly areas • *Selected pilot cities*: Xiamen, Jiaxing, Shanghai, Ningbo, Fuzhou, Zhuhai, Shenzhen, Sanya in Region 1 and Zhenjiang, Pingxiang, Suining, Wuhan, Changde, Nanning, Gui'an New District in Region 2	• Heavy rainfall, long rainy season, frequent occurrence of flooding • Well-developed natural water systems but poor connectivity of waterways for many existing water systems • High population density, large sewage discharges, and relatively serious problems of black and odorous urban waters • High groundwater level, high soil moisture, and poor soil infiltration	*Strategies* ○ Waterlogging control as the main SCC objective and recommendable use of gray and green sponge facilities and/or structures with the functions of 'stagnation', 'storage' and 'drainage' ○ Region 1 (river delta or plain river network areas): Provision of sufficient storage capacity and improvement of connectivity between water bodies, attention paid to runoff pollution reduction and remediation of black and odorous water bodies ○ Region 2 (mountainous, piedmont, and hilly areas): Safety as the top priority, attention paid to good organization of

(*Continued*)

Table 11.5 Classification of SCC strategies and demonstration technologies based on natural factors (*Continued*).

Regional Characteristics	Major Common Problems	SCC Strategy and Technical Measures
		sponge facilities and structures with a focus on smooth drainage of surface runoff *Main technical measures* ○ Construction of any kind of sponge facilities as needed ○ Avoidance of using infiltration facilities in urban areas of mountainous terraces ○ Recommended technologies for black and odorous remediation: dredging, chemical or biological agent addition, water reoxygenation, bypass treatment, ecological floating beds, sand sedimentation, etc.

effectiveness of the individual technologies adapted to the local conditions, and another is to contribute to achieving the SCC goal set for the selected area.

11.3.3.3 Implementation of SCC pilot projects

The total number of SCC projects completed in selected areas of the 30 pilot cities has amounted to 5221, including implementation of sponge facilities and/or structures associated with the following construction and renovation works:

- urban blocks and residential quarters;
- urban roads;
- parks and green spaces;
- urban water bodies (including rivers and lakes);
- drainage and waterlogging prevention projects (including pipeline network construction and renovation).

The total investment cost for these projects amounted to about 181 billion CHY. This included financial aid from the central government (roughly 20%), local governmental investment (30–40%), and the remaining from other sectors through Public-Private Partnerships (PPP).

11.3.3.4 Establishment of SCC management systems

Along with the implementation of SCC pilot projects, each pilot city is mobilized to establish an SCC management system under the leadership of the municipal government and with the collaborative participation of different departments and related sectors. The SCC management system is not merely for the unified management of the SCC pilot projects but more importantly for the promotion of SCC in long-term urban development. Many pilot cities have established specialized agencies for coordinating the SCC work and put forward new policies on land use, project planning, engineering work permits, project construction, and completion acceptance, etc., so that a green channel can be provided to speed up SCC projects.

Some cities also incorporate the SCC requirements into the governmental approval of all the proposals of construction projects. This is important for the transition of SCC activities from the first 'top-down' stage (pilot stage with central governmental aid) to the second 'bottom-up' stage (local projects and invests responding to the central governmental call), which is the original intention of the government when SCC was proposed.

11.3.3.5 Localized development of SCC technologies

Another important progress is in the localization of SCC technologies in these pilot cities. As shown in Table 11.5, the pilot cities are widely distributed in four typical regions with varied natural conditions and major distinct problems to solve. Therefore, the implementation of all the planned pilot projects is also a process of 'learning while doing' with attention paid to the adaptation of existing technologies and importation of successful experiences from other regions to local conditions and development of new technological measures for solving specific local problems. For example, in the northwest loess plateau area, collapsible soil is the main obstacle in the building of sponge facilities using conventional ways and soil erosion control is a prerequisite for projects implementation. To cope with such topographic and geological difficulties, a comprehensive technical route has been proposed in Qingyang city (No. 12 in Table 11.4). The so-called 'Three-in-One' SCC strategy provides a good combination of various sponge measures with local engineering experiences in water conservation and ecological restoration. Recent findings on local soil permeability and pavement structures have also been incorporated into the technological scheme for runoff flow reduction and pollution control in this city. Another city, Xining (No. 13 in Table 11.4) in the same area has successfully incorporated the method of 'Fish Scale Pits' for construction into SCC measures. This is a method for soil and water conservation on mountain slopes by digging half-moon-shaped pits along the contour lines from top to the bottom, forming a shape of fish scales.

For some cities, such as Shenzhen (No. 9 in Table 11.4) in South China, SCC projects are combined with the reformation of the so-called 'Urban Villages', which are villages that appear on both the outskirts and downtown segments of some Chinese cities as a result of rapid urbanization. This has led to a standardized technological scheme for incorporating various sponge measures into the renovation of shantytowns and dilapidated houses, construction of new residential blocks, road repair, sewer improvement, and elimination of black and odorous waters.

The localized development and integration of SCC technologies and their successful application in pilot SCC projects have led to the formulation of many regional and local standards and specifications, covering the entire process of sponge city planning, design, construction, operation, and maintenance.

11.4 MAJOR ACHIEVEMENTS OF NATIONAL PILOT PROJECTS

11.4.1 Urban ecological space protection and environmental improvement

To enlarge the environmental carrying capacity of cities, the improvement of eco-environmental settings is an important task of SCC. Several pilot cities have achieved much in this aspect.

11.4.1.1 Shanghai Lingang SCC pilot area

Shanghai (No. 3 in Table 11.4) is the largest metropolis in China. For this city, the Lin-Gang Special Area covering about 79 km^2 was selected as a pilot project area for demonstrating SCC technologies. In Chinese language, 'Lin-gang' means 'adjacent to the harbor', which indicates its characteristic location in this coastal city. As this area is within the China (Shanghai) Pilot Free Trade Zone, it is targeted that the Lin-Gang Special Area will be built in a 'CASE of Future', where CASE is the combination of the first letters of Creative, Active, Smart, and Ecology. Following the general plan of urban development, the ecological spatial structure (Figure 11.3) is in the form of 'One Belt, Three Rings, and Multiple Corridors'. The 'One belt' refers to the riverside and coast ecological belt, the three rings are the core activity ring, urban green ring, and ecological leisure ring, while the multiple corridors include a series of green corridors.

Following the abovementioned general target, the protection and improvement of ecological spaces have been set as the main objectives of SCC projects in this area. The largest water body in this area is the 'Di-Shui Lake', which is an artificial lake of round shape built at the beginning of this century, with a water surface area of about 5.66 km^2 and storage volume of about 16.2 million m^3. Many of the SCC projects implemented are associated with Di-Shui Lake protection for point and non-point source reduction and pollution control. For the lake itself, a series of special

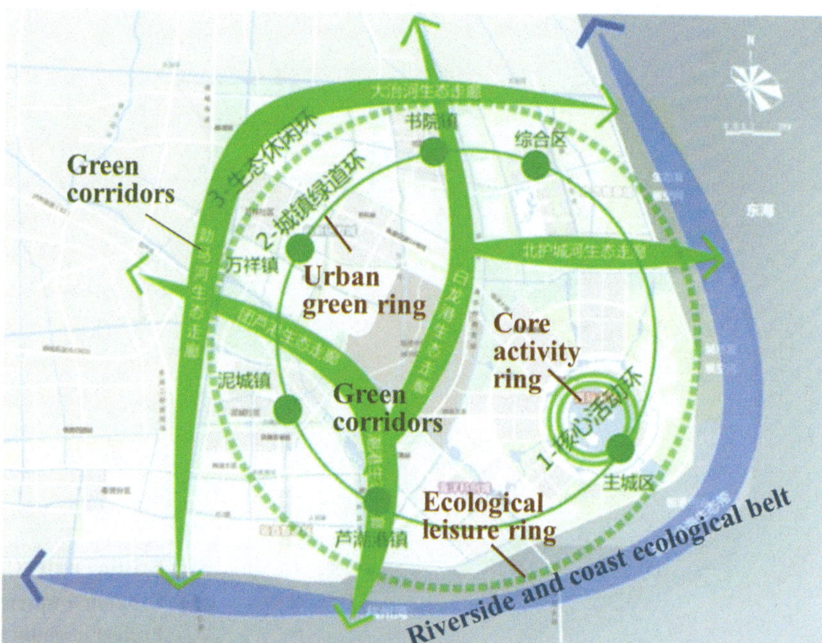

Figure 11.3 Ecological spatial structure of the Lin-Gang Special Area (adapted from Shanghai Sponge City Pilot Construction Self-Assessment Report (unpublished), 2019).

technologies have been developed and adopted for lake ecosystem improvement, such as aquatic animal and plant regulations. In addition, many projects have been implemented for the restoration of ecological shorelines, including the construction of ecological banks for rivers and streams, eco-reformation of existing shorelines for rivers in the main city area, and eco-reformation of soil slopes for rivers in the old city area. After three years of the pilot project implementation, the percentage of ecological shoreline restoration in the main city area has reached 83.6%, while that in the old city area has reached 89.5%, all over the original target of 80%.

SCC projects implementations have also been combined with other projects for the restoration of local streams and construction of new river channels. Figure 11.4 shows the water network in the Lin-Gang Special Area after the pilot SCC projects. At present, the water coverage is 12.06% in the pilot project area.

The protection and enlargement of ecological spaces have notably improved the ecological setting of this area and significantly improved the local aquatic environmental quality. According to the water quality monitoring results in the recent 12 months at 16 locations, including designated river sections and lake sampling points, waters in this area all meet the requirements of the surface water

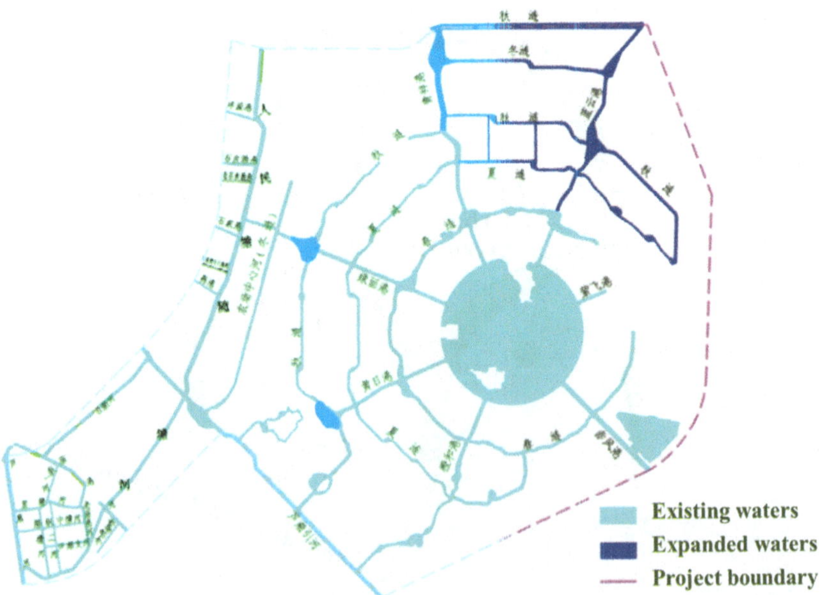

Figure 11.4 Water network in the Lin-Gang Special Area after the SCC projects (adapted from Shanghai Sponge City Pilot Construction Self-Assessment Report (unpublished), 2019).

quality standards (GB 3838-2002) of Grade IV (suitable for industrial water use and recreational waters with no direct human contact) or above.

11.4.1.2 Shenzhen SCC pilot area in Guangming district

In the southern metropolis of Shenzhen (No. 9 in Table 11.4), a pilot project area was selected in Guangming district as shown in Figure 11.5. This area is located in the northwest of the city and almost covers two sub-basins of the branch streams of the Maozhou river, namely, the Dongkeng watershed and Ejing watershed, with a total area of 24.6 km^2. The areas shaded in blue in Figure 11.5 are protected areas designated by the local government for which the waterfront ecological spaces should be well protected and controlled. They are also the targets of ecological protection and aquatic environmental improvement for the pilot SCC projects.

The basic strategy for achieving this target is the integration of gray-green technologies with emphasis on a complete restoration of ecological shorelines of water bodies and recovery of a healthy aquatic environment. In the Ejing watershed, for example, the stream water is used occasionally due to heavy pollution from, first, the high pollutant loading of the watershed, uncontrolled discharge of wastewater into the stream channel, and second the very low regional ecological capacity. After a full investigation and analysis of the existing problems, it is decided that 100% restoration of the ecological shorelines is the main measure for the direct protection of the stream, while clear zonation of land reserves and areas for ecological protection is the main measure for optimizing the local ecological setting (Figure 11.6). Of course, all these are associated with a series of measures for point and non-point source pollution reduction and control.

Due to the pilot project implementation, the water environmental condition of the area has been significantly improved, including the stream water and the upstream reservoirs.

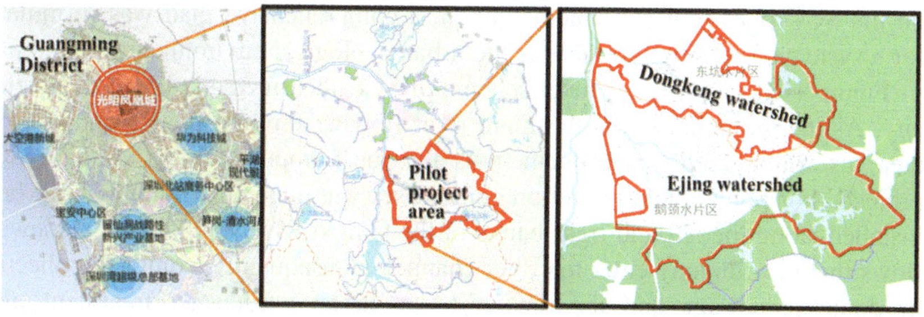

Figure 11.5 Pilot project area in Guangming District, Shenzhen City (adapted from Shenzhen Sponge City Pilot Construction Self-Assessment Report (unpublished) 2019).

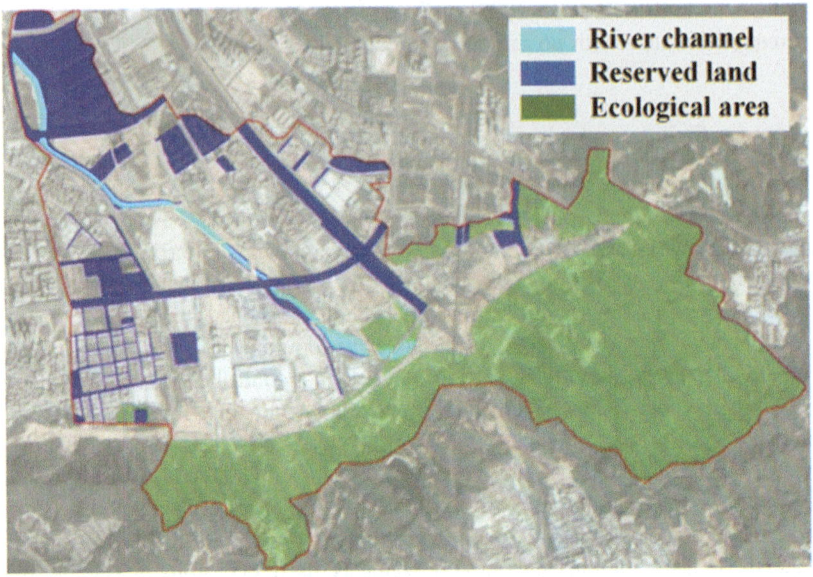

Figure 11.6 Local zonation of reserved land and ecological area in the Ejing watershed (adapted from Shenzhen Sponge City Pilot Construction Self-Assessment Report (unpublished) 2019).

11.4.1.3 Chizhou city as an SCC pilot area

Chizhou (No. 8 in Table 11.3) is a city located in a hilly area adjacent to Changjiang (Yangtze) River. It has limited land available for urban development. As shown in Figure 11.7, the SCC pilot projects have targeted the whole city for the protection of its unique ecological setting characterized by a series of river flows in mountain valleys surrounding the city. However, due to rapid urban development with increasing population and economic activities, the water environment and ecological quality have considerably deteriorated in the past decades. Capitalizing on the SCC pilot project implementation, a comprehensive plan was formulated for a systematic solution to recover the urban ecological environment.

Figure 11.8 shows the general scheme of SCC measures employed in Chizhou city. As the Qingxi River is the main river flowing through the expanded urban area, its water quality improvement is the most important target of the SCC projects. For point source pollution control, sewage interception mains have been installed along the left and right banks of the Qingxi River to intercept all sewage flows that may directly enter the river channel. Consequently, complete collection of domestic sewage and its transfer to an urban wastewater treatment plant (WWTP) downstream of the city could be achieved. To prevent pollution of the Yangtze River, the final receiving water of this city, the WWTP effluent is not discharged directly into the river but further purified by a wetland system before

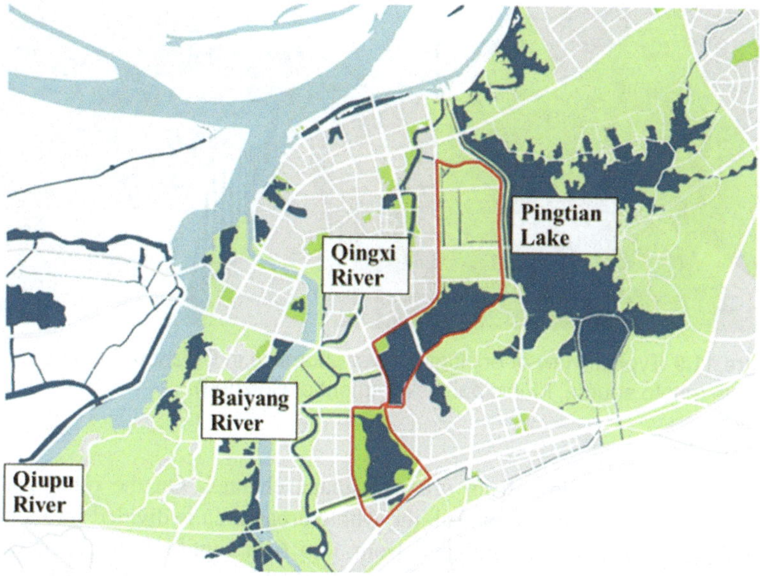

Figure 11.7 Ecological setting of Chizhou characterized by a series of river flows in maintain valleys surrounding the city (adapted from the Sponge City of Small and Medium-Sized Cities Along the River with Ecological Resources Construction – Chizhou Pilot Summary Report (unpublished) 2018).

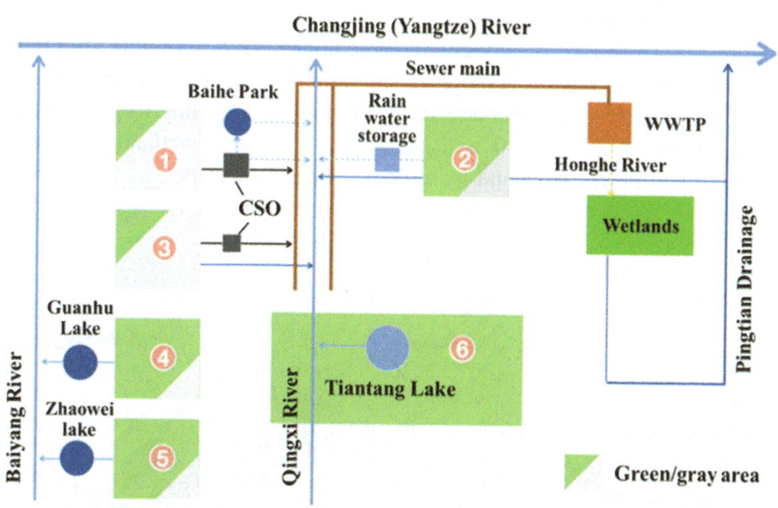

Figure 11.8 Ecological setting of Chizhou characterized by a series of river flows in maintain valleys surrounding the city (adapted from the Sponge City of Small and Medium-Sized Cities Along the River with Ecological Resources Construction – Chizhou Pilot Summary Report (unpublished) 2018).

Figure 11.9 Typical view of ecological river shoreline in Chizhou City after the SCC projects (photo source: Municipal Government of Chizhou).

the final discharge through an urban drainage channel. On the other hand, for non-point source pollution control, the whole urban area is divided into a number of drainage zones (areas numbered from 1 to 6 in Figure 11.8), each with the implementation of green-gray sponge facilities that function for surface runoff reduction as well. It can be seen that the large area surrounding the Tiantang Lake (area No. 6 in Figure 11.8) is almost completely covered by green space or served with other green measures. Such a systematic solution has effectively reduced pollutant loading to rivers and, overall, improved the urban ecological and environmental conditions.

Figure 11.9 shows a typical view of the ecological river shoreline after the SCC projects. About 9.5 km of ecological shoreline has been reconstructed and/or restored, and a percentage increase in the ecological shoreline for the whole city of up to 70.4%. Figure 11.10 is a panoramic view of a wetland park for both WWTP effluent purification and its reuse for urban landscape.

Figure 11.10 The panoramic view of a wetland park in Chizhou City for WWTP effluent purification and urban landscape (photo source: Municipal Government of Chizhou).

11.4.2 Urban flood control and waterlogging prevention

11.4.2.1 Significant improvement of urban flood control and waterlogging prevention capabilities

The implementation of SCC projects in all the pilot cities has significantly improved the capabilities for urban flood control and waterlogging prevention. Based on comprehensive analyses of historical data and urban flooding risks under varied conditions, many of the cities have upgraded their targets of flood control and set detailed countermeasures in different areas.

Figure 11.11 shows the example of Jiaxing City (No. 11 in Table 11.3), where the whole city area has been divided into first-level catchment zones according to the waterways (natural streams and/or artificial drainage channels), and further into 38 second-level catchment zones related to the layout of the urban drainage network. Within each catchment zone, sponge facilities have been introduced for runoff reduction and local waterlogging prevention. Through the SCC projects, the flood control area has been increased by 9.43 km^2 (about 8.7% of the constructed area in the central city), and the flood protection embankment has been increased by 8.84 km. All these measures have collectively increased the capability for smooth drainage of stormwater from the whole urban area. An evaluation of these measures demonstrates that the city can now withstand floods with up to a 100-years return period.

Another example is Pingxiang City (No. 7 in Table 11.3) located in the mountainous Jiangxi Province and with average annual rainfall over 1600 mm. Two major rivers flowing through the urban area should have provided favorable drainage conditions for the city, but with large mountain river catchment areas in the upstream, the river channels are often full due to the large volume of

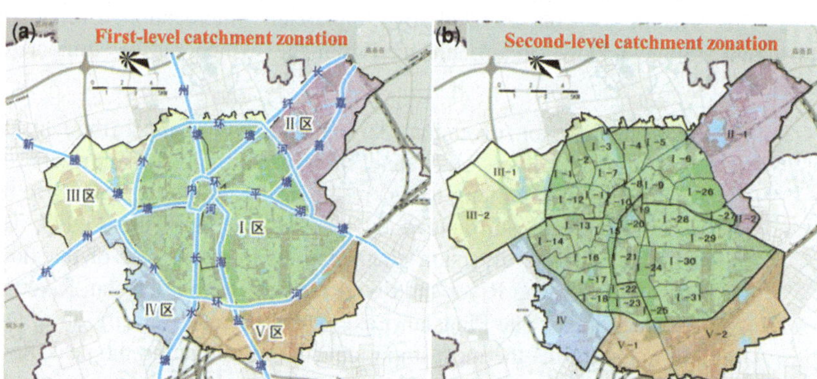

Figure 11.11 Zonation of first-level catchment second-level catchment areas in Jiaxing City (adapted from Jiaxing Sponge City Pilot Construction Self-Assessment Report (unpublished) 2018).

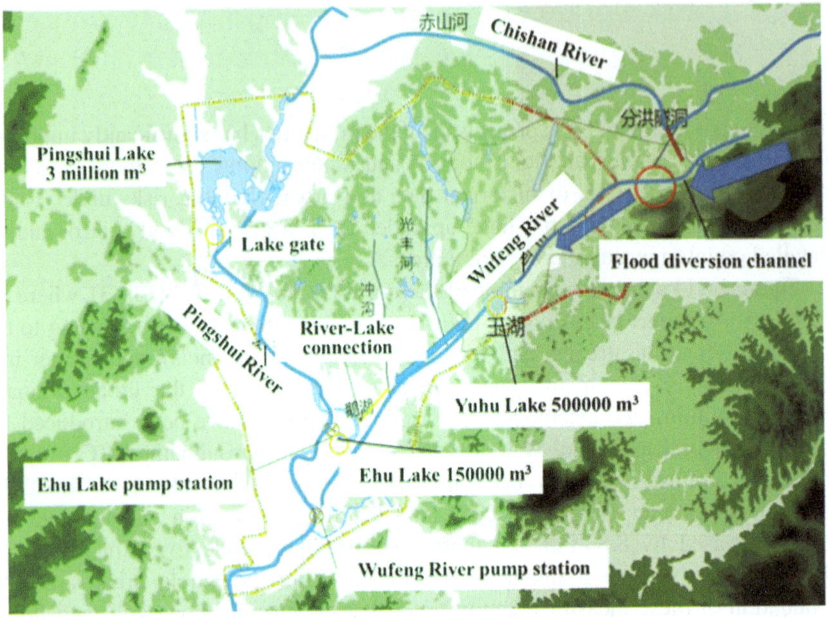

Figure 11.12 Layout of the large-scale drainage system for Pingxiang City (adapted from Pingxiang Sponge City Pilot Construction Summary Report (unpublished) 2018).

incoming flood flows and are no longer able to receive local surface runoff smoothly from the city area. Therefore, in addition to various sponge measures to reduce urban runoff, a large-scale drainage system has been implemented.

Figure 11.12 shows the layout of the large-scale drainage system for Pingxiang City. The system is characterized by 'Upstream interception/diversion + Midstream storage + Downstream enhanced drainage' which can be summarized as below:

(1) Upstream interception/diversion. An existing river channel (the Chishan River) is used as a diversion channel in between the two main rivers (the Pingshan River and Wufeng River) in the upstream before flowing to the urban area (the area surrounded by the red line in Figure 11.11). The main function of the diversion channel is to intercept part of the flood volume to the Wufeng River and divert the flow to the Pingshui River.

(2) Midstream storage. The Pingshui Lake with a storage capacity of up to 3 million m^3 performs the most important role in mitigating the peak flood flow from the upstream of the Pingshui River and that diverted from the upstream of the Wufeng River. Several smaller lakes, such as the Yuhu Lake (500 000 m^3) and the Ehu Lake (150 000 m^3) in the urban area also provide considerable storage volume.

(3) Downstream enhanced drainage. Several pumping stations that could be put into operation in flooding seasons to smoothly discharge local runoff into the river channels downstream of the urban area.

The implementation of the large-scale drainage system, along with various sponge measures and facilities for runoff reduction in the urban area, has significantly increased the capability for flood control and waterlogging prevention in Pingxiang City.

Other SCC pilot cities have formulated similar plans for combating flooding and waterlogging and taken various measures according to local conditions.

11.4.2.2 Effects of flood control and waterlogging prevention in SCC pilot cities

In August 2019, Typhoon Lekima attacked the east coastal area of mainland China through a track of heavy storms, as shown in Figure 11.13. A number of the pilot cities, such as Qingdao (No. 7 in Table 11.4), Shanghai (No. 4 in Table 11.4), Ningbo (No. 5 in Table 11.4), and Jiaxing (No. 11 in Table 11.3), are within the affected area, and thus underwent a test of the effects of flood control

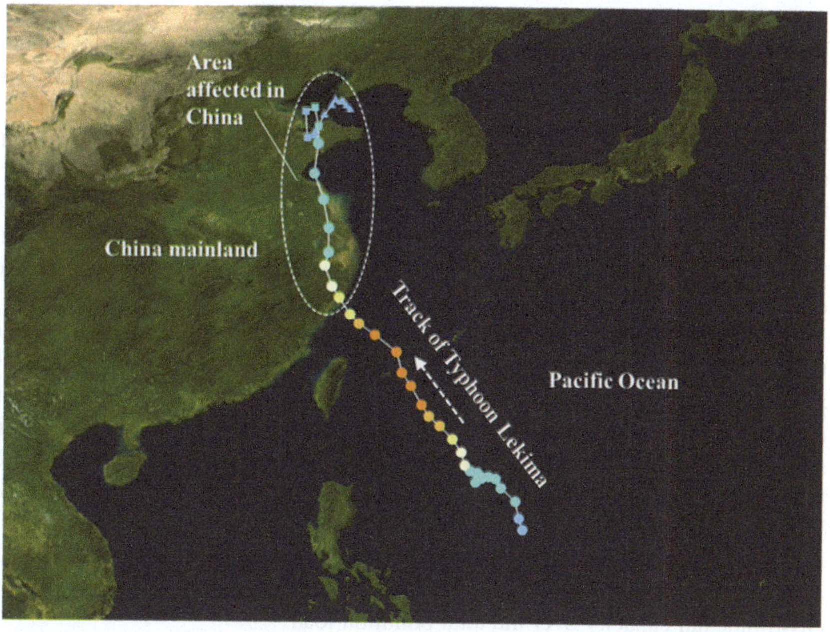

Figure 11.13 The track of Typhoon Lekima towards the coastal area in China (adapted from Wikipedia, https://en.wikipedia.org/wiki/Typhoon_Lekima).

immediately after the SCC projects. In some of these cities, the daily rainfall was as high as 269 mm but none of them were heavily flooded in contrast to what had happened before the SCC projects under similar storm conditions. In Jiaxing City, for example, the total volume of runoff flow reduction and/or storage through all the sponge facilities was equivalent to 100 mm of rainfall, which significantly mitigated the peak flood flow and enabled the urban drainage system to function well for smooth discharge of surface runoff.

Pingxiang City, introduced in section 11.4.2.1, also underwent a test of its capability to combat heavy storms in the same year. From the 7th to 9th of July 2019, the total rainfall in over 24 consecutive hours was as high as 256.8 mm. Thanks to the efficient functioning of the drainage system shown in Figure 11.12, all the storage facilities accommodated about 400 million m^3 of runoff flow and protected the city area well from severe flooding.

In 2020, although the vast area of south China, especially the Yangtze River basin, experienced unexpected heavy rainfall and flooding, most of the SCC pilot cities showed much better conditions due to the implementation of various sponge facilities and innovative urban drainage systems.

11.4.3 Restoration of urban black and odorous rivers

11.4.3.1 Water environmental restoration of Nakao River in Nanning City

The Nakao River, with a total length of 6.6 km and flowing through the Nanning urban area, used to be heavily polluted and had no sufficient natural flow. However, it received urban drainage from various sources. According to long-term monitoring data, before the SCC projects, the overall river water quality had always been worse than the lowest permissible level, namely, worse than Class V of the national environmental standard for surface water (GB 3838-2002). The river became black and odorous, the dirtiest water in the city. Therefore, a systematic plan was formulated (Figure 11.14) for its pollution control and water quality improvement within the SCC pilot project scheme.

The main pollutants that severely deteriorated the river water quality include COD, NH_3-N, TN, and TP. Taking COD as the representative pollutant for pollution analysis, as shown in Figure 11.14, it has been demonstrated that domestic pollutants from urban sewer and drainage systems are the sole point sources (2.12×10^7 kg COD). Moreover, non-point sources (6.15×10^5 kg COD) from urban runoff cannot be neglected due to the difficulty of effective control. Furthermore, the significant contribution of the internal pollution from the river sediments (3.1×10^6 kg COD) to the river pollution problem is notable. To eliminate the pollution, measures should be taken principally to provide an environmental capacity over the total pollution load.

For point source pollution control, various measures have been adopted for the effective reduction of domestic pollution loads, such CSO overflow interception and storage, improvement of the urban drainage system, sewer interception, and

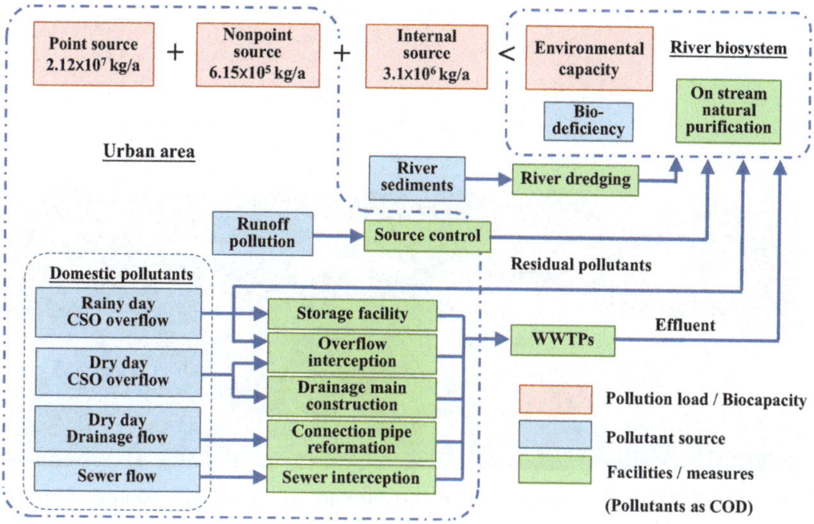

Figure 11.14 Systematic plan for Nakao River pollution control and water quality improvement (adapted from Nanning Sponge City Pilot Construction Self-Assessment Report (unpublished) 2018).

so on. The intercepted flows are then sent to the WWTPs for secondary and advanced treatment. In the whole urban area, various recommended sponge measures have been taken for non-point source reduction from urban surface runoff on rainy days. The internal pollution source, namely the river bottom sediments, has been removed by river dredging.

In addition to the control and reduction of pollution sources, measures have also been taken to improve the ecological condition of the river channel. One such measure is to increase the frequency of river water replenishment by using the WWTP effluent, which can supply a steady flow of 50 000–70 000 m^3/d to significantly improve the river's hydraulic conditions. Advanced treatment is conducted in the WWTP to upgrade the effluent quality to a suitable quality standard for river water replenishment. Another measure is to improve the environmental conditions of the river channel by the restoration of ecological shorelines along the whole river length.

For the success of the project, the Nanning municipal government has introduced a PPP engineering model for project design, construction, operation, and maintenance. All the related construction works, including river improvement, pollution control, ecological restoration, treatment facility construction, landscape improvement, and information management have been put into a package of 1.19 billion CHY investment. This has provided a new engineering model for SCC projects in China.

As a result, the Nakao River has been completely changed from a typical black and odorous urban drainage channel into a beautiful stream and landscape attraction

Figure 11.15 Ecological shoreline of the Nakao River (photo source: Municipal Government of Nanning).

in Nanning City. Figure 11.15 shows an overview of the ecological shoreline, and Figure 11.16 shows an overview of the Nakao River after the SCC projects implementation.

Figure 11.16 Appearance of the Nakao River after the SCC projects implementation (photo source: Municipal Government of Nanning).

11.4.3.2 Water environmental restoration of Chuanzi River in Changde City

Changde (No. 15 in Table 11.3) is a water city in south China. The Chuanzi River is one of the streams flowing through the central city area. With a total length of 17.3 km, the riverbank area is home to more than 300 000 residents and inevitably receives pollutants due to the concentrated human activities. Before the SCC projects, the two banks of the river were steep hard surfaces in most sections with 118 drainage outlets discharging not only rainwater but also sewage directly into the river channel. The direct waste discharge turned the river into an urban drainage channel.

Targeting at a complete change of the water environment of the Chuanzi River, the following measures were taken for this river within the scheme of the SCC projects in Changde City.

(1) River channel improvement. The narrow river channel in many sections has been broadened. Also, the original high flood dike which completely isolated the city from the river was lowered to a more natural level without affecting the function of the river channel for the discharge of flood flow smoothly downstream. Incorporated with many sponge measures in the surrounding area for the reduction and storage of surface runoff, ecological shorelines have replaced the hard river banks, which not only made the river more natural but also increased the ability to prevent waterlogging.

(2) Supplement of river water flow. With very limited natural replenishment in the dry season, the river did not have sufficient water flow most of the time. This was one important reason why the river became black and odorous occasionally. To overcome this problem, clean water is pumped from a nearby larger river (at a flow rate of 1 m^3/s or about 86 000 m^3/d) to enlarge the river flow. Through many onsite facilities for runoff storage, additional water can also be supplied for supplementing the river flow. This has effectively provided a necessary base flow for the river and prevented water stagnation in the river channel.

(3) Water pollution control. For point source reduction, all the rainwater pumping stations along the river have been reformed so that no overflow may occur in dry days. Also, CSO during moderate rainfall can be treated to a certain extent before entering the river. Direct discharge of sewage and the polluted first flush rainwater into the river channel through the drainage outlets have also been prevented by flow interception. The newly constructed WWTP (50 000 m^3/d) has sufficient capacity to treat the sewage and intercepted wastes. Non-point source pollution control could also be achieved by the introduction of various sponge measures in the urban area.

(4) River ecological improvement. To improve the river ecological condition, over 370 000 m^3 of sludge was dredged out of the river channel to eliminate internal pollution sources. Along with the river channel improvement, the cross-section of the river bank was remodeled into a double trapezoidal shape. Between the design normal water level and low water level, palm mats and plant rollers were installed, on which moisture- and drought-tolerant plants were planted for bank protection and provision of habitat for waterfowls and other amphibians. Ecological floating islands were also installed to increase self-purification.

(5) Waterfront space creation. Scenic footpaths, waterfront platforms, and bicycle lanes were built along the embankments of the river to provide waterfront spaces and parks for residents to enjoy the restored waterscape. Integrated with the leisure facilities in the surrounding area, the Chuanzi River has become a famous attraction in Changde City.

Figure 11.17 shows the waterscape of the Chuanzi River in Changde City after the SCC projects implementation. This project has provided a successful case to

Figure 11.17 Chuanzi River in Changde City after the SCC projects (photo source: Municipal Government of Changde).

transform a once black and odorous river into a multifunctional open river park with diversified waterfronts and entertainment spaces.

11.4.3.3 Water environmental restoration of Meishe River in Haikou City

Haikou is the capital city of Hainan Province. As shown in Figure 11.18, the Meishe River, with a total length of 23.86 km and a catchment area of about 53 km^2, flows through the central urban area and finally discharges to the Qiongzhou Strait of the Beibu Gulf in the South China Sea. Therefore, the water quality of this river strongly affects the living environment of the city as well as the coastal environment. Unfortunately, the river had long been an urban drainage channel receiving treated and untreated sewage from about 130 outlets. The total volume of sewage

Figure 11.18 Project area for the restoration of Meishe River in Haikou City (figure by authors).

flowing into the river could amount to 50 000 m^3/d and formed the major external pollutant source. The water quality of the Meishe River reportedly used to be very poor, with a DO concentration as low as 0.87 mg/L, and NH_3-N and COD_{Cr} concentrations as high as 15.2 and 61.5 mg/L, respectively. The river had been entered on the 'Black List' for black and odorous water and its treatment and remediation was urgently required.

Although Haikou City was not selected as an SCC pilot city, the municipal government implemented a project for the water environmental restoration of the Meishe River by an integration of the following measures:

(1) Sewage system upgrading and improvement. Based on a comprehensive survey and diagnosis of the problem within the entire catchment area of the river, priority was given to the upgrading and improvement of the domestic sewage system, including the construction of new sewer interception pipelines, modification of the sewer collection network, optimization of sewage treatment schemes by combining decentralized treatment and centralized treatment on an as-needed basis, and upgrading and/or enhancement of wastewater treatment processes. To effectively prevent the direct river water pollution from CSOs, eight storage facilities with a total capacity of 150 000 m^3 were built for storing large volumes of the highly polluted first flush rainwater and/or mixed wastewater and then gradually sending the stored water to the wastewater treatment facilities for pollutant removal.
(2) Non-point source pollution control by various sponge measures. Within the entire catchment area of the Meishe River, various sponge facilities were built for onsite infiltration, runoff reduction, rainwater harvesting, storage, purification, and reuse, covering residential communities, public buildings, roads, squares, and parks. Non-point source pollution has thus been effectively reduced, and the harvested and treated water could be used for replenishing the river flow.
(3) Restoration of aquatic ecosystems. The river channel was largely modified under the concept of ecological design. The whole length of the shoreline was ecologically reshaped, waterfront spaces were created, blue-green corridors were built, and specific habitat zones were provided. Under the subtropical local climate conditions, recovery of natural wetlands and building constructed wetlands were proven to be most effective for the restoration of healthy river ecosystems.

After the project implementation, the Meishe River turned from a severe black and odorous water body into a beautiful urban waterfront area. The water quality steadily meets Class V of the Chinese surface water standard. Figure 11.19 is a panoramic view of the current Meishe River flowing through the central city.

Figure 11.19 A panoramic view of the current Meishe River flowing through the central city of Haikou (photo source: Municipal Goverment of Haikou).

REFERENCES

Brown R. and Clarke J. (2007). The Transition Towards Water Sensitive Urban Design: the Story of Melbourne, Australia, Report No. 07/01. Facility for Advancing Water Biofiltration, Monash University, Clayton, VIC.

General Office of the State Council. (2015). Instruction on Promoting Sponge City Construction (in Chinese), General Office of the State Council, People's Republic of China, Beijing.

Hoang L. (2016). System interactions of stormwater management using sustainable urban drainage systems and green infrastructure. *Urban Water Journal*, **13**(7), 739–758.

MOHURD. (2014). Technical Guidelines (Tentative) for Sponge City Construction – Low-Impact Development Rainwater System Construction (in Chinese). Ministry of Housing and Urban-Rural Development, Beijing, People's Republic of China.

MOHURD. (2015). Action Plan for Prevention and Control of Water Pollution (in Chinese). Ministry of Housing and Urban-Rural Development, Beijing, People's Republic of China.

MOHURD. (2018). Standard for Assessment of Sponge City Construction (GB/T51345-2018, in Chinese). Ministry of Housing and Urban-Rural Development, Beijing, People's Republic of China.

SEPA and GAQSIQ. (2002). Environmental Quality Standards for Surface Water (GB3838-2002, in Chinese). State Environmental Protection Administration and General Administration of Quality Supervision, Inspection and Quarantine, Beijing, People's Republic of China.

Shanghai Municipal Housing and Urban-Rural Construction Management Committee. (2019). Technical Specification for Construction of Sponge City (in Chinese). Tongji University Press, Shanghai.

State Council. (2015). Action Plan for Prevention and Control of Water Pollution (in Chinese). The State Council, Beijing, People's Republic of China.

US EPA. (2000). Low-Impact Development (LID): A Literature Review (EPA-841-B-00-005). United States Environmental Protection Agency, Office of Water, Washington, DC.

Chapter 12

LID-BMPs for urban runoff pollutant source control

Mingyi Jia[1] and Haifeng Jia[2]

[1]Environmental Technology, Wageningen University & Research, Wageningen, 6708GA, The Netherlands
[2]School of Environment, Tsinghua University, Beijing 100084, China

12.1 INTRODUCTION

In recent decades, urbanization has become prevalent globally. Since 1950, the urban population has increased from 751 million to 4.2 billion. In 1950, 30% of the world's population lived in urban areas, and by 2018, it had increased to 55%. It is predicted that 68% of the world's population, around 6.7 billion, will be residing in the urban area by 2050. Although the level of urbanization, in terms of speed and magnitude, differ regionally, it is one of the most urgent challenges around the world for its strong environmental, social, and economic implications. In particular, the rapid development of urban areas and increasing anthropogenic activities have greatly altered the natural landscape. Forests, farmlands, and grasslands are gradually displaced by streets, buildings, and parking lots, transforming the previous natural surfaces to impervious surfaces. This ongoing urbanization has brought adverse impacts on the environment, which will, in turn, affect the health and wellbeing of urban residents.

The marked alteration of natural landscapes, natural processes, and natural resource consumption by urbanization directly affects the four spheres of the Earth: atmosphere, lithosphere, biosphere, and hydrosphere, corresponding to the

Editors: Xiaochang C. Wang and Guangtao Fu
doi: 10.2166/9781789060768_0325

environmental resources of air, soils, water, and biodiversity. In this chapter, only the impact on water environments is discussed.

12.1.1 Urban water environment problems and pollution sources

12.1.1.1 Urban water environment problems

From the water environment perspective, urbanization has induced the following problems: change of hydrological dynamics and water quality degradation. The increase in impervious surfaces can alter surface runoff dynamics and runoff generating processes by reducing infiltration, reducing surface storage capacity, decreasing evapotranspiration, increasing runoff velocity, increasing runoff volume, and increasing discharge rates and flood peaks (Walsh *et al.*, 2012). Consequently, the subsurface flow dynamics are also affected. Less water enters the groundwater aquifers due to impervious surfaces, while water supply or sewage infrastructure contributes to water recharge (Jacobson, 2011). The degradation of water quality is another outcome of urbanization. Different materials and structures used in the built environment give rise to multiple sources and types of pollution. These diffuse contaminants and emerging priority pollutants deteriorate the stormwater quality and further intensify the problem with runoff. When discharged into the water cycle, they can have severe impacts on aquatic animals and human health (McGrane, 2016).

Many urban water bodies are heavily polluted by the discharge of polluted water from different sources due to the lack of effective urban water management and control, especially in developing countries. In China, highly polluted water bodies that contain excessive amounts of nutrients and wastes are referred to as 'black and odorous water' (Chen *et al.*, 2018). The degraded water quality not only impairs the aquatic ecology but also threatens public health. Hence, water quality restoration and ecological rehabilitation of urban rivers have become one of the main concerns for city managers and planners.

12.1.1.2 Urban water environment pollution sources

The sources of pollutants in the inflow can be categorized into point and non-point sources. In the Clean Water Act section 502, a point source is defined as 'any discernible, confined and discrete conveyance, including but not limited to any pipe, ditch, channel, tunnel, conduit, well, discrete fissure, container, rolling stock, concentrated animal feeding operation, or vessel or other floating craft, from which pollutants are or may be discharged. This term does not include agricultural stormwater discharges and return flows from irrigated agriculture'. In short, point source pollution refers to pollution from a single, identifiable source where pollutants are collected and discharged. Examples of point sources of water pollution include sewage treatment plants, factories, oil refineries, paper, and pulp mills, and all kinds of manufacturing. In urban areas, municipal sewage treatment plants and industrial plants are the two main point sources. The discharges from these plants can transport pollutants, such as heavy metals, pathogens, nutrients, oxygen-depleting substances, and toxic chemicals, into waterways causing severe water quality degradation and health hazards. With the process of urbanization, point source pollution has been increasingly monitored and regulated. In the US, the effluent limit is established and managed by EPA in two ways. The first is technology-based control, which is based on available technologies and preventative measures. The other is water quality-based control, which is based on the protection of receiving water bodies. Similarly, many countries around the world have set strict effluent limits to ensure a certain degree of pollutant removal before sewage enters the water bodies. However, under certain circumstances, the discharges can still be problematic. Currently, the most commonly used sewer system is the combined system, which carries domestic wastewater, industrial wastewater, and stormwater together to the treatment plant. However, with the combined sewer system, problems could occur during the rainy and snow-melting season as excessive volumes of inflow can cause sewer overflow, resulting in excess inflow being discharged directly into the river without any treatment. According to US EPA, the combined sewer overflows are considered major sources of water pollution for around 772 cities in the US that have combined sewer systems (Tibbetts, 2005).

In comparison, non-point sources (NPS) of pollutants come from diffuse sources and cannot be attributed to a single source. According to the EPA, NPS pollution results from rainfall or snowmelt moving over and through the ground. When stormwater or snowmelt moves across the underlying surface, it picks up and carries away the natural or man-made pollutants accumulated during the dry season, and eventually brings these pollutants into waterways, such as lakes, rivers, and groundwater. The increase of impervious surfaces has severely exacerbated the problems as the runoff will stay on the ground longer, carrying more pollutants. Also, with urbanization, industrialization, and technological development, more sources and a variety of pollutants have emerged. Some common non-point source pollution in urban areas includes:

- excess fertilizers, herbicides, and insecticides from urban green areas, such as parks, golf courses, and grasslands in residential areas;
- oil, grease, and toxic chemicals from urban runoff and energy production;
- sediment from improperly managed construction sites and eroding streambanks;
- salt from irrigation practices;
- bacteria and nutrients from pet wastes and faulty septic systems;
- atmospheric deposition and hydromodification.

Nowadays, NPS pollution has become the largest source of water impairment in many cities and countries, which is mainly due to the difficulty of controlling and monitoring diffuse pollution sources. The Montana Department of Environment

Quality in the US (2012) claimed that 75% of Montana's assessed rivers and streams and 45% of its lakes, reservoirs, and wetlands fail to meet state water quality standards largely due to the impacts of NPS pollution. Urban water runoff can degrade waterways by altering flow quality, volume, and pattern. With urban runoff that contains high pollutant loadings entering the waterways, the surface water quality is severely affected due to the inflow of a variety of pollutants. A previous study by Mallin *et al.* (2009) reported that impervious surface coverage is positively correlated with the pollutant concentration in the runoff. Runoff from areas with greater imperviousness carries a higher amount of biochemical oxygen demand (BOD), bacteria, orthophosphate, and surfactants into the surface waters, all of which can be attributed largely to anthropogenic sources. Similarly, after monitoring the water quality parameters of a river passing through five towns, Glińska-Lewczuk *et al.* (2016) also concluded that urbanization is likely to be the primary cause for surface water quality deterioration due to significant loads of pollutants and the increased impervious cover. The deteriorated water quality can endanger aquatic species. With a high level of nutrients inflow, such as nitrogen and phosphate present in the water bodies, eutrophication can occur, triggering explosive algae growth that can deplete oxygen in the water. Moreover, pathogens and toxic chemicals can spread diseases, causing the death of aquatic species.

12.1.2 Urban runoff pollutants characteristics

12.1.2.1 Pollutants in stormwater runoff

To tackle non-point source pollution, the European Commission has set forth a control policy in the European Water Framework Directive 2000/60/EC (WFD), which requires the monitoring of surface water quality throughout Europe. The WFD established a list of priority pollutants that need to be reduced or eliminated. However, not all the pollutants in the list are relevant to stormwater because of their specific industrial sources. Moreover, the pollutants exist in both the water phase and solid phase, but WFD focused primarily on the water phase, so the priority pollutants identified in WFD are not sufficient for monitoring and evaluating stormwater quality (Eriksson *et al.*, 2007). Hence, different approaches were used to select priority pollutants in stormwater, including a scientific approach CHIAT.

The Chemical Hazard Identification and Assessment Tool (CHIAT) is a risk assessment tool for chemicals advocated by the European technical guidance document (TGD) and is used widely by governments, industries, etc. It consists of five steps: (1) source characterization, (2) recipient, exposure targets, and criteria identification, (3) hazard and problem identification, (4) hazard assessment and (5) stakeholder involvement.

Using this method, a sample list of priority pollutants is selected by Eriksson *et al.* (2007), which include the following categories: (1) common water quality parameters, (2) metals, (3) PAHs, (4) herbicides, and (5) miscellaneous.

As shown in Table 12.1, common water quality parameters include biological oxygen demand (BOD), chemical oxygen demand (COD), suspended solids (SS), nitrate, phosphate, and bacteria (i.e. fecal coliforms/pathogens). These pollutants are generally very common and high in concentration, but are easier to treat. The rest of the pollutants are lower in concentration but are either difficult to treat, which means long-term accumulation, or toxic and, hence, have harmful impacts.

The list could not include all the priority pollutants in stormwater, and an increasing number of new pollutants are emerging and threatening the ecological value of the water cycle. Aiming for better efficiency of priority pollutants removal in stormwater, the utilization of conventional ways of urban stormwater management can no longer meet the need for addressing the current water issues.

12.1.2.2 Runoff water quality spatiotemporal variability

Runoff quality monitoring and sampling are required for better control and management. However, due to the dynamic nature of runoff flow, it is challenging to qualify the runoff, which varies both spatially and temporally.

Temporally, the pollutant concentrations in the runoff are typically higher with the first few rainfall events following a drying period (among storm variability) as well as at the start of a rainfall event (within storm variability), a process referred to as the 'First Flush' (Bertrand-Krajewski *et al.*, 1998). The accumulated pollutants on impervious surfaces during the dry season are quickly washed off at the beginning of a rainfall event, resulting in high pollutant concentrations in the runoff generated. By contrast, in the middle/end of a rainfall event or a wet season, the runoff carries low levels of pollutants as the majority have been washed away by the first flush. This results in a disproportionately large discharge of pollutants in proportion to the volume during the first flush. This phenomenon is significant for practice for it indicates that by capturing the initial runoff with only a small storage tank, the majority of the pollutants can be removed, while the remaining runoff contains fewer pollutants and could be drained directly. Thus, the 'Half-Inch' rule, which requires that the first half-inch of runoff should be treated by LID-BMPs, has been adapted in many places. However, as the concept of first flush is highly site-specific, the first flush-based sizing criteria vary from place to place. Chang *et al.* (1990) demonstrated that the first flush effect is correlated with impervious surfaces and that the effect is more evident at highly impervious sites but is weak at sites with low imperviousness. In addition to the first flush effect, previous research also indicates that the pollution level in runoff is inversely correlated with rainfall intensity and duration. At a higher rainfall duration or intensity, the average concentration of pollutant constituents is significantly lower. Moreover, at increased rainfall intensity, the magnitude of the first flush effect is decreased (Schiff *et al.*, 2016).

Table 12.1 List of common and priority pollutants in stormwater with their concentration and justification (adapted from Eriksson *et al.*, 2007 and Birch, 2012).

Type	Constituent	Concentration in Stormwater (μg/L)	Justification
Common water quality parameters	BOD/COD	–	General water parameter
	Suspended solids	–	General water parameter
	Total Nitrate (TN)	–	General water parameter
	Total Phosphate (TP)	–	General water parameter
	Faecal/Pathogens	–	General water parameter
Heavy metals	Zn	0–38,061	*P*
	Cd	<0.1–700	*P, CMR*
	Cr	<0.5–4200	*P*, occurs widely as anionic form in natural waters
	Cu	<0.5–6800	*P, T*
	Ni	5–580	*P, CMR*
	Pb	<0.5–2764	*P, CMR*
	Pt	–	*P, CMR*
PAHs	Naphthalene	0.006–49	Representative indicator of PAH (two ring compound), *P, B, CMR*
	Pyrene	0.0001–120	Representative indicator of PAH (four ring compound), *P, B*
	Benzo(a)pyrene	0.00015–300	Representative indicator of PAH (five ring compound), *P, B, CMR*
Herbicides	Glyphosate	<0.05–1.92	*B, T*
	Pendimethalin	–	*P, B*
	Phenmedipham	–	*P, B*
	Terbuthylazine	0.017–0.16	Representative indicator of triazines, *P, B*
Other substances	Di (2-ethylhexyl) phthalate (DEHP)	3.0–44	*P, B*
	Methyl tert-butyl ether	–	Technical problems (odour)
	Nonylphenolethoxylates	<0.04–23	CMR
	Pentachlorophenol	–	*P, B*
	Polychlorinated biphenyl 28	–	*P, B*, Water solubility

Note: CMR= Carcinogenic/mutagenic/hazardous to reproduction and/or endocrine-disrupting. P = Persistent, B= Bioaccumulating, T = High acute aquatic toxicity.

Table 12.2 Typical pollutant concentrations in urban runoff for different land uses (adapted from US EPA, 1999 and Jia *et al.*, 2013a, 2013b).

Pollutant (units)	Median Event Mean Concentration for Land Use		
	Residential	Mixed	Commercial
Biological oxygen demand (mg/L)	10	7.8	9.3
Chemical oxygen demand (mg/L)	73	65	57
Total suspended solids (mg/L)	101	67	69
Total Kjeldahl nitrogen (μg/L)	1900	1288	1179
Nitrate and nitrite (mg/L)	0.736	0.558	0.572
Total phosphorus (mg/L)	0.383	0.263	0.201
Soluble phosphorus (mg/L)	0.143	0.056	0.080
Total lead (μg/L)	144	114	104
Total copper (μg/L)	33	27	29
Total zinc (μg/L)	135	154	226

The spatial variation of runoff results from different land covers, such as highways, parking lots, residential areas, commercial areas, parks, open land, rangelands, etc. Pollution levels and pollutant types of runoff vary between different surface covers. For example, on roads, highways, and bridges, the level of pollution is usually high, containing contaminants such as oils and grease, heavy metals, and road salts. In parks and residential areas, the pollution level is typically lower, with common pollutants such as fertilizers, pesticides, and herbicides, etc. Table 12.2 shows a summary of some typical pollutant median event mean concentrations in urban runoff from residential, mixed, and commercial land use.

12.2 LID-BMPs TECHNOLOGY FOR URBAN RUNOFF CONTROL

12.2.1 Concepts and evolution of LID-BMPs techniques

The urban stormwater runoff management systems in use are often designed with a performance standard based on historical data. However, as urbanization and climate change introduce new emerging pollutants as well as alter rainfall-runoff regimes in urban areas, the current stormwater pattern no longer conforms with the pattern projected based on historical data. Consequently, conventional stormwater management systems can no longer effectively and sustainably handle current and future urban runoff. Urban stormwater management should take into account and be able to handle the impacts of climate change on stormwater runoff peak flows and volumes as well as rainfall intensities (Zahmatkesh *et al.*,

2014). Therefore, new land development methodology and engineering should be proposed and applied to relieve the stress on the environment and local ecology.

Currently, countries around the world are exploring new ways of stormwater management. Along with that, many new concepts are developed and adopted. Some widely known concepts include Water Sensitive Urban Design (WSUD) in Australia, Sustainable Urban Drainage Systems (SUDS) in the UK, Low Impact Urban Design and Development (LIUDD) in New Zealand, Sponge City in China, and Low Impact Development Best Management Practice (LID-BMPs) in the USA.

As a widely used and well recognized term internationally, LID-BMPs is considered an innovative stormwater management scheme that could be the potential solution for the current urban water issues. Low Impact Development (LID) is a philosophy for development that focuses on specific sustainable water conservation goals. The goal of LID is to use and manage developed lands ecologically to minimize negative environmental impacts. LID designs attempt to mimic pre-development hydrologic conditions as closely as possible. The desired result is not only a reduction in stormwater runoff from the site but improvements in water quality as well. The objectives of LID design are to use natural approaches, such as preserving and recreating natural landscape features, and incorporates as little impervious surface as possible to create functional and aesthetic sites that retain and treat runoff. LID consider rainwater as an important source of surface and underground flows and, thus, should be retained in the city and treated properly, instead of a waste product that needs to be drained down from the city as quickly as possible, as it does with the traditional urban sewerage system (US EPA, n.d.).

Best Management Practices (BMPs) are often used to describe a type of practice or structural approach to pollution prevention. In the context of stormwater control, BMPs are a 'technique, measure, or structural control that is used for a given set of conditions to manage the quantity and improve the quality of stormwater runoff in the most cost-effective manner' (US EPA, 2004). It represents the practices determined to be the most efficient, practical, and cost-effective measures. The term was first used in the US to describe a type of water pollution control. In the field of stormwater management, practitioners have used the term BMPs to describe both structural or engineered control devices and systems (e.g. retention ponds) to treat polluted stormwater, as well as operational or procedural practices (e.g. minimizing the use of chemical fertilizers and pesticides). There are a variety of BMPs available, depending on pollutant removal capabilities.

Integrating the two concepts, LID-BMPs represent all the BMPs for urban stormwater runoff control using the LID strategy. The basic principle of LID-BMPs design is to manage rainfall at the source using uniformly distributed decentralized micro-scale controls. As an integrated approach for comprehensive stormwater management planning, it aims at prevention first, and mitigation second (Jia *et al.*, 2013a, 2013b). Comparing with conventional stormwater design, the LID-BMPs approach advocates for more careful site design in the planning phases. It addresses stormwater management through small, cost-effective landscape features located at the lot level, including open space, rooftops, streetscapes, parking lots, sidewalks, and medians. The purpose of the site design is to preserve as much of the site in an undisturbed condition, and when a disturbance is unavoidable, reduce the on-site impacts to the soils, vegetation, and aquatic systems. While LID-BMPs primarily focus on the design characteristics at the individual lot level, improvements are expected as cumulative impacts are integrated over the entire developed areas (Jia *et al.*, 2012).

Incorporating simple LID-BMPs into a site design can significantly reduce runoff flows and pollutant loads. A study by Liu *et al.* (2015) has found, after modelling 16 LID-BMPs scenarios, that the various levels and combinations of LID-BMPs practices can reduce runoff volumes by 0–26.47%, total nitrogen by 0.3–34.2%, total phosphorus by 0.27–47.41%, total suspended solids by 0.33–53.59%, lead by 0.3–60.98%, BOD by 0–26.7%, and COD by 0–27.52%. LID-BMPs also have beneficial functions on enhancing groundwater recharge, reducing stormwater and combined sewer system expenditure, reducing energy use, improving air quality, enhancing aesthetics and property values, and increasing educational opportunities (Jia *et al.*, 2013a, 2013b). LID-BMPs may also provide increased resilience to future climate change. Studies have found LID practice to be effective in moderating potential climate change impacts such as extreme temperatures and increased surface runoff (Pyke *et al.*, 2011).

LID-BMPs is a versatile and flexible approach that can be applied equally well to new developments, urban retrofits, and redevelopment/revitalization projects.

12.2.2 Functionality and characteristics of typical LID-BMPs techniques

LID-BMPs techniques can be, in general, categorized into non-structural LID-BMPs and structural LID-BMPs. Non-structural LID-BMPs are management-based practices, which focus on broader planning, policies, and principles. It refers to stormwater runoff management techniques that use natural measures to not only mitigate stormwater runoff impacts by reducing pollution levels but also prevent runoff generation through developing lands with non-conventional practices. Non-structural practices do not require extensive construction efforts as they do not involve fixed, permanent facilities.

Structural LID-BMPs are engineering-based practices with explicit physical forms. Structural LID-BMPs practices are implemented to control water quality and quantity through the combination of a series of physical, biological, and chemical processes, including detention/retention, settling, absorption, infiltration, flocculation, and/or biological uptake.

The integration of engineering- and management-based techniques can minimize the overall stormwater-related impacts and provide an entire technical system for urban stormwater management. The following subsections will discuss the types of structural and non-structural LID-BMPs.

12.2.2.1 Structural LID-BMPs

As LID-BMPs infrastructures often have multiple functions in terms of stormwater management, there is no defined way to classify and conceptualize these infrastructures. Currently, LID-BMPs techniques are often classified according to their functionalities, forms, or types. The main functionalities of LID-BMPs include volume/peak rate reduction and quality improvement by infiltration/filtration, detention and retention, storage, and reuse. The forms of LID-BMPs can be categorized based on whether they can be purchased and put into use directly, such as rain cisterns, or they require the growth of vegetation before they can serve their purposes, such as constructed wetlands. The type-based classification can be divided into point LID-BMPs (e.g. bioretention, rain barrel), linear LID-BMPs (e.g. vegetated filter strip, grass swale), and area LID-BMPs (e.g. green roof, porous pavement). In the following subsection, structural LID-BMPs techniques are clustered based on their dominant functional and/or performance characteristics and then explained in detail. Overall, there are five major classes of techniques, namely detention/retention, infiltration, filtration, vegetated, and storage and reuse (Table 12.3).

In Table 12.4 and Figure 12.1, some primary pollutant removal mechanisms are displayed.

Table 12.3 List of main structural LID-BMPs categorized by their major functions (summarized by authors)

Major Function	Structural LID-BMPs
Detention/retention	Dry detention ponds Dry extended detention ponds Wet retention ponds
Infiltration	Infiltration trench Infiltration basin Porous pavement
Filtration	Surface sand filters Media filter Underground vault filters
Vegetated	Grass swales Filter strip/buffer Bioretention cells Stormwater wetlands Green roof
Storage and reuse	Rain barrels and cisterns

Table 12.4 Primary removal mechanisms in structural LID-BMPs (adapted from Scholes *et al.*, 2005)

Removal Processes	Key Measurements
Sedimentation	Settling velocity
Adsorption	Adsorption coefficient
Microbial degradation	Biodegradation rate
Precipitation	Solubility
Filtration	Function of adsorption and precipitation
Volatilization	Henry's Law constant
Photolysis	Photodegradation rate
Plant uptake	Bioaccumulation

12.2.2.1.1 Detention/retention LID-BMPs

In this class, the three widely used LID-BMPs techniques are dry detention ponds, dry-extended detention ponds, and wet retention ponds. Stormwater pond refers to an engineered structure to collect rainfall and surface runoff, and treat water. Detention systems are designed to collect a specified volume of runoff, temporarily retain it for a short period, and slowly release it. In comparison, retention systems are designed to capture the runoff and retain them for on-site use until displaced in part or total by the next runoff event. Both systems are used for flood control and the settling of suspended particles in the runoff.

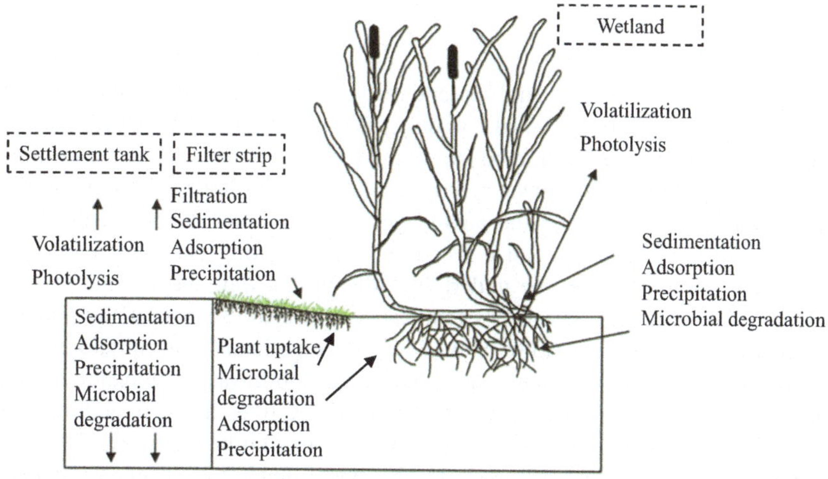

Figure 12.1 Primary stormwater pollutant removal processes in structural LID-BMPs (adapted from Scholes *et al.*, 2005).

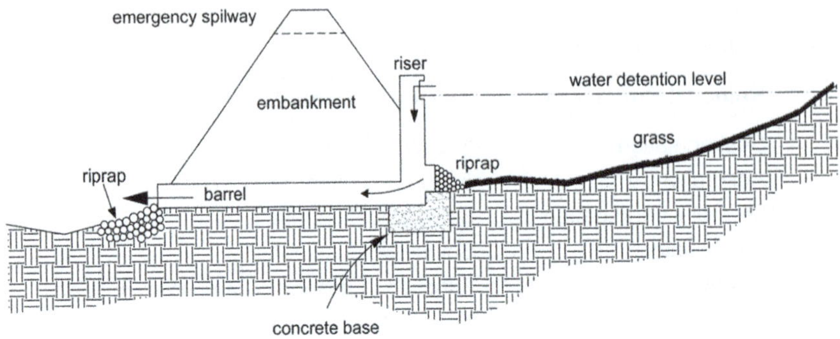

Figure 12.2 A schematic overview of a detention pond (adapted from Rivanna-Stormwater at https://rivanna-stormwater.org/what-happens-to-the-rain/bmp-photos-descriptions/#BMP1).

Detention systems are temporary holding spaces for stormwater that store peak flows and, subsequently, release it to lessen the demand for treatment facilities, especially during large storm events and, thus, prevent flooding. Detention ponds are also known as dry ponds because they are completely drained between storms. Figure 12.2 illustrates the typical profile of a detention pond. According to Stormwater Design Guidelines, detention ponds are designed to release the runoff within 48 hours, with no more than 50% of the total volume draining in the first 16 hours. It can help improve the water quality by allowing the particles to settle to the bottom. Detention systems have low construction and maintenance costs and are good for sites where infiltration is not possible.

12.2.2.1.2 Infiltration LID-BMPs

An infiltration system has a high potential for stormwater runoff control through disposal at the local site level. Widely used infiltration LID-BMPs include infiltration trenches, infiltration basins, and porous pavements. They are designed to capture a volume of stormwater runoff, retain, and infiltrate it into the ground. Infiltration LID-BMPs have the advantages of both water quantity and quality control. Water quantity control is done by capturing surface runoff and infiltrating this runoff into the underlying soil. In this way, the volume of water discharged to receiving streams is reduced, which can also reduce the potential impacts caused by excess flows as well as reduce pollutant concentrations in the receiving stream. The pollutants are removed as stormwater percolates through the various soil layers, during which the particles are filtered out. In addition, soil microbial degradation of organic pollutants in the infiltrated stormwater also occurs during the process. The system also has a number of secondary benefits including increased recharge of underlying aquifers, increased base flow levels of nearby streams, and water quality treatment. Despite these advantages, infiltration systems also display some drawbacks. First, infiltration LID-BMPs may not be suitable in areas where groundwater is a primary source of drinking water due to the possibility of pollutant migration. Second, infiltration LID-BMPs can experience reduced infiltration capacity and even clogging due to excessive sediment accumulation. Moreover, system performance is limited in areas with poorly permeable soils.

Porous pavement is an infiltration system that can help reduce runoff by infiltrating stormwater runoff to the ground through a permeable layer of pavement. It is considered an alternative to the traditional pavement where little runoff can permeate to the ground. The alternative materials can be pervious asphalt, pervious concrete, interlocking pavers, and plastic grid pavers, all of which allow runoff to seep through the surface down to the underlying layers of soil and gravel. In addition to reducing runoff, porous pavement can also help filter out pollutants in the stormwater. It can also reduce the need for road salt as well as decrease the costs for residential and commercial construction by reducing the need for some conventional drainage features.

12.2.2.1.3 Filtration LID-BMPs

A filtration system is primarily a water quality control device that uses a combination of granular filtration media, such as sand, gravel, organic material (e.g. peat, compost), or other acceptable treatment media to remove a part of the pollutants in runoff. There are a wide range of filtration LID-BMPs in use (Claytor & Schueler, 1996). Some frequently used ones are surface sand filters, media filters, and underground vault filters. These systems are commonly installed in areas with high pollution potentials, such as industrial areas and parking lots, or highly urbanized areas where land availability and costs limit the use of other LID-BMPs techniques. Filters should be placed off-line and are sometimes designed to intercept and treat only the first flush while bypassing larger storm flows. The benefit of using a filtration system in highly urbanized areas is that it can be placed under parking lots or in the basement of buildings, which could largely reduce or even eliminate the costly land requirement. However, an implication for this kind of installation could be the difficulty for maintenance. The main function of the filtration system is quality control, hence, to also control the runoff quantity, additional LID-BMPs techniques should be incorporated, for example by providing additional storage volume in a pond or basin, providing vertical storage volume above the filter bed, or allowing the water to be temporarily detained in a place before being discharged to the filter.

12.2.2.1.4 Vegetated LID-BMPs

Typical vegetated systems include grass swales, filter strip/buffer, bioretention cells, stormwater wetlands, and green roofs. A vegetated system is designed explicitly to capture and treat the stormwater runoff within dry or wet cells formed by check dams or other means, as well as systems designed to convey and treat either shallow flow (swales) or sheet flow (filter strips) runoff. Such a system is also called 'biofilter' because the grasses and vegetation filter the stormwater as it flows over them (Field & Tafuri, 2006). Besides treating and retaining runoff, vegetated systems also have some additional functions such as to enhance vegetation diversity and wildlife habitat in urban areas as well as increase site aesthetics.

Constructed stormwater wetlands are engineered systems designed and built to mimic natural wetlands for water quality improvement. They treat and contain surface water runoff pollutants and decrease loadings to surface water. Pollutants are removed through sedimentation, adsorption to soil particles, and infiltration through the soil and other media, microbial decomposition, and plant uptake. Constructed stormwater wetlands incorporate the natural functions of wetland to aid in pollutant removal from runoff, while also allowing quantity control by providing a significant volume of ponded water above the permanent pool elevation. In constructed wetlands, flow is controlled so that the water is spread evenly around the wetland vegetation. They provide simple and effective treatment for domestic and agricultural wastewater. On the other hand, construction wetlands also have limitations in their application. A water balance must be performed to determine the availability of water to sustain the aquatic vegetation during dry periods.

Green roof is another typical engineering system placed on the top of conventional waterproofed roof surfaces of buildings. Green roofs have been used in urban areas for centuries to retain water for plant uptake while reducing runoff volumes, peak runoff rates, and pollutant loads from rooftops. Green roofs use natural sedimentation to filter pollutants. Despite the number of forms and types of green roofs available, usually, a distinction is made between extensive, intensive, and biodiversity/wildlife roofs. Extensive green roofs are designed to have minimal maintenance requirements and, thus, vegetations do not need irrigation or fertilization. The composition of extensive green roofs is lightweight layers of free-draining materials that support low-growing, hardy, drought-tolerant vegetation. In comparison, intensive green roofs have a relatively thicker substrate and, thus, support a wider range of plant types, such as trees, shrubs, and grasses. With increasing attention on biodiversity issues, biodiversity/wildlife roofs emerged and gaining more popularity. The design of biodiversity/wildlife roofs either aims at replicating specific habitat needs of a single or a small number of species or creating a range of habitats that can maximize the number of species that inhabit and use the roof.

12.2.2.1.5 Storage and reuse LID-BMPs

In an urban community, rain barrels and cisterns are widely used facilities for storage and reuse. Rain barrels are used to prevent runoff from roofs entering the storm drain system. The collected runoff can be used by homeowners to provide water for gardens, lawns, and flower beds. Businesses usually build cisterns to trap runoff. These methods do not remove pollutants so sediments will have to be removed periodically from the barrel or cistern.

12.2.2.2 Non-structural LID-BMPs

According to the Pennsylvania stormwater Best Management Practice Manual (Department of Environmental Protection, 2006), some common non-structural practices include:

(1) Protect sensitive and special value features. Before site development, it is a priority to prevent disturbance to and alteration of natural features that provide important stormwater functions, such as wetlands, riparian habitats, and natural flow pathways.
(2) Minimize disturbance and maintenance. During construction, it is important to reduce site grading and vegetation removal as well as minimizing soil compaction and ensuring soil quality. When disturbances are unavoidable, amendments to restore permeability should be considered. Vegetation selected for revegetated areas should require little maintenance with fertilizers, herbicides, and pesticides.
(3) Reduce impervious surfaces. When drafting development planning, an important principle is to reduce impervious surfaces as much as possible, especially to avoid the direct connection of impervious surfaces. For example, we can try to reduce street and square areas by minimizing its widths and lengths.
(4) Concentrate and cluster. Use of 'smart growth' planning techniques to plan and zone for concentrated development patterns that can accommodate reasonable growth and development; construction clustered on the smallest area possible to reduce overall disturbed area.
(5) Decentralize, disconnect, and distribute. Minimize stormwater volumes by disconnecting rooftop leader and disconnecting roads and driveways from stormwater collection points. Redirecting runoff to vegetated areas or non-vegetated structural BMPs.
(6) Source control. Regular street sweeping to remove large pollutant particles, preventing large debris from clogging stormwater management systems.

12.2.3 Case study: An LID-BMP treatment train system

As described above, LID-BMPs are effective to control urban runoff. Several individual LID-BMPs arranged in a series is called an LID-BMP treatment train, which is believed to enhance the treatment effects (Mid-America Regional Council/American Public Works Association, 2012; Minnesota Pollution Control Agency, 2014). The field tests of a full-scale LID-BMP treatment train system can be very useful in the planning and design phase when implementing

LID-BMPs. However, these field data are currently lacking. Thus, a field monitoring project for the LID-BMP treatment system was carried out to test and evaluate the effectiveness of the treatment train system on urban runoff control (Jia *et al.*, 2015a). The project is based on the campus of the Guangdong Vocational College of Environmental Protection Engineering in the city of Foshan in Guangdong province, China, where a demonstration project of the LID-BMP treatment train system was constructed in 2012.

12.2.3.1 Site and field monitoring

The campus is located in a humid subtropical monsoon climate zone with annual averages of 1567.1 sunshine hours, 23.2°C, and 350 frost-free days. The constructed LID-BMP treatment train system comprises three grassed swales, a buffer strip, a bioretention cell, two infiltration pits, and a constructed wetland, which are connected in series (Figure 12.3). The influent stormwater runoff originates from four tennis courts with a total area of 2808 m^2 and eight basketball courts with a total area of 4864 m^2. These sites were chosen because of their high surface imperviousness and heavy uses.

For this field monitoring, a total of 19 storm events were monitored in two years (2012 and 2013), of which 10 produced runoff and nine did not. Hence, only the ten complete storm events were used for performance evaluation. However, this method, to some extent, underestimates the effectiveness of the LID-BMPs as smaller rainfall events with total control of runoff is not taken into account. The sampled rainfall events varied from small (19.3 mm) to very heavy (61.6 mm). The average rainfall intensity ranged from 10.6 to over 30 mm/h. This corresponds to a rainfall return period of about 3–5 months for the city of Foshan.

12.2.3.2 Results and discussion

In the study, runoff quantity control and pollutant removal performance of LID-BMPs are assessed.

For quantity control, the findings show the bioretention cell to be efficient in reducing the runoff peak and volume, whereby 50–84% of the runoff volume and 47–80% of the peak flow was reduced. In comparison, the swale was less effective for relatively large storm events as only 17–79% of the runoff volume and 9–74% of the peak flow was reduced, on average. It is believed that the rectangular shape and long length of the bioretention cell are the determinants of its excellent volume reduction. It was also observed that the flushing of unsettled media material posed a problem for the bioretention cell as the outflows contained a high level of sediments. Hence, a settlement period is required after the construction of a bioretention cell. Moreover, the 'flower soil', a nutrient-rich planting soil, used in the bioretention cell also increased the phosphorus content in the media and, therefore, the effluent phosphorus concentrations from the bioretention cell.

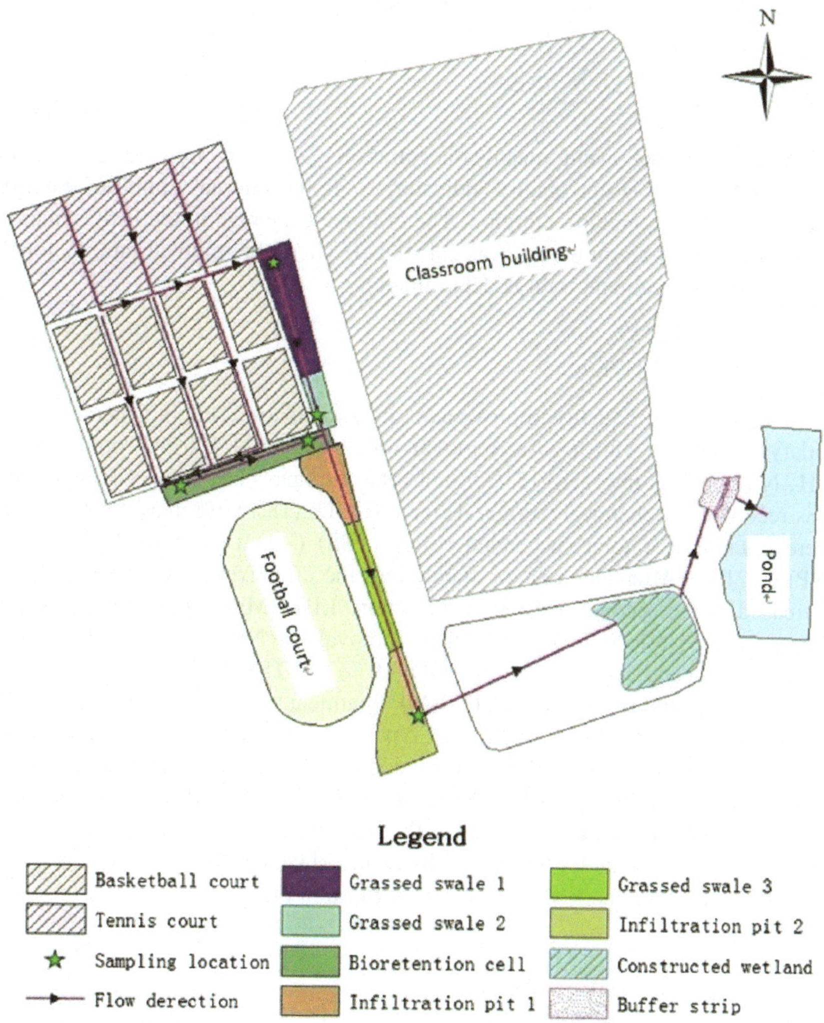

Figure 12.3 Placement of the LID-BMPs treatment train system at the campus site (reprinted by permission from Springer Nature: Jia *et al.*, 2015a).

The pollutant removal efficiency was calculated using the 'sum of loads or SOL' and the 'efficiency ratio or ER' methods, based on the event mean concentration or EMC. The SQL method is based on a mass balance computation comparing the input pollutant loads to the LID-BMP and the loads leaving the LID-BMP during the sampling period. The removal rates are calculated as follows:

$$R = 1 - \frac{SL.out}{SL.in} \tag{12.1}$$

The EMC method for an individual storm event is calculated by:

$$EMC = \sum_{t=0}^{T} C_t Q_t \Delta t / \sum_{t=0}^{T} Q_t \Delta t \qquad (12.2)$$

where EMC is the event mean concentration, Q_t is the flow rate, C_t is the concentration of a water quality indicator, Δt is the internal sampling duration, and T is the total sampling duration. The arithmetic average of the EMCs, namely *EMCAVE*, for all storm events was calculated. Then, the removal rate, R, for the LID-BMP or a treatment train is determined by:

$$R = 1 - \frac{EMCAVE.out}{EMCAVE.in} \qquad (12.3)$$

In terms of water quality improvement, the bioretention cell, in general, showed satisfying results, with a good removal for zinc (nearly 100%), copper (69%), NH_3-N (ammonia-nitrogen) (51%), and total nitrogen (TN) (49%). Fair removal was recorded for chemical oxygen demand (COD) (18%), whereas poor removals were obtained for total suspended solids (TSS) (−11%) and total phosphorus (TP) (−21%). After one year of stabilizing, the effectiveness of the LID-BMPs increased in the second year. As to the entire LID-BMP treatment train system, the aggregated effect showed excellent removal for NH_3-N (73%), TN (74%), and TP (95%) and fair removal for COD (19%) and TSS (35%).

The assessment results of the LID-BMP treatment train system provide valuable information on how to link the different types of LID-BMP facilities and maximize the integrated effects on urban runoff control. Therefore, it is concluded that compared with individual LID-BMPs, the use of an LID-BMP treatment train design would significantly enhance the runoff quantity and quality control efficiency. In this study, good results were obtained for even a sub-system including only bioretention, swales, and infiltration pits.

12.3 LID-BMPS PLANNING

12.3.1 Planning methodology

12.3.1.1 Principle

The main principles of LID-BMPs planning to be taken into account include the following (Jia *et al.*, 2015b):

(1) Preserve the original terrain. Construction of LID-BMPs should avoid affecting and disturbing areas with important natural stormwater functional values, such as floodplains, wetlands, riparian areas, etc., as well as areas with stormwater impact sensitivities (steep slopes, adjoining properties). This measure should be considered on a site-by-site and area-wide basis. New developments should not encroach upon areas with

special and sensitive resources so that their natural stormwater functions will not be lost and, therefore, not further intensify the stormwater impacts. Natural flow pathways should be identified, protected, and utilized in the planning. Therefore, before the LID-BMPs planning, the local terrain should be examined and weighted according to their functional values to avoid any disturbance to special and sensitive natural features.

(2) Controlling stormwater at the source. For a more effective treatment, one of the main goals is to mitigate the hydrological impacts of the various anthropogenic activities close to the source of generation. Particularly at sites with high anthropogenic activities, incorporating hydrological functions such as interception, depression, and infiltration is essential.

(3) Thinking of micro-management. LID-BMPs are often designed to be small-scale, distributed in residential areas, parking lots, shopping malls, and roads. Although the effect of each individual LID-BMP is limited, the synergetic effects are expected to have significant impacts on the control at the watershed-scale.

(4) Using simplistic, natural practices combined with traditional stormwater controls. Although the use of only natural practices has the potential to be more effective in restoring natural hydrologic functions at the sites when dealing with large volumes and high peak flows, the function of the conventional system is also indispensable. Thus, incorporating natural practices with conventional stormwater control systems is a more practical approach, and by doing so, the scale of the conventional system can be decreased.

(5) Using better site design and creating a multifunctional landscape. A good site design aims to incorporate multiple functions into the planning and, thus, can simultaneously mitigate the stormwater issues by mitigating flooding, reducing pollutant loads, preserving natural areas, increasing property values, and saving costs.

12.3.1.2 Technical route

In general, the route of planning analysis would include the following steps, which is illustrated in Figure 12.4 (Jia *et al.*, 2015b):

(1) Set runoff control goals. The first step in the LID-BMPs planning is to set the runoff control goal, which should include targets for runoff volume control, peak flow control, water quality control, and ecology and landscape expectations. The runoff control goals should be coordinated with the overall planning objectives of the development.

(2) Baseline analysis. The basic data of the development site should be collected and analyzed. The related data include weather conditions (historical rainfall, temperature, etc.), river and hydrology (geometry of

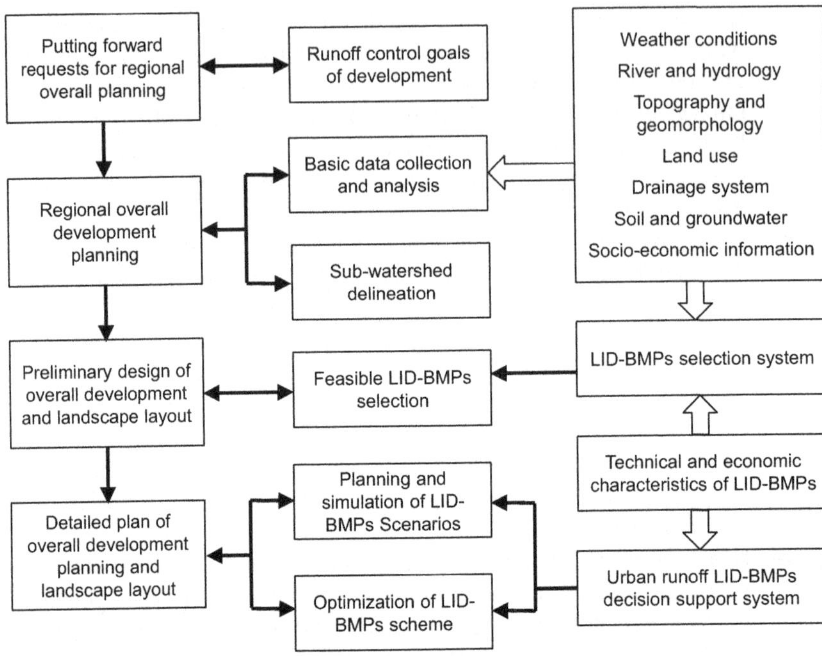

Figure 12.4 LID-BMPs planning process schematic for urban runoff control (figure by authors).

receiving waters, recorded flow and water quality), topography, land use, natural and manmade drainage systems, soil and groundwater, relevant socio-economic information (e.g. population, zoning, regulations, and ordinances), and potential pollution sources.

(3) Sub-watershed delineation. A sub-watershed is the basic unit for runoff control analysis. The delineation should be done according to the overall regional planning, topography, the current and future land use, and the underlying surface characteristics. These characteristics for each sub-watershed are then analyzed to help identify feasible sites for different LID-BMPs.

(4) Feasible LID-BMPs selection. After analyzing the basic data of the development site and considering the runoff control goals, the overall planning, and landscape planning, the candidate LID-BMPs can be screened and a set of suitable ones are selected for the specific development site. For the selection process of feasible LID-BMPs, a multi-criteria scheme was developed as discussed in the latter section.

(5) LID-BMPs scenario planning and simulation. Based on the selection results of runoff control goals and feasible LID-BMPs, a variety of scenarios for LID-BMPs implementation can then be devised under the umbrella of

the overall site planning schemes. Next, scenario analysis could be conducted to evaluate runoff control effects. The LID-BMPs planning layout schemes which comply with the urban runoff control goals can thus be selected.

(6) Optimization of the implementation plan. To optimize the LID-BMPs planning option with respect to control efficiency and associated costs, the key parameters of each LID-BMP (such as the geometric size, media composition, etc.) need to be optimized with respect to goals of best runoff controls benefits and/or least costs. Optimization modules such as the SUSTAIN framework can be used to provide the decision-makers with choice options in terms of the most efficient and/or the least cost implementation of the LID-BMPs at the site. However, it should be integrated and implemented with related planning, such as overall development and landscape planning.

12.3.2 Feasible LID-BMPs selection

12.3.2.1 The importance of LID-BMPs selection

The selection of site-specific LID-BMPs techniques is a crucial step in planning and design. Currently, there are many available LID-BMPs techniques, and each has its own intrinsic technical and/or economic characteristics and limitations for implementation. Thus, at different installation sites, the selection for feasible LID-BMPs is complex and intricate given the variability of site conditions, performance, and costs of LID-BMPs. With the increasing need for comprehensive LID-BMPs selection in recent years, a number of BMP planning/design support systems have been developed. Among them, GIS-based Best Management Practice Decision Support System (BMPDSS) (Cheng *et al.*, 2009; Jia *et al.*, 2012) and the System for Urban Stormwater Treatment and Analysis Integration (SUSTAIN) are the two most notable systems. Both systems provide multiple-scale (site, region, and watershed) applications, detailed BMP process simulations, and cost optimization analyses. More recently, a front-end BMP tool was also developed that includes process-based BMP and multi-criteria analysis for BMP selection (Viavattene *et al.*, 2010).

All of the abovementioned planning systems are comprehensive and highly technical analysis tools, which are very helpful for selecting the final BMP planning strategy prior to design and implementation. However, under some circumstances, a simpler system is sufficient, much like using a one-dimensional as opposed to a two- or three-dimensional model. In addition, a simple analysis tool can also be used as a preliminary or 'screening' tool which can facilitate the complex modeling process involving the use of BMPDSS, SUSTAIN, etc. An example of such a simple system would be an Analytical Hierarchal Process (AHP) (Young *et al.*, 2009). AHP can be used as an analytical framework for

BMP selection through priority ranking of stormwater control objectives and BMP performance metrics.

12.3.2.2 Multi-Criteria index ranking system for LID-BMPs selection

LID-BMPs selection is a multi-objective decision problem. It involves the performance effects (such as runoff quantity control, runoff quality control, and additional ecological benefits) and the costs (such as construction cost and operation cost). Here, a simple system for LID-BMPs selection known as Multi-Criteria Index Ranking System (MCIS) is developed (Jia *et al.*, 2013a, 2013b), which will be described below.

MCIS can either be used as a simple and comprehensive screening tool for preliminary siting and BMP implementation plan or an adequate tool for LID-BMPs planning and implementation under situations where a simpler approach is acceptable or even preferred. For the latter objective, three key considering factors were taken into account.

(1) Installation site or regional suitability: area, soil characteristics, topography, depth of groundwater table, space available for LID-BMPs, etc.
(2) LID-BMP runoff control effectiveness: both quantities, e.g. peak flow and volume reduction, and quality, e.g. sediment, nutrient, and metal removal.
(3) Cost: capital, operation, and maintenance considerations.

The site suitability category is aimed at screening the various LID-BMPs for appropriateness at the specific site. The runoff control and cost consideration categories are both assessment tools for the comparison of applicable LID-BMPs to choose one or more cost-effective LID-BMPs. These basic criteria allow MCIS to be used as an adequate stand-alone tool for selecting the most cost-effective LID-BMPs.

In the site suitability category, a total of six first-level and nine second-level index factors, as shown in Figure 12.5, were established for the screening of available LID-BMPs.

In the runoff control category, the performance of runoff control is assessed, which refers to the mitigating effects of LID-BMPs on runoff water quantity and quality. In this category, three first-level performance indicators are presented, namely runoff quantity control, runoff quality control, and additional benefits. These indicators are then further divided into 12 second-level categories, as shown in Figure 12.6.

In the cost and maintenance category, three impacting factors considered are capital investment cost, operational/maintenance, and system reliability (Figure 12.7). In the capital costs category, construction cost is the index factor. In many situations, land costs can be an important factor, especially in highly urbanized areas. In operation and maintenance, the operation costs include the spending on personnel, material, and operation, and the maintenance costs

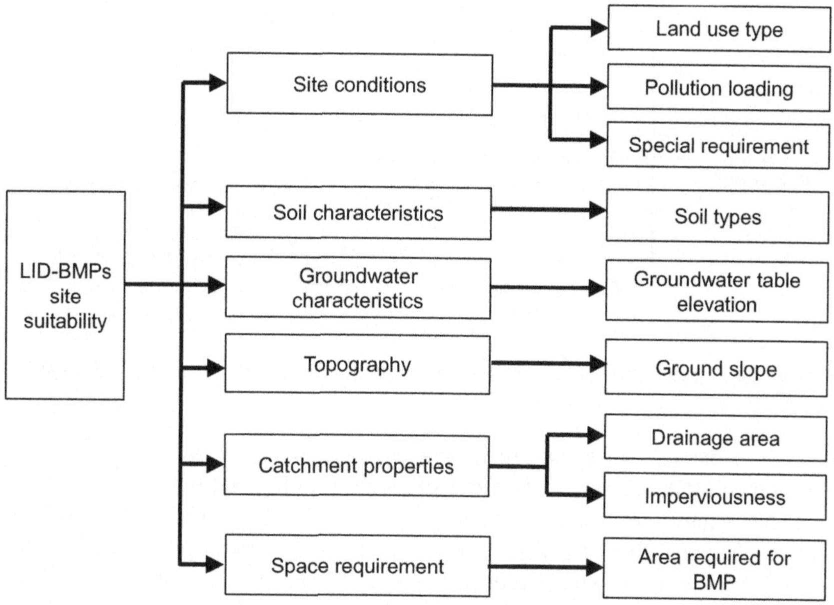

Figure 12.5 Indicators of LID-BMPs site suitability (figure by authors).

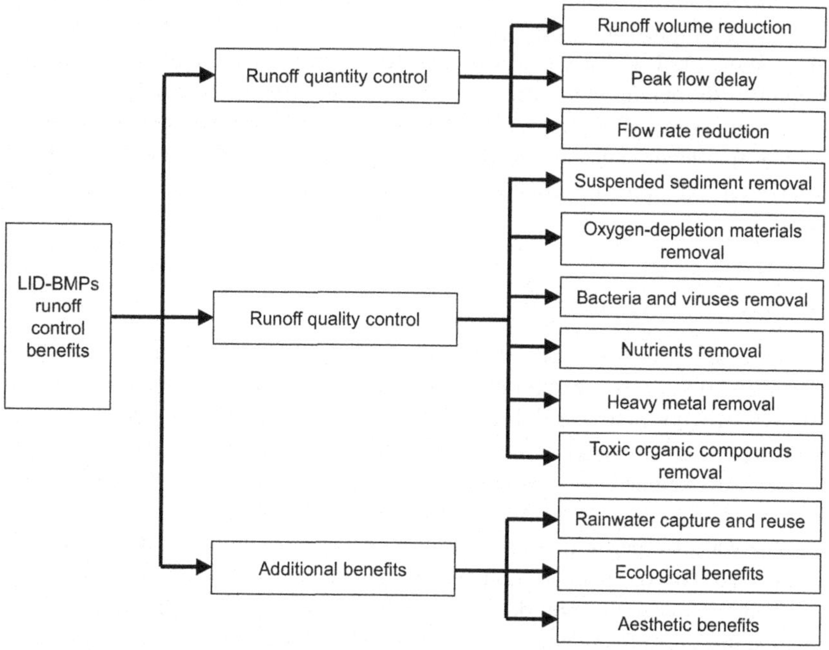

Figure 12.6 Indicators of LID-BMPs runoff control benefit (figure by authors).

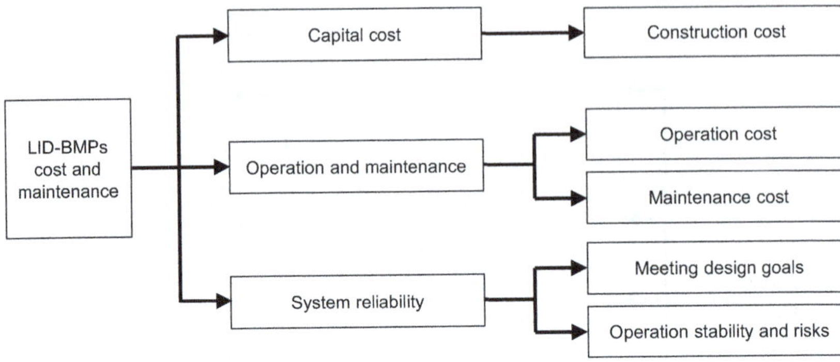

Figure 12.7 Indicators of LID-BMPs cost and maintenance (figure by authors).

include personnel, material, and replacement part costs. System reliability is determined by the design goals and operational stability and risks.

Thus, a multi-criteria selection index system (MCIS) for LID-BMP planning was developed. The selection indexes include 12 first-level indices and 26 second-level indices which reflect the specific installation site characteristics pertaining to site suitability, runoff control performance, and economics of implementation.

A mechanism for ranking the LID-BMPs was devised. First, each second level index was assigned a numeric value that was based on site characteristics and information on LID-BMPs. The quantified indices were normalized and then integrated to obtain the score for each of the first-level indexes. The final evaluation scores of each LID-BMP were then calculated based on the scores for the first-level indices. Finally, the appropriate BMP types for a specific installation site were determined according to the rank of the final evaluation scores.

12.3.3 Case study: Lay-out optimization of LID-BMPs

A case study was conducted (Jia *et al.*, 2015a, 2015b) to optimize LID-BMPs implementation. By examining information obtained, potential feasible LID-BMPs were first selected. SUSTAIN was then used to analyze four runoff control scenarios, namely: pre-development scenario; base scenario (existing campus development plan without BMP control); Scenario 1 (least-cost BMPs implementation); and, Scenario 2 (maximized BMPs performance). A sensitivity analysis was also performed to assess the impact of hydrologic and water quality parameters. The optimal solution for each of the two LID-BMPs scenarios was obtained by using the non-dominated sorting genetic algorithm-II (NSGA-II). Finally, the cost-effectiveness of the LID-BMPs implementation scenarios was examined by determining the incremental cost for a unit improvement of control.

12.3.3.1 Case study site and model set-up

The case study was conducted at the campus of the Guangdong College of Environmental Protection in Foshan City, Guangdong Province, China. Figure 12.8 depicts the location of the case study and associated land uses.

The climate of the site is sub-tropical with an average annual rainfall of 1614 mm and an average temperature of 21.9°C. The rainy season is between April and September, which accounts for 80% of the annual rainfall. The college is undergoing an expansion project; construction was scheduled in two phases. Phase I construction is finished and phase II is to be conducted. According to the construction plan, the total area of the campus is about 30 ha with an imperviousness ratio of 59%.

All the relevant data were collected and then processed in accordance with the requirements of the SUSTAIN model. The collected data include a Digital Elevation Model (DEM), land-use map and its attribute data, impervious surface map, soil type map and its attribute data, natural water system map, and drainage pipe system map.

Based on the basic data of the study site, a preliminary set of LID-BMPs was selected using the above Multi-Criteria Index Ranking System (MCIS). The selected LID-BMPs include bioretention cells, wet ponds, green roofs, porous pavements, infiltration trenches, and rainwater barrels. The SUSTAIN modeling framework was then set up in the ArcGIS platform. SUSTAIN was developed by the US EPA to evaluate alternative plans for stormwater runoff management in urban and developing areas. It provides a decision-support tool capable of evaluating the optimal location, type, and cost of LID-BMPs needed to meet water quality and quantity goals (Lee *et al.*, 2012).

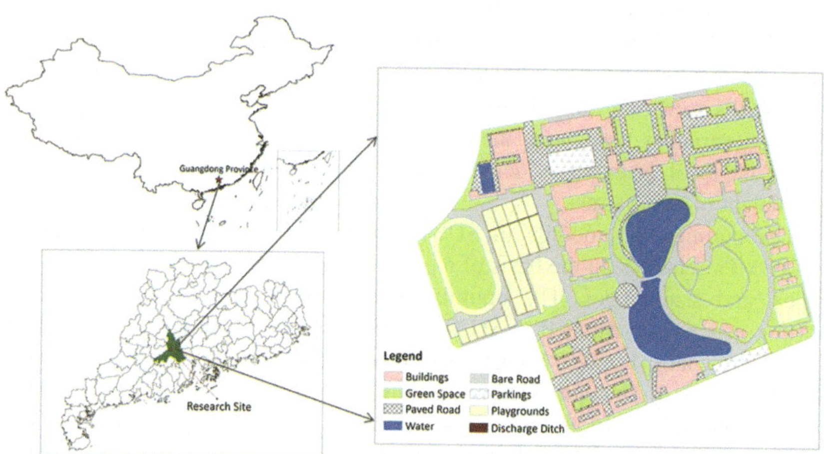

Figure 12.8 Location of the study site and associated land uses (reprinted by permission from Elsevier: Jia *et al.*, 2015b).

The required data for SUSTAIN include various types of digital maps and its attribute data, meteorological data, hydrological data, and cost data of different LID-BMPs, etc.

After the analysis of recent meteorological data, the hourly precipitation, air temperature, and evaporation data in the typical year of 2008 were used. The hydrological parameters (such as roughness coefficient and surface depression storage of different land covers) and pollutant characteristic parameters (such as maximum buildup, buildup rate constant, wash off coefficient, and wash off exponent) were set by referencing the calibrated parameters reported in the literature.

12.3.3.2 LID-BMPs planning scenario simulation and optimization

For the analysis of runoff control performance of the LID-BMPs planning options, four scenarios were designed, namely: predevelopment scenario; base scenario (existing campus development plan without BMP control); Scenario 1 (least-cost LID-BMPs implementation); and, Scenario 2 (maximized LID-BMPs performance). Then the scenarios were simulated using SUSTAIN to evaluate the expected runoff quantity (runoff volume and peak flow) and quality (SS, COD, TN, TP) control effectiveness.

The results showed that the runoff volume, peak flow, and the pollutant load were much larger under the base scenario when compared to the pre-development scenario. The results also show that the two LID-BMPs scenarios could effectively cut the runoff quantity and pollutant loads. Under Scenario 1, total runoff volume, peak flow, and pollutant loads were reduced by 14.5, 13.8, and 17–21%, respectively, compared to those under the base scenario. Under Scenario 2, total runoff volume, peak flow, and pollutant loads were reduced by 40, 46.8, and 46–51%, respectively.

LID-BMPs planning scenario optimization was conducted to obtain the most cost-effective scheme. The reduction rate of total runoff volume under different LID-BMPs scenarios was selected as the optimization objective. The optimization decision variables used were the LID-BMPs design configurations, such as the media depth of green roofs, length and depth of bioretention, and the diameter and height of rain barrels.

The optimization module in SUSTAIN was used to develop cost-effective LID-BMPs schemes on the basis of a pre-selected list of potential sites, LID-BMP types, and size ranges. The optimization algorithm used is a non-dominated sorting genetic algorithm-II (NSGA-II). Ten thousand iterations were conducted for both Scenario 1 and 2. Figure 12.9 shows the optimization curves of Scenario 1 and Scenario 2. The optimized best solution for Scenario 1 showed a 15.1% total runoff volume reduction rate, with a total cost of about 11.7 million Yuan. Compared with the original scheme of Scenario 1, the optimized one increases the reduction rate by 0.6% and reduce the costs by 2.6 million Yuan. For Scenario 2, the optimized plan has a 40.6% total runoff volume reduction rate, with a cost of

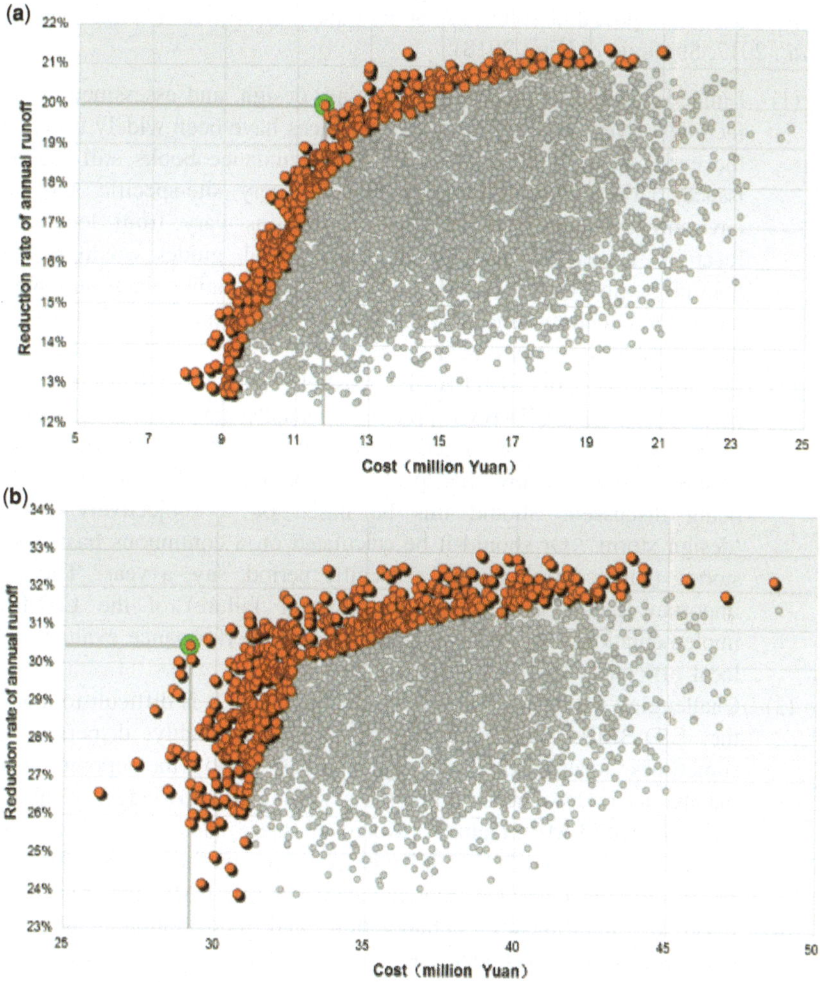

Figure 12.9 Optimization curves of Scenario 1 and Scenario 2. Orange dots are the optimized points (reprinted by permission from Elsevier: Jia *et al.*, 2015b).

about 29.1 million Yuan. It realizes an increase of 1.1% in runoff volume reduction rate while saving 5.0 million Yuan compared with the original Scenario 2 scheme.

12.4 SUMMARY AND FUTURE PERSPECTIVES

12.4.1 Limitation of current LID-BMPs techniques implimentations

There are still many existing challenges that hinder the wide application of LID-BMPs practices. Different studies have been carried out to identify these barriers and offer potential solutions. Some other key barriers identified are (Jia *et al.*, 2017; Shafique & Kim, 2018):

(1) Limited technical guidance on planning, design, and assessment of LID facilities. Even though LID-BMPs practices have been widely used since the early 2000s, technical manuals and guidance books still are been issued. Urban runoff characteristics are very site-specific and local environmental and socioeconomic conditions vary from location to location. Therefore, local, or at least regional, guides would be most helpful. A case in point is the need for a list of native plants suitable for use in bioretention cells. Currently, localized technical guidance is still not available for many places.

(2) Quantification of LID-BMP cost-effectiveness. Currently, the performance of a LID-BMP practice is usually measured as 'percentage of runoff volume' or 'fraction of pollutant load' removed by the practice. However, how this 'percentage of fraction' is calculated is still being discussed. Should this be based on a subjectively selected 'design storm'? Or should it be calculated on a continuous basis for all storms that occurred during a specific period, say, a year? The other important factor is how the success (or failure) of the LID-BMP implementation can be incorporated into the performance evaluation by local officials.

(3) Challenging to operate and maintain LID-BMPs. It is difficult to manage the LID-BMPs systematically as they are highly dispersed and sometimes located on private properties. Also, the operation and maintenance crew needs to possess sufficient knowledge of how to maintain LID-BMPs features. Moreover, there has been a discrepancy when it comes to whether private sectors or public sectors should be responsible for the long-term maintenance of LID-BMPs features.

(4) Lack of inter-agency collaboration and cooperation. Stormwater management is not under the sole discretion of a single sector but requires collaboration and coordination between multiple sectors and departments. Hence, it is important to understand how LID-BMPs fit within the codes and ordinances of each department, as well as the role each department plays for the successful implementation of LID-BMPs.

(5) Education and training do not provide skills to design and implement LID-BMPs. To successfully implement an LID-BMPs construction project, knowledge from many disciplines is required. For example, the planning/design of LID-BMPs facilities would need skills in stormwater management, urban hydrology and hydraulics (scales from site to region to watershed), water quality modeling, optimization techniques, etc. However, a specific system of education and training programs are still lacking.

12.4.2 Future technical research direction

Urbanization will still be the development trend globally. Along with the rapid urbanization process, urban water problems (flooding, water pollution, and ecological degradation) have become obstacles to urban sustainability. The implementation of LID-BMPs is believed to be the solution.

However, research still lags in all respects, and we still need a holistic system of theory, technology, and management for LID-BMPs. The required research includes studies of the effects of individual LID BMPs (structures, media, etc.), LCA analysis of individual/aggregate LID BMPs, urban runoff simulation models for LID-BMPs, LID-BMPs planning, and the necessary guidelines and technical codes.

Moreover, questions are frequently raised in regard to the suitability of LID BMPs for all sites, groundwater contamination, and the winter performance of LID BMPs practices. In some situations, LID BMP concepts are not allowed or are highly restricted under many current regulatory statutes. Concrete curbs and gutters may be required to rapidly convey runoff because of safety concerns. Runoff collection and ponding may be discouraged because of public health concerns over mosquito breeding. Therefore, continued research, development, and case study results are needed to address these issues.

REFERENCES

Bertrand-Krajewski J., Chebbo G. and Saget A. (1998). Distribution of pollutant mass vs volume in stormwater discharges and the first flush phenomenon. *Water Research*, **32**, 2341–2356.

Birch H. (2012). Monitoring of Priority Pollutants in Dynamic Stormwater Discharges from Urban Areas. PhD dissertation. Department of Environment, Technical University of Denmark, Denmark.

Chang G., Parrish J. and Souer C. (1990). The First Flush of Runoff and Its Effect on Control Structure Design, report, 33 pp., Environ. Resource Mgt. Div. Dept. of Environ. and Conservation Services. City of Austin, Austin, Tex.

Chen G., Luo J., Zhang C., Jiang L., Tian L. and Chen G. (2018). Characteristics and influencing factors of spatial differentiation of urban black and odorous waters in china. *Sustainability*, **10**(12), 4747. doi: 10.3390/su10124747

Cheng M. S., Zhen J. and Shoemaker L. (2009). BMP decision support system for evaluating stormwater management alternatives. *Frontiers of Environmental Science & Engineering in China*, **3**(4), 453–463.

Claytor R. A. and Schueler T. R. (1996). Design of Stormwater Filtering Systems. Chesapeake Research Consortium, Edgewater, MD, USA.

Department of Environmental Protection (2006). Pennsylvania stormwater Best Management Practices Manual. Documment Number:363-0300-002, PA, USA.

Eriksson E., Baun A., Scholes L., Ledin A., Ahlman S., Revitt M., Noutsopoulos C. and Mikkelsen P. S. (2007). Selected stormwater priority pollutants – a European perspective. *Science of the Total Environment*, **383**, 1–3.

Field R. and Tafuri A. N. (eds.) (2006). The Use of Best Management Practices (BMPs) in Urban Watersheds. DEStech Publications, Inc., Lancaster, PA, USA.

Glińska-Lewczuk K., Gołaś I., Koc J., Gotkowska-Płachta A., Harnisz M. and Rochwerger A. (2016). The impact of urban areas on the water quality gradient along a lowland river. *Environmental Monitoring and Assessment*, **188**(11), 624. doi: 10.1007/s10661-016-5638-z

Jacobson C. R. (2011). Identification and quantification of the hydrological impacts of imperviousness in urban catchments: a review. *Journal of Environmental Management*, **92**, 1438–1448. doi: 10.1016/j.jenvman.2011.01.018

Jia H., Lu Y. W., Zhen X. and Yu S. (2012). Planning of LID-BMPs for urban runoff control: the case of Beijing Olympic village. *Separation and Purification Technology*, **84**, 112–119.

Jia H., Yao H. and Yu S. (2013a). Advances in LID BMPs research and practice for urban runoff control in China. *Frontiers of Environmental Science & Engineering*, **7**(5), 709–720.

Jia H., Yao H., Tang Y., Yu S., Zhen J. and Lu Y. (2013b). Development of a multi-criteria index ranking system for urban stormwater best management practices (BMPs) selection. *Environmental Monitoring and Assessment*, **185**(9), 7915–7933.

Jia H., Wang X., Ti C., Zhai Y., Field R., Tafuri A. N., Cai H. and Yu S. (2015a). Field monitoring of a LID-BMP treatment train system in China. *Environmental Monitoring and Assessment*, **187**(6), 373. doi: 10.1007/s10661-015-4595-2

Jia H., Yao H., Tang Y., Yu S., Field R. and Tafuri A. (2015b). LID-BMPs planning for urban runoff control and the case study in China. *Journal of Environmental Management*, **149**, 65–76.

Jia H., Wang Z., Zhen X., Clar M. and Yu S. (2017). China's Sponge City Construction: A Discussion on Technical Approaches. *Frontiers of Environmental Science and Engineering*, **11**(4), 18. doi: 10.1007/s11783-017-0984-9

Lee J., Selvakumar A., Alvi K., Riverson J., Zhen X., Shoemaker L. and Lai F. (2012). A watershed-scale design optimization model for stormwater best management practices. *Environmental Modelling & Software*, **37**, 6–18.

Liu J. and Niyogi D. (2019). Meta-analysis of Urbanization Impact on Rainfall Modification. *Scientific Reports*, **9**(1), 1–14.

Liu Y., Bralts V. F. and Engel B. A. (2015). Evaluating the effectiveness of management practices on hydrology and water quality at watershed scale with a rainfall-runoff model. *Science of The Total Environment*, **511**, 298–308.

Mallin M. A., Johnson V. L. and Ensign S. H. (2009). Comparative impacts of stormwater runoff on water quality of an urban, a suburban, and a rural stream. *Environmental Monitoring and Assessment*, **159**(1–4), 475–191.

McGrane S. J. (2016). Impacts of urbanisation on hydrological and water quality dynamics, and urban water management: A review. *Hydrological Sciences Journal*, **61**(13), 2295–2311.

Mid-America Regional Council/American Public Works Association (2012). Manual of Best Management Practices for Stormwater Quality. Kansas City, MO.

Minnesota Pollution Control Agency (2014). Minnesota Stormwater Manual: Stormwater Management for Lake Protection and Restoration. Duluth, MN, USA. http://stormwater.pca.state.mn.us

Pyke C., Warren M. P., Johnson T., LaGro J., Jr., Scharfenberg J., Groth P., Freed R., Schroeer W. and Main E. (2011). Assessment of low impact development for managing stormwater with changing precipitation due to climate change. *Landscape and Urban Planning*, **103**(2), 166–173.

Rivanna Stormwater (n.d.). https://rivanna-stormwater.org/what-happens-to-the-rain/bmp-photos-descriptions/#BMP1, (accessed 7 June 2020).

Schiff K. C., Tiefenthaler L. L., Bay S. M. and Greenstein D. J. (2016). Effects of rainfall intensity and duration on the first flush from parking lots. *Water*, **10**(4), 435.

Scholes L., Revitt D. M. and Ellis J. B. (2005). The Fate of Stormwater Priority Pollutants in BMPs. Public Report. DayWater Project, London, UK.

Shafique M. and Kim R. (2018). Recent progress in low-impact development in South Korea: Water-management policies, challenges and opportunities. *Water*, **10**(4), 435.

Tibbetts J. (2005). Combined sewer systems: Down, dirty, and out of date. *Environmental Health Perspectives*, **113**(7), 465–467.

US EPA (1999). Preliminary Data Summary of Urban Storm Water Best Management Practices. EPA-821-R-99-012. US Environmental Protection Agency, Washington.

US EPA (2004). The Use of Best Management Practices (BMPs) in Urban Watersheds. EPA/600/R-04/184. US Environmental Protection Agency, Washington.

US EPA (n.d.). Polluted Runoff: Nonpoint Source (NPS) Pollution. http://www.epa.gov/nps/urban-runoff-low-impact-development, (accessed 24 October 2019).

Viavattene C., Ellis J. B., Revitt D. M., Seiker H. and Peters C. (2010). The application of a GIS-based BMP selection tool for the evaluation of hydrologic performance and storm flow reduction. Proceeding of 7th International Conference on Sustainable Techniques and Strategies for Urban Water Management. NOVATECH2010, Lyon, France.

Walsh C. J., Fletcher T. D. and Burns M. J. (2012). Urban stormwater runoff: A new class of environmental flow problem. *PLoS One*, **7**(9), e45814.

Young K. D., Kibler D. F., Benham B. L. and Loganathan G. V. (2009). Application of the analytical hierarchical process for improved selection of storm-water BMPs. *Journal of Water Resources Planning and Management*, **135**(4), 264–275.

Zahmatkesh Z., Karamouz M., Burian S. J., Tavakol-Davani H. and Goharian E. (2014). LID implementation to mitigate climate change impacts on urban runoff. In: World Environmental and Water Resources Congress 2014: Water without Borders, W. C. Huber (ed.), American Society of Civil Engineers, Reston, VA, USA, pp. 952–965.

Chapter 13

Water quality improvement for upgrading urban landscapes

Qiang He, Caihong Liu, Ziwei Chen, Riuxing Huang and Xiaoliu Huangfu

Key Laboratory of Eco-environments in the Three Gorges Reservoir Region, Ministry of Education, Chongqing University, Chongqing 400044, China

13.1 INTRODUCTION

The urban water environment is mainly composed of urban rivers and urban lakes, which differ from natural rivers because of the direct connection they have with human habitats (Huang *et al.*, 2019). Urban rivers and lakes are important water resources for urban dwellers (Wu *et al.*, 2019) due to their close relation to socioeconomic development and the improvement of living environments. According to the United Nations Department of Economic and Social Affairs (UN, 2018), 55% of the world's population currently lives in urban areas, and by 2050, this proportion is expected to increase to 68%. This ever-increasing urbanization process and global population growth have caused the number of urban residents in various countries to surge from 750 million in 1950 to 4.2 billion in 2018, especially in developing countries and regions. However, as an important indicator of a country's development level, urbanization has increased the interaction between human society and the ecological environment, to a certain extent, which has led to a series of society-environment-ecology problems (Wan *et al.*, 2015).

Water pollution is a crucial problem of urbanization, which causes a series of water environmental problems (Wen *et al.*, 2006). Pollutant input to urban rivers

Editors: Xiaochang C. Wang and Guangtao Fu
doi: 10.2166/9781789060768_0357

occurs through point source discharges, agricultural irrigation, surface runoff, etc., causing a series of water environmental problems (Zhou, 2014). A survey of river water quality in Beijing in 2019 showed that 28% of 104 rivers in Beijing were below the requirement of Class IV of the Chinese Environmental Quality Standards for Surface Water (GB 3838-2002). The water quality deterioration is mainly attributed to over-exploitation of rivers, which results in changes to the river morphology, riparian vegetations, weakened self-purification capacity, and accumulation of pollutants (Deng *et al.*, 2015; Yousefi *et al.*, 2019). Furthermore, the weakening of the self-purification capacity and accumulation of pollutants lead to water body eutrophication, thus, producing algal blooms and malodor (Li *et al.*, 2018). Water pollution causes the loss of river landscape value, thus reducing landscape benefits significantly. Consequently, the improvement of water quality and urban river landscapes has become an important topic of discussion in academic circles.

13.2 CURRENT PROBLEMS AND NEEDS FOR URBAN WATER QUALITY IMPROVEMENT

Urban expansion causes a rapid increase in impervious areas, which affects the processes in the urban water cycle, leading to increases in the runoff coefficient and runoff volumes and, thereby, extreme precipitation events and risks of urban flooding (Fang *et al.*, 2018). Second, increases in domestic sewage and industrial wastewater production causes environmental problems such as water quality deterioration and water ecosystem degradation (Ilyas *et al.*, 2019).

13.2.1 Water pollution

The main problem of the urban water environment is the deterioration of water quality caused by water pollution. With the increase of urban population and density as well as economic development, the discharge of domestic and industrial sewage has increased exponentially. The rate of pollutant discharges far exceeds the self-purification capacity of urban water bodies, which significantly affects the quality of the urban water environment. According to the annual report of the national environmental statistics in China, in 2015, there was a total of 73.53 billion m^3 of wastewater discharge, including 19.95 billion m^3 of industrial wastewater and 53.52 billion m^3 of urban domestic wastewater. Domestic and industrial wastewater are important factors that lead to urban water environment pollution.

The increasing deterioration in the quality of urban water bodies leads to severe eutrophication, black and odorous waters, and the disappearance of aquatic life. The degradation of the urban water ecosystem, including that of its biological composition from advanced flora and fauna to germs and other bacteria, has caused urban water to lose its ecological service functions that are vital to humankind (Arthington *et al.*, 2010). Pollutants are also noted to increasingly interfere with human endocrine formation. This is the direct harm to human beings caused by the degradation of the urban aquatic ecosystem. Overall, the degradation of urban water ecosystems leads to the degeneration and destruction of cities.

13.2.2 River morphology and landscapes destruction

Rivers in China have been fundamentally channelized for water conservancy. River damming plays a role in the control and drainage of urban floods and water resource allocation. However, river damming leads to irregularities in its hydrology, causing continuous changes to the river water level. Such changes could affect the way a river functions and its normal flow rate, resulting in erosion of the river bed and banks downstream. In addition, with the continuous development of urbanization, people have channelized urban rivers for urban development, turning urban rivers into canals and ditches. These changes in river morphology bring a series of problems:

- Loss of meander. Using concrete materials to manually bend and straighten the originally winding river could be important for the restoration and reformation of the river (Wolff & Sutton, 2010). However, the reformed river no longer has its natural forms and characteristics and loses the original characteristics of deep shoals. The water flow rate of the reformed river is accelerated, resulting in the loss of habitat for organisms originally living in the river. Thus, the ecological environment changes.
- Blockage by water devices. To regulate the water level some water conservancy devices, such as sluice gates, are artificially set up in some urban river sections. As the river is blocked, fish and other organisms cannot pass through the water conservancy devices normally, which leads to changes in natural behaviors such as migration and spawning, and eventually, the number of species gets reduced (Trifu *et al.*, 2011).
- Channelization (Qi & Luo, 2006). Since urban river courses are mostly composed of narrow linear green spaces, in the process of river course restoration, simple hard stone or brick paving and monoculture plant configurations are often used, resulting in boring river landscape designs. The river landscape loses its original natural form, and the river's landscape benefit is greatly reduced.

13.2.3 Pollution characteristics of urban waters

13.2.3.1 Pollution sources

The sources of pollution to the urban water environment are mainly classified into three categories: point source pollution, non-point source pollution, and internal pollution (Schaffner *et al.*, 2009).

Point source pollution mainly comes from domestic sewage. Pollution areas expand with the growth of residential areas. Urban rivers usually have their upstream channels and/or branch streams flowing through the outskirts of rural areas. Due to the scattered population and less developed economic conditions, it is almost impossible to develop centralized treatment systems for domestic sewage. Consequently, in some areas, about 95% of the sewage is directly discharged to the environment, causing pollution of water bodies. Direct pollution from small township enterprises also causes pollution in the upper reaches of rivers. The discharge of domestic and industrial sewage not only causes a source of pollution for the city itself but also an external source of pollution for cities downstream.

Non-point source pollution is mainly caused by the encroachment of river revetment by farmlands, where fertilizers from the farmlands flow into the river (Nie *et al.*, 2012). The direct discharge of the initial runoff of a rainfall event also has a certain impact on water quality. The nutrients carried in the runoff may lead to the eutrophication of the water body, especially in areas where the river flow is low, to induce the growth of water hyacinth. Water and soil losses in most rural areas occur in the upstream of rivers, and the pollution from chemical fertilizers and pesticides become the main water pollution sources. Agricultural non-point source pollution is an important factor affecting the water environmental quality.

Internal pollution mainly refers to the pollutants contained in the sediment of urban water bodies and the degradation products formed from various floating matters, garbage, and aquatic plants or algae in the water body (Stimson & Larned, 2000). When a water body receives sewage discharge, the accumulation of sediment may lead to the formation of a thick black and odorous sediment layer which releases pollutants and results in the deterioration of the overlying water quality (Testa *et al.*, 2013). The origin of the internal pollutants may be mainly from urban domestic sewage (Bohlin *et al.*, 2006), and also suburban agriculture and aquaculture backwater, industrial wastewater, catering wastewater, sediment, etc., with complex components, including inorganic pollutants, organic pollutants, toxic pollutants, etc.

The pollutants from all the above-mentioned sources are composed of inorganic and organic substances. The inorganic pollutants include acids, alkalis, and inorganic salts, which mainly come from industrial wastewater and atmospheric sedimentation. The organic pollutants include hydrocarbons, proteins, fats, cellulose, and others, which mainly come from domestic sewage, aquaculture, and industrial wastewater. Organic pollutants may also contain nutrients such as nitrogen, phosphorus, and potassium. When these pollutants enter the water body, they are decomposed by microorganisms into nutrient elements for use by aquatic plants, resulting in eutrophication of water bodies. Many toxic pollutants such as phenol, cyanide, arsenic, mercury, cadmium, and copper may also be discharged into water bodies and cause damage to aquatic flora and fauna.

13.2.3.2 Reduction of self-purification capacity

Healthy water bodies should be capable of assimilating pollutants through a number of physiochemical and biological actions, which are called self-purification processes. However, for many urban water bodies, their self-purification capacities are declining due to artificial reasons. With the rapid urban development, many urban water bodies have lost their ecological shorelines and have come to be directly adjacent to increasingly impervious surfaces. Their hydrological conditions have also been largely disturbed by improper engineering structures. All these have resulted in a loss of the water body's self-purification capacity. The increase of impervious surface area also significantly increases surface runoff in rainy seasons. This also negatively impacts the natural purification processes not only for the water bodies but also for the whole urban ecological system.

Another factor impacting on the self-purification capacity of urban waters is the channelization of urban rivers. In many cases, for urban river management, attention is traditionally mainly paid to the capacity of the river channel for flooding flow. Many urban rivers are cut straight and flood control walls are built on the river banks, thus, transforming the natural streams into concrete drainage channels that have little ecological function as well as landscape value.

In some cities, rapid urbanization has transformed existing wetlands and even lakes into lands for infrastructure construction. This has resulted in the loss of urban water and green areas and the deterioration of urban water environment quality.

13.2.3.3 Characteristics of the water environment in Chongqing

A case study will be presented in this chapter regarding Chongqing, a mountainous megacity in southwest China. In this sub-section, we present a short introduction to the characteristics of the water environment of this city.

Chongqing is situated in the transitional area between the Tibetan Plateau and the plain on the middle and lower reaches of the Yangtze River in a sub-tropical climate zone often swept by moist monsoons. It often rains at night in late spring and early summer and, thus, the city is famous for its 'night rain in the Ba Mountains'. Chongqing covers a large area crisscrossed by rivers and mountains. The Daba Mountains stand to the north, the Wu Mountains to the east, the Wuling Mountains to the southeast, and the Dalou Mountains to the south. The whole area slopes down from north and south towards the Yangtze River valley, with sharp rises and falls. The area is featured by a large geological massif, of mountains and hills, with large sloping areas at different heights.

Specifically, the central urban area is located on a huge folding area. This makes Chongqing a city of unique features. Built on mountains and partially surrounded by the Yangtze and Jialing rivers, it is known as a 'mountain city' and a 'city on rivers'. Therefore, the water environment in Chongqing has the following characteristics.

- Variable hydrological conditions. Affected by the mountainous topography, geology, and climate, the hydrological characteristics of the urban watersheds are large slopes, short duration of confluence, and obvious seasonal changes in discharge and water level. Due to the uneven spatial and temporal distribution of rainfall, in July, the flood flow into the river is large, the inflow rate is fast and the peak flow is high, and the river section is sharply widened. Conversely, in April, the flow in the river is small and the water level is low. This highly variable hydrological characteristic not only jeopardizes the safety of the city but also destroys the stability of the river ecosystem.
- Complex channel structure. Due to the complex mountainous topography and landforms, mountainous urban rivers exhibit large relief, varying widths, narrow depths, and diverse river morphologies. According to the degree of artificial modification, rivers in mountainous cities are mainly divided into three types, namely (1) rivers with natural morphologies, (2) cascade reservoir-type channels, and (3) artificial canalized channels. Among them, the rivers with natural morphologies have a high flow rate with continuous vegetation distribution along the river bank. Also, the river bed is mostly formed from pebbles, gravel, or sand. On the other hand, the cascade reservoir channel has a multi-level reservoir structure, the water system structure is fragmented, the water system is longitudinal, the connectivity is poor and the flow velocity is generally small. Meandering in the channel of the artificial canalized river is minimal, mainly manifested by the hardening of the bottom of the channel, and the water depth is shallow.

13.3 URBAN RUNOFF CONTROL AND POLLUTANTS REDUCTION FOR URBAN WATER QUALITY IMPROVEMENT

Urban waters, including rivers/streams, lakes, and wetlands, basically have two important functions in urban aquatic environments. One is their ecological function, and the other is the landscape effect. Both require a favorable water quality. As discussed in the former sections, pollutants that may enter urban waters and deteriorate water quality originate from point sources, non-point sources, and internal sources within a water body. Point source control is usually by the construction and upgrading of urban sewage and drainage systems, while non-point source control and internal source elimination cannot be solely dependent on engineering measures but a good combination of engineering and ecological measures. Nowadays in China, such measures have been incorporated into the scheme of Sponge City Construction, where the multi-functional effects are stressed, including surface runoff reduction (Opher *et al.*, 2009), soil erosion control (Gromairemertz *et al.*, 1999), non-point source reduction (Mikkelsen *et al.*, 1996), riparian ecosystem recovery (Ahiablame *et al.*, 2012), and landscape improvement (Gobel & Coldewey, 2013; Karnatz *et al.*, 2019). In this section, several practical technologies for urban runoff control and pollutant reduction are discussed.

13.3.1 Urban surface runoff control and pollutants reduction by an innovative sand filter

In many Chinese cities, especially a mountainous city like Chongqing, it is found that many organic and inorganic pollutants are attached to particulates that accumulate on various surfaces. These particulates, subsequently, can be easily carried into urban waters by stormwater runoff and become a major non-point source of water pollution (Gromaire *et al.*, 2001; Li *et al.*, 2010b). Building roofs, roads, and parking lots are the main impervious surfaces in urban areas that accumulate particulates as carriers of heavy metals, nutrients, and toxic organic pollutants (Lin *et al.*, 2007). Therefore, the effective reduction of particulate pollutants from surface runoff is key to reducing the impact of non-point source pollution (Wang *et al.*, 2017). Due to the shortage of land and traffic pressure on urban roads and parking lots, the space available for building facilities is limited. Therefore, how to effectively remove the particulate pollutants from surface runoff with limited space and without affecting the normal function of the surface is a problem to be solved (Seo *et al.*, 2017; Zhang *et al.*, 2017).

The core component of the technology for the effective removal of particulate pollutants from surface runoff is an innovative sand filter (Hou *et al.*, 2013). The main filter medium is sand and an appropriate proportion of peat or granular activated carbon for enhancing organic pollutants and heavy metals removal. As shown in Figure 13.1, the filter system is mainly composed of a sedimentation chamber, a filter chamber, and a drainage chamber. The sand bed in the filter chamber comprises a gravel grid layer, coarse sand filter layer, and gravel drainage layer from top to bottom. The filter system can be buried into impervious surfaces such as roads and parking lots.

In the filter chamber, from the top to the bottom, the gravel grid layer is filled with gravels of 1–2 cm in media size and 5 cm in depth. Also, the coarse sand filter layer is filled with sand of 0.5–2 mm in media size and 45–60 cm depth, and the gravel drainage layer is filled with gravels of 2–4 cm in media size and 15–30 cm in depth. The sand filter system is very effective and stable for the removal of TSS from urban surface runoff. The removal rates of TSS, TN, NH_4^+, NO_3^-, TP, and PO_4^{3-} average about 95.3, 7.8, 18.7, 3.9, 37.7 and 18.5%, respectively. TSS is the main pollutant of urban surface runoff pollution. It can be seen that the sand filter system can effectively remove TSS from the heavily polluted runoff under high hydraulic loading. Also, it can remove TP in the urban surface runoff to a certain extent.

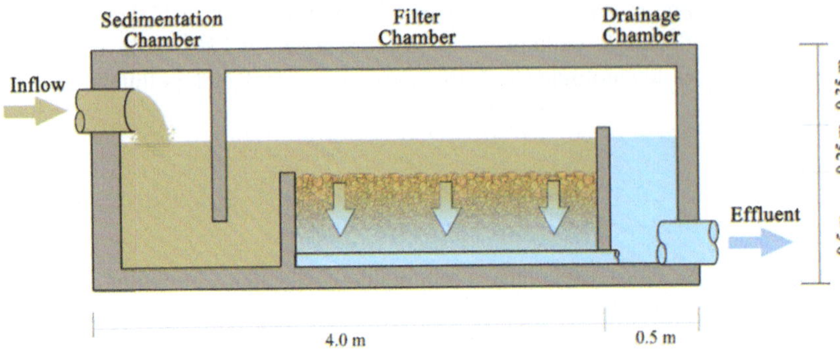

Figure 13.1 The innovative sand filter system (figure by authors).

According to the site conditions, the sand filter system can either be installed on the ground surface or below the ground surface. The surface runoff firstly flows through the sedimentation chamber to remove large particles, and then enters the filter bed to remove fine particles and other pollutants attached to the particle surfaces. When peat or granular activated carbon is added to the filter bed, higher removal of organics, nutrients, and heavy metals can be achieved. Water collection pipes are set at the bottom of the filter bed to discharge the treated effluent through the drainage chamber.

13.3.2 Urban surface runoff control and pollutants reduction by an enhanced ecological filter and flow reduction system

Wetland systems are usually used for receiving urban surface runoff to reduce both runoff volume and pollutants. However, for mountainous cities, the initial runoff usually contains a large amount of sediment and high pollutant loadings (Chai *et al.*, 2019). Under a continuous heavy rainfall, the wetlands receiving the increasing runoff will soon become flooded. When the rainfall intensity gradually decreases, most of the sediments that entered the wetlands will settle and accumulate, forming a dense sediment layer that reduces the ecological purification effect (Ahmed *et al.*, 2020; Knowles *et al.*, 2011).

To overcome this problem, an enhanced ecological infiltration and flow reduction system is designed by utilizing the landscape green space adjacent to the impervious surface to reduce the urban surface runoff and control the initial runoff pollution. As shown in Figure 13.2, this system is composed of a sand bed with sufficient depth and water permeability and also cultivated plants. When it receives surface runoff, water can quickly be filtered and infiltrate into the void volume of the whole sand bed. In the process of runoff filtration, pollutants can be effectively removed by interception, adsorption, and biological actions (Roy-Poirier *et al.*, 2010).

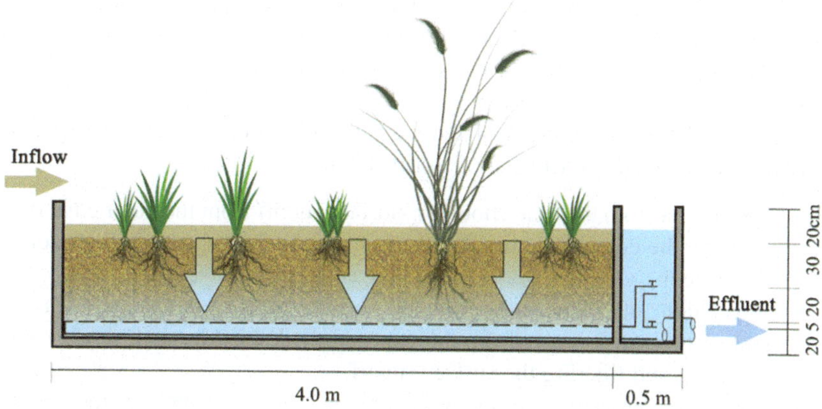

Figure 13.2 Diagram of the enhanced ecological filter and flow reduction system (figure by authors).

This system can perform the functions of hydraulic detention and enhanced ecological filtration. At the bottom of the sand bed, the effluent can be discharged directly to nearby drainage or receiving water as the bottom outlet is open. Otherwise, water can be stored in the sand bed as the outlet is closed. Therefore, the detention and filtration processes are adjustable according to the inflow volume and requirement of pollutant removal. Usually, at the beginning of a rainfall event, the initial runoff with high pollutant loading can be detained in the system for a longer residence time so that pollutant removal can be enhanced. When rainfall continues, the system may receive larger volumes of surface runoff than the sand bed can accommodate. As the continuous inflow no longer contains high concentrations of pollutants, the hydraulic residence time can be shortened in correspondence with the inflow rate while the effluent quality is not deteriorated.

The cultivated plants also perform important roles in the enhanced ecological filtration. It is preferable to select local plants with developed root systems. The system is suitable for small-scale installations for onsite urban surface runoff reduction and pollution control, such as in residential areas, commercial areas, and industrial areas. However, it may not be suitable for installation in areas with shallow groundwater levels.

13.3.3 Cascade infiltration system for urban road shoulder runoff control

Urban roads are typical impermeable surfaces attributed to runoff pollution (Dorchin & Shanas, 2010). When space conditions allow, grass planting ditches can partially replace drainage pipes for surface runoff pollution control (Shafique & Kim, 2017). However, in mountainous cities such as Chongqing, the road slope is usually steep, so grass planting ditches may not be applicable because of the strong scouring effect of runoff flow. To meet the needs for road shoulder runoff control in mountainous cities, a cascade infiltration system is developed (Zhou *et al.*, 2018).

Figure 13.3 shows a section of the cascade infiltration system. The system's composition and characteristics are described below:

- Water inlet from the road shoulder. Surface runoff from the outer edge of the lane and the edge of the roadbed is dispersed to the vegetation filter zone of the system.
- Vegetation filter zone. It is arranged on both sides of the cascade filtration system to receive water from the exposed road shoulder and play a role of buffering and filtering the surface runoff.
- Detention percolation zone. It is arranged in the center of the cascade percolation system where part of the surface runoff is retained to percolate downward.

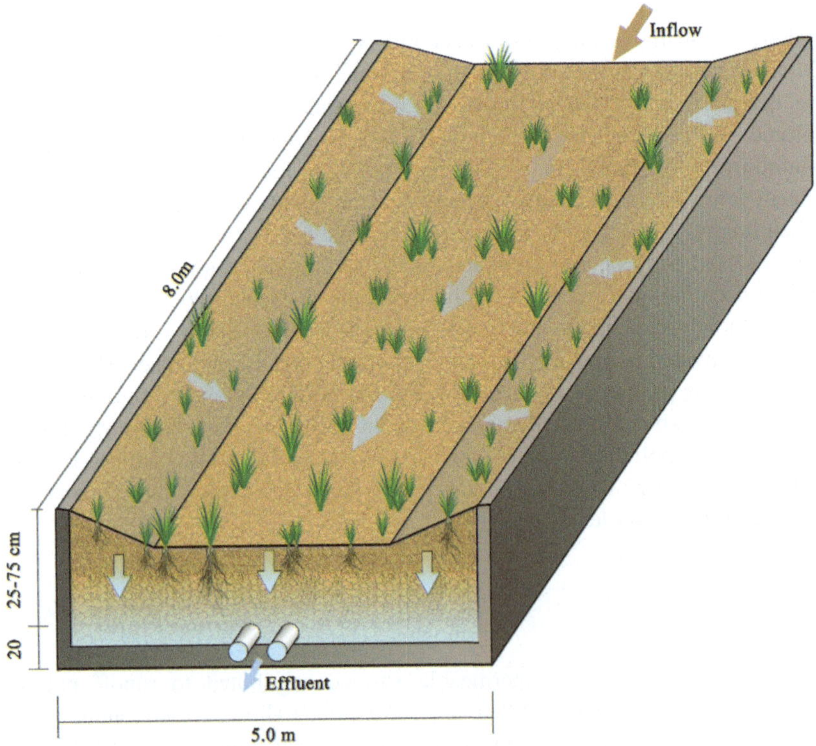

Figure 13.3 Diagram of the cascade infiltration system (figure by authors).

- Cascade weir. It is set at a certain distance in the longitudinal slope of the detention percolation zone and increases the residence time of surface runoff in the detention percolation zone. It can also slow down the flow rate and allow water to overflow to the next unit.
- Stagnant water space. It provides temporary storage and retention of surface runoff as well as deposition of particulate pollutants.
- Plants. They assist with water transpiration and purification.
- Substrate layer. It supports plant growth and maintains the rhizosphere community for water purification, as well as other functions, such as physical and chemical interception, filtration, and adsorption for pollutant removal.
- Gravel layer. It is composed of larger size gravels (20–40 mm) and has a depth of 45–60 cm. It can provide temporary storage for infiltration from the upper layer, prolong the infiltration time, and discharge the infiltrated water through the opening of drainage pipes.
- Overflow facility. It is arranged at the end of the cascade filtration system, mainly including an overflow discharge channel, overflow pipe, and discharge pipe.

The cascade infiltration system can be applied in mountainous cities with up to 10% slope for road shoulder runoff control through the setting of cascade weirs.

13.3.4 Enhancement of water circulation and oxygen enrichment in urban lakes

Many urban lakes are insufficiently replenished due to relatively small catchment areas to receive natural inflows and/or shortage of source water for artificial replenishment. Therefore, they are usually closed, slow-flowing water bodies (Li *et al.*, 2013; Zhou *et al.*, 2001), and are characterized by limited water environmental capacity, non-diversified aquatic ecology, and fragmented landscapes. Water quality may easily be deteriorated or even become black and odorous due to external and/or internal pollution, and seasonal algae outbreak (Lawson & Anderson, 2007; Li *et al.*, 2010a).

To improve water quality in urban lakes, artificial measures are widely taken for water circulation enhancement and oxygen enrichment. Water fountains and artificial waterfalls can well perform such kinds of roles along with water landscape improvement (Takyi & Lence, 1999). Mechanical devices, such as submersible thrusters, can also be installed in dead water areas to assist water flow and convection, and increase dissolved oxygen content (Mitrovic *et al.*, 2003).

Figure 13.4 is a diagram of a novel floating island equipped with a submerged thruster driven by solar power. The floating island itself functions as an ecological system for water purification, and also for the improvement of the lake landscape (Kwon *et al.*, 2004). The submersible thruster can effectively promote

Figure 13.4 Diagram of a novel floating island equipped with a submerged thruster for water circulation and oxygen enrichment (figure by authors).

water flow and enhance water reoxygenation. The installation of this device can be very flexible to meet the requirement of water quality improvement.

13.4 CASE STUDY

13.4.1 Case introduction

Chongqing Garden Expo Park, the main venue for the Eighth China International Garden Expo, is located at the center of Longjing Lake district in the New North Zone of Chongqing. The park covers an area of 2.2 km^2 and is characterized by the integration of ecological landscape, local culture, leisure entertainment, and science and education into a feature zone.

Longjing Lake in the central area of the park is the most important scenic spot. It is an artificial lake formed by constructing a dam over a mountain valley. Upstream of the dam, the water surface area is 0.53 km^2. With the original diverse geographic features of the valley, many bays and peninsulas are naturally formed when water from the 11.8 km^2 catchment area is stored in the lake. The total storage capacity is 663×10^4 m^3, and the water depth, on average, is over 10 m, with a maximum depth of about 22 m. Upstream of the Longjiang Lake, there are two mountain rivers, namely Zhaojia River and Longjing River, both with short river channels, narrow widths, and small flow rates on dry days. After the construction of the lake, the local climate and ecological environment have improved.

With the irregular shape of the lake area and the existence of bays and peninsulas of varied sizes and water depths, the Longjing Lake consists of a number of semi-closed water bodies in addition to the main lake. Most of these semi-closed water bodies have very low water flows and poor water circulation and, therefore, poor self-purification capacities. Monitoring data shows that the water quality of Longjing Lake tended to deteriorate after the Chongqing Garden Expo Park was opened. The TN and TP concentrations cannot always meet the prescribed level of Class IV or above according to the National Environmental Quality Standards for Surface Water (GB3838-2002). Sometimes, the water quality is even worse than the lowest requirement for surface waters (Class V of GB3838-2002). The comprehensive trophic state index (TLI) in some of the semi-closed water bodies is occasionally higher than the guideline value for eutrophication control.

To improve the Longjing Lake water quality, a project was implemented to study countermeasures to solve the current problem for guaranteeing the water quality of Longjing Lake and increasing its ecological and landscape values.

13.4.2 Diagnosis of water environment problems

Based on long-term water quality monitoring regarding the main parameters specified in the National Environmental Quality Standards for Surface Water (GB3838-2002), such as DO, COD, BOD, NH_3-N, TN, and TP, the characteristics of water quality in the Longjing Lake and its catchment area were determined, in association with mathematical modeling. On average, the water quality could meet the requirement of Class IV, but seasonal and spatial variations were apparent. In February, the lake water quality was the most unfavorable due to the increased number of visitors during the Chinese New Year holidays. Conversely, in the summer season of June and July, the water quality was between Class IV and Class V. The quality of water flowing in the upstream rivers was usually good, while the quality of water in the lake area, especially in several bays with poor hydrodynamic conditions, was poor. The lake water was evaluated to be moderately eutrophic.

In the whole catchment area of the Longjing Lake, including the two rivers in the upstream, there is no centralized point source of pollution such as effluent discharge from any large-scale wastewater treatment plant, except for the treated effluents from the scattered facilities in the Garden Expo Park area, including hotels, public buildings, restaurants, and toilets. Therefore, water pollution of the Longjing Lake is mainly due to non-point sources from the park, as well as certain pollutants from the upstream rivers. The endogenous release of pollutants from the lake sediments is also a pollution source.

By pollution analysis, it has been demonstrated that the pollution loads of four typical pollutants, namely COD, TN, NH_3-N and TP entering the Longjing Lake are 2.747×10^5, 1.227×10^4, -3.47×10^3, and 3.832×10^3 kg per year, respectively. The negative pollution load of NH_3-N is an indication of a good

Table 13.1 Comparison of pollution load and water environmental capacity regarding typical pollutants (Unit: kg/a).

Pollutants	Pollution Load	Water Environmental Capacity	Capacity Surplus
COD	2.747×10^5	7.097×10^5	4.35×10^5
TN	1.227×10^4	2.813×10^4	1.586×10^4
TP	3.832×10^3	2.065×10^3	-1.767×10^3

nitrification condition in the water catchment to convert ammonia-nitrogen into more oxidized forms.

The water environmental capacity of the Longjiang Lake has also been evaluated by dividing the lake area into a number of calculation units (Zhou *et al.*, 2014) and using the established relationship between pollution sources and water quality (Li *et al.*, 2010c). As shown in Table 13.1, regarding COD, TN, and TP, the Longjing Lake has sufficient capacity to accommodate the pollution loads of COD and TP, but an insufficient capacity to accommodate the pollution load of TP. It has thus been concluded that TP can be taken as a priority pollutant for setting the target of pollutant load reduction.

13.4.3 Technology integration and demonstration

As non-point source pollution control is the main task for water quality improvement in the Longjing Lake, a demonstration project was implemented for technology integration in the Jiangnan park area of the Chongqing Expo Park. Figure 13.5 shows an aerial view of the project location. According to calculations, the non-point source pollution load in the project area accounts for 36% of the total non-point source pollution load of the Chongqing Expo Park.

The demonstration project is a centralized exhibition area for non-point source control by 'infiltration, stagnation, storage, purification, utilization and drainage', which are the measures recommended in the Technical Guidelines for Sponge City Construction. These basic measures are integrated into a landscape design according to the local conditions, including a number of engineering technologies for non-point source pollution control in the ultra-fine sand area, infiltration enhancement under complex geological conditions, road shoulder runoff control, roadside green space for runoff reduction, modular filtration in steep-slope area, micro waterscape detention, enhanced lateral flow biofilter and so on.

Non-point source control for the demonstration project is mainly by rainwater treatment. Figure 13.6 shows the technology framework. Rainwater runoff first enters a sedimentation and energy dissipation pool, where the water flow is slowed down and particulate pollutants are partially precipitated. The following

Figure 13.5 Aerial view of the integrated technology demonstrative project (photo by authors).

treatment facilities include the combined modular large gradient runoff control filter system and the micro waterscape system. The combined modular large gradient runoff control filter system is mainly used in mountainous areas for water treatment by filtration, while the micro waterscape system is mainly used for runoff retention and ecological treatment under the combined actions of plants, substrates, and microorganisms. For long-duration rainfall events, the amount of water entering the micro waterscape system may overflow. The overflow water can enter the infiltration area of the road shoulder or the permeation system for subsequent treatment before entering the Longjing Lake.

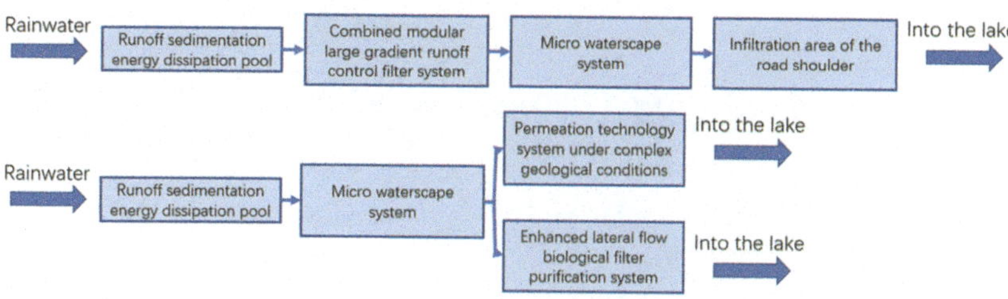

Figure 13.6 Technology framework of rainwater treatment for the demonstrative project (figure by authors).

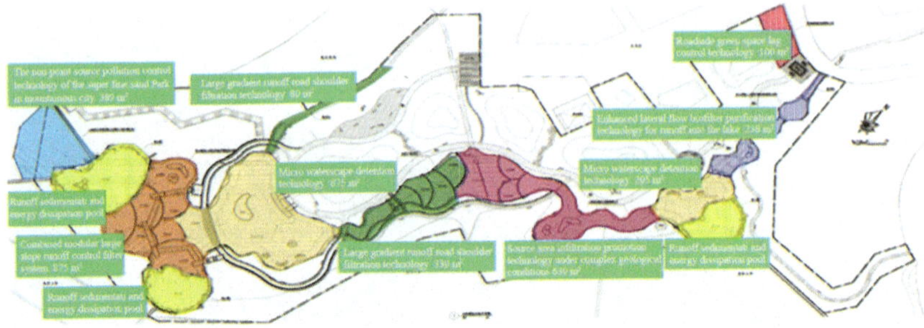

Figure 13.7 Layout of major facilities implemented in the demonstrative project area for non-point source pollution control (figure by authors).

Figure 13.7 shows the layout of the facilities implemented in the demonstration project area with adaptations to the topographic condition. Figure 13.8 demonstrates the direction of runoff water flow, and Figure 13.9 shows some photos of the rainwater treatment facilities.

13.4.4 Upstream river water quality control

Although the water quality of the upstream rivers is usually good, its further improvement is still important for guaranteeing the downstream Longjiang Lake

Figure 13.8 Rainwater flow direction in the demonstrative project area (figure by authors).

Figure 13.9 Rainwater treatment facilities integrated into a non-point source control system (photos by authors).

water quality. To protect the river water quality of the Zhaojia River, which is one of the upstream tributary rivers, Zhaojia River Ecological Park is built as a buffer zone of the river catchment. As shown in Figure 13.10, the ecological park receives water from the Huangjue reservoir on the Zhaojia River. It was built mainly by vegetation restoration to strengthen the ecological conservation capacity of the catchment and provide time and space for natural purification.

In addition to vegetation restoration, artificial ecological filters were also implemented in the Zhaojia River Ecological Park for effective removal of the

Figure 13.10 Aerial view of the Zhaojia River Ecological Park (photos by authors).

Figure 13.11 Location of ecological filters in the Zhaojia River Ecological Park (photos by authors).

occasional high concentration of solid particles. Figure 13.11 shows the location of the ecological filters. Figure 13.12 shows an overview of the ecological filters after their construction. It can be seen that the construction of these filters well utilized the steep slope of the mountainous area and natural flow direction.

13.4.5 Lake inflow quality control

The quality of direct inflow to the Longjing Lake also needs to be controlled for the reduction of pollutants. The facilities implemented for this purpose include enhanced lateral-flow biological filters and cascade pool chain systems.

13.4.5.1 Enhanced lateral-flow biological filter

The enhanced lateral-flow biological filter is a multi-stage filter system integrating sedimentation, filtration, adsorption, and biotransformation, and with radical inflow into the lake (Figure 13.13). The first stage is a filter with zeolite as filter media which can effectively remove organic substances and nutrients by adsorption as the influent flows downward through the filter layer. The second stage is an

Figure 13.12 Constructed ecological filters in the Zhaojia River Ecological Park (photos by authors).

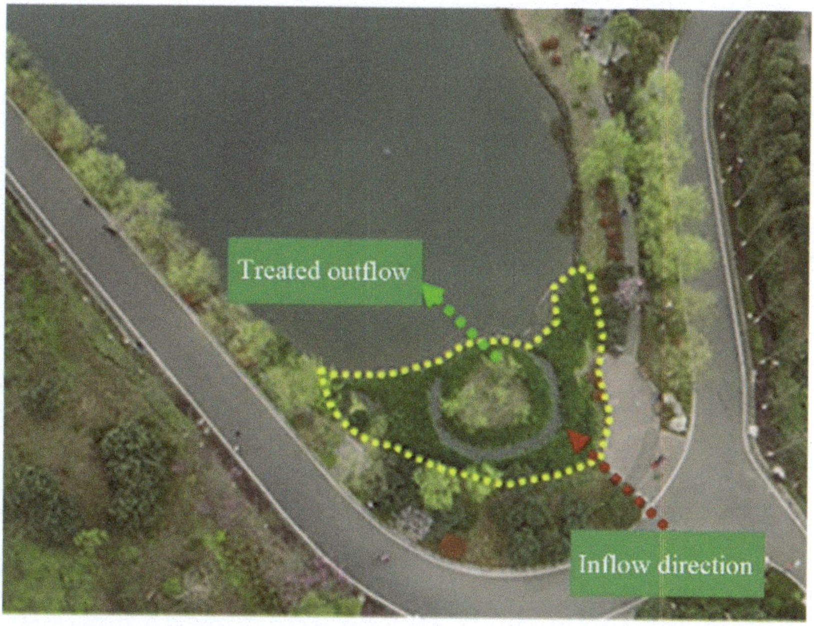

Figure 13.13 Aerial view of enhanced lateral-flow biological filter for lake inflow quality control (photo by authors).

upward-flow filter with ceramsite as the filter media. In addition to filtration, the ceramisite media are also biological barriers for microbial growth and perform the function of denitrification under anoxic or anaerobic conditions. The third stage is, again, a downward-flow filter with iron ore as filter media to enhance the adsorptive removal of phosphorous.

13.4.5.2 Cascade pool chain system

The cascade pool chain system consists of a series of deep pools by utilizing the local diverse topographic feature. Each deep pool provides a unique hydraulic and biological environment according to its geometric shape and flow pattern, and the cascade of different deep pools in series collectively perform multiple functions for enhanced water purification, such as water reaeration, extension of hydraulic retention time, and other physicochemical and biological processes. Figures 13.14 and 13.15 show the aerial view and real scene of the cascade pool chain system, respectively.

13.4.6 Pollution control in the main lake area

The main lake area is the central area of Chongqing Garden Expo Park where various services are provided to visitors. Therefore, local pollution control is also required for the assurance of lake water quality and landscape value.

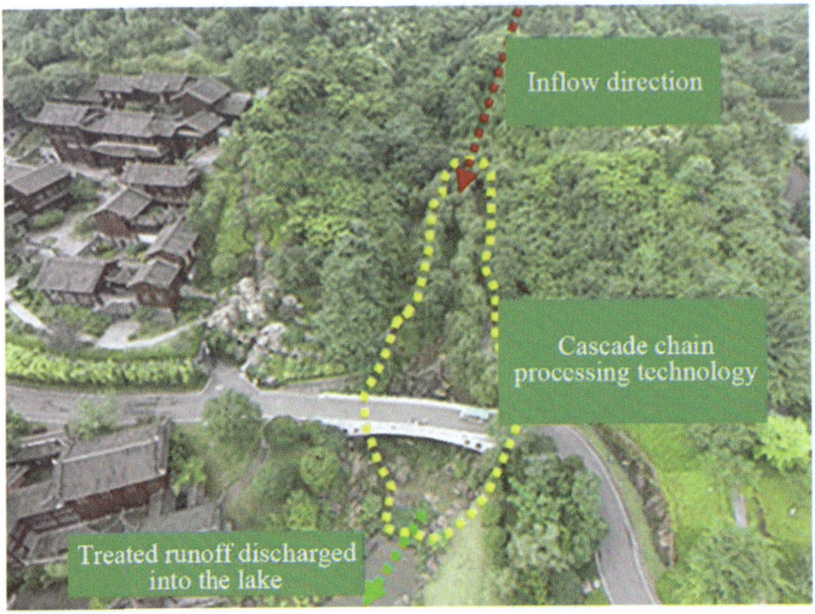

Figure 13.14 Aerial view of the cascade pool chain system for lake inflow quality control (photo by authors).

13.4.6.1 Elimination of point source pollution by ecological measures

In the central area of the Expo Park, although onsite treatment facilities are installed at each location where sewage or wastewater is generated, direct discharge of the treated effluent into the Longjing Lake or any waterways connected to the lake may still become a pollutant source. To eliminate all the point sources, various ecological measures have been taken for the treatment and onsite reuse of the

Figure 13.15 Real scene of water flow through the cascade pool chain system (photos by authors).

effluent from public toilets, restaurants, and other service buildings. According to the available space and local natural and scenic conditions, ecological ponds or constructed wetlands of different types are selected as the main treatment facilities. The final treated water is reused mostly for onsite irrigation of the green belt so that almost no pollutants from these scattered point sources are allowed to enter the lake water area.

13.4.6.2 Integrated measures for local non-point source pollution reduction

Scouring of super fine sand particles from the sandy soil by surface runoff is always a major problem in the mountainous city of Chongqing. For the reduction of local non-point pollutant sources in the central park area, a special method has been worked out to improve the soil structure by spreading selected organic polymer powders onto the sandy soil. This is proved to be a very effective measure because the polymeric molecules can easily react with the superfine soil particles and make them agglomerate with each other. As a result, the sandy soil layer becomes very stable and its permeability is much improved.

In addition to the chemical modification of the sandy soil layers, other measures have also been taken for local runoff control to further reduce non-point source pollution, such as the installation of infiltration beds and multi-functional filters at locations where runoff water is easy to accumulate or flow through. Local conditions have been well considered in designing and installation of these facilities to formulate an integrated system for effective reduction of local runoff flow and pollutant removal.

13.4.6.3 Enhancement of self-purification in the lake

To guarantee the water quality in each location of the Longjing Lake, water quality simulation was conducted by taking into account the hydraulic condition and pollution load under different circumstances. Various engineering and ecological measures have been adopted for enhancing the self-purification capacity of the lake water, especially in the bays where water tends to be stagnant. Artificial hydraulic circulation has been introduced as needed, and novel floating islands equipped with submerged thrusters have been widely applied for water circulation and oxygen enrichment, in addition to their function of water purification.

13.4.7 Overall effects of water quality control and landscape improvement for the Longjing Lake

The lake water quality has been significantly improved by the implementation of a systematic plan for water pollution control, including various engineering and ecological facilities and measures in the upstream tributary rivers, at the inflow sections of the lake, and within the lake. Since the completion of the project at

Figure 13.16 Real scene of part of the Longjing Lake in the mountain valley (photo by authors).

the end of 2014, the Longjin Lake water quality has steadily improved to reach the requirement of Class IV of the national surface water standard, even in the hottest summer seasons. Abnormal algae growth has well been controlled and the unique landscape of water and green spaces has been sustained. Figures 13.16 and 13.17 show the landscapes of the Longjing Lake at Chongqing Garden Expo Park.

Figure 13.17 Real scene of part of the Longjing Lake surrounded by restored traditional houses (photo by authors).

REFERENCES

Ahiablame L., Engel B. A. and Chaubey I. (2012). Effectiveness of low impact development practices. Literature review and suggestions for future research. Water, Air and Soil Pollution, **223**(7), 4253–4273.

Ahmed S. M., Zhou B. X., Zhao H., Zheng Y. P., Wang Y. and Xia S. B. (2020). Developing a composite vertical flow constructed wetlands for rainwater treatment. *Membrane and Water Treatment*, **11**(2), 87–95.

Arthington A. H., Naiman R. J., Mcclain M. E. and Nilsson C. J. (2010). Preserving the biodiversity and ecological services of rivers: new challenges and research opportunities. *Freshwater Biology*, **55**(1), 1–16.

Bohlin H. S., Mörth C. M. and Holm N. G. (2006). Point source influences on the carbon and nitrogen geochemistry of sediments in the Stockholm inner archipelago, Sweden. *Science of the Total Environment*, **366**(1), 337–349.

Chai H., Chen Z., Shao Z., Deng S., Li L., Xiang Y., Li L., Hu X. and He Q. (2019). Long-term pollutant removal performance and mitigation of rainwater quality deterioration with ceramsite and Cyperus alternifolius in mountainous cities of China. *Environmental Science and Pollution Research*, **26**(32), 32,993–33,003.

Deng X. J., Xu Y. P., Han L. F., Song S., Yang L., Li G. and Wang Y. F. (2015). Impacts of urbanization on river systems in the Taihu Region, China. *Water*, **7**(4), 1340–1358.

Dorchin A. and Shanas U. (2010). Assessment of pollution in road runoff using a Bufo viridis biological assay. *Environmental Pollution*, **158**(12), 3626–3633.

Fang G. H., Yuan Y., Gao Y. Q., Huang X. F. and Guo Y. X. (2018). Assessing the effects of urbanization on flood events with urban agglomeration polders type of flood control pattern using the HEC-HMS model in the Qinhuai River Basin, China. *Water*, **10**(8), 1003.

Gobel P. and Coldewey W. G. (2013). Concept for near-natural storm water control in urban areas. *Environmental Earth Sciences*, **70**(5), 1945–1950.

Gromaire M., Garnaud S., Saad M. and Chebbo G. (2001). Contribution of different sources to the pollution of wet weather flows in combined sewers. *Water Research*, **35**(2), 521–533.

Gromairemertz M. C., Garnaud S., Gonzalez A. and Chebbo G. (1999). Characterisation of urban runoff pollution in Paris. *Water Science & Technology*, **39**(2), 1–8.

Hou L., Liu F., Feng C. and Wan L. (2013). Efficiencies of multilayer infiltration systems for the removal of urban runoff pollutants. *Water Science & Technology*, **67**(8), 1851–1858.

Huang H., Wan F., Gao Y., Zhong Z. X., Annanurov S. and Zeng X. G. (2019). Study on measures to improve water quality in urban lakes: casing in Lake Nanhu in Wuhan. *Fresenius Environmental Bulletin*, **28**(4), 2625–2632.

Ilyas M., Ahmad W., Khan H., Yousaf S., Yasir M. and Khan A. (2019). Environmental and health impacts of industrial wastewater effluents in Pakistan: a review. *Reviews on Environmental Health*, **34**(2), 171–186.

Karnatz C., Thompson J. R. and Logsdon S. (2019). Capture of stormwater runoff and pollutants by three types of urban best management practices. *Journal of Soil & Water Conservation*, **74**(5), 487–499.

Knowles P., Dotro G., Nivala J. and Garcia J. (2011). Clogging in subsurface-flow treatment wetlands: occurrence and contributing factors. *Ecological Engineering*, **37**(2), 99–112.

Kwon S. B., Ahn H. W., Ahn C. J. and Wang C. K. (2004). A case study of dissolved air flotation for seasonal high turbidity water in Korea. *Water science & Technology A Journal of the International Association on Water Pollution Research*, **50**(12), 245–253.

Lawson R. and Anderson M. A. (2007). Stratification and mixing in Lake Elsinore, California: an assessment of axial flow pumps for improving water quality in a shallow eutrophic lake. *Water Research*, **41**(19), 4457–4467.

Li F. P., Zhang H. P., Zhu Y. P., Ling C. and Zhao J. J. (2010a). Spatial and temporal dynamics in the relationship of phytoplankton biomass and limnological variables in a small artificial lake. *AIP Conference Proceedings*, **1251**(1), 29–32.

Li L., Zhu R. X., Guo S. G. and Yin C. Q. (2010b). Research on spatial differentiation of urban stormwater runoff quality by source area monitoring. *Environmental Sciences*, **31**(12), 2896–2904.

Li Y., Qiu R., Yang Z., Li C. and Yu J. (2010c). Parameter determination to calculate water environmental capacity in Zhangweinan Canal Sub-basin in China. *Journal of Environmental Sciences*, **2010**(6), 904–907.

Li F. P., Zhang H. P., Zhu Y. P., Xiao Y. and Chen L. J. (2013). Effect of flow velocity on phytoplankton biomass and composition in a freshwater lake. *Science of the Total Environment*, **447**, 64–71.

Li A. D., Zhang Y., Zhou B. H. and Lu X. Q. (2018). Influence of algae blooms on DOM characteristic in water bodies in urban landscape river. *Spectroscopy and Spectral Analysis*, **38**(1), 188–193.

Lin L. F., Li T. and Li H. (2007). Characteristics of surface runoff pollution of Shanghai urban area. *Environmental Sciences*, **28**(7), 1430–1434.

Mikkelsen P. S., Jacobson P. and Fujita S. (1996). Infiltration practice for control of urban storm water. *Journal of Hydraulic Research*, **34**(6), 827–840.

Mitrovic S. M., Oliver R. L., Rees C., Bowling L. C. and Buckney R. T. (2003). Critical flow velocities for the growth and dominance of Anabaena circinalis in some turbid freshwater rivers. *Freshwater Biology*, **48**(1), 164–167.

Nie J., Gang D. D., Benson B. C. and Zappi M. E. (2012). Nonpoint source pollution. *Water Environment Research*, **84**(10), 1642–1657.

Opher T., Ostfeld A. and Friedler E. (2009). Modeling highway runoff pollutant levels using a data driven model. *Water Science & Technology*, **60**(1), 19–28.

Qi S. Z. and Luo F. (2006). Land-use change and its environmental impact in the Heihe River Basin, arid northwestern China. *Environmental Geology*, **50**(4), 535–540.

Roy-Poirier A., Champagne P. and Filion Y. (2010). Review of bioretention system research and design: Past, present, and future. *Journal of Environmental Engineering*, **136**(9), 878–889.

Schaffner M., Bader H. P. and Scheidegger R. (2009). Modeling the contribution of point sources and non-point sources to Thachin River water pollution. *Science of the Total Environment*, **407**(17), 4902–4915.

Seo M., Jaber F. H. and Srinivasan R. (2017). Evaluating various low-impact development scenarios for optimal design criteria development. *Water*, **9**(4), 270.

Shafique M. and Kim R. (2017). Green stormwater infrastructure with low impact development concept: a review of current research. *Desalination and Water Treatment*, **83**, 16–29.

Stimson J. and Larned S. T. (2000). Nitrogen efflux from the sediments of a subtropical bay and the potential contribution to macroalgal nutrient requirements. *Journal of Experimental Marine Biology & Ecology*, **252**(2), 159–180.

Takyi A. K. and Lence B. J. (1999). Surface water quality management using a multiple-realization chance constraint method. *Water Resources Research*, **35**(5), 1657–1670.

Testa J. M., Brady D. C., Toro D. M. D., Boynton W. R., Cornwell J. C. and Kemp W. M. (2013). Sediment flux modeling: simulating nitrogen, phosphorus, and silica cycles. *Estuarine Coastal Shelf Science*, **131**, 245–263.

Trifu M. C., Luca E., Daradici V. and Sgem R. (2011). Fish passes – an ecological measure for the river and habitat continuity preservation in the Cris River Basin. 11th International Multidisciplinary Scientific Geoconference, Int Scientific Conference Sgem, Sofia, pp. 359–366.

UN. (2018). 2018 Revision of World Urbanization Prospects, United Nations Department of Economic and Social Affairs. Available from: www.un.org/development/desa/publications/2018-revision-of-world-urbanization-prospects.html

Wan L. L., Ye X. Y., Lee J., Lu X. Q., Zheng L. and Wu K. Y. (2015). Effects of urbanization on ecosystem service values in a mineral resource-based city. *Habitat International*, **46**, 54–63.

Wang Q., Zhang Q. H., Wu Y. D. and Wang X. C. (2017). Physicochemical conditions and properties of particles in urban runoff and rivers: implications for runoff pollution. *Chemosphere*, **173**, 318–325.

Wen Y., James P. and Kai Y. J. (2006). Impact of urbanization on structure and function of river system – case study of Shanghai, China. *Chinese Geographical Science*, **16**(2), 102–108.

Wolff G. and Sutton S. (2010). Low Impact Development 2010 – Using the Bay-Friendly Landscape Standards to Implement Low Impact. American Society of Civil Engineers Low Impact Development International Conference (LID) 2010, April 11–14, 2010, San Francisco, California, United States, pp. 1056–1066.

Wu J., Luo J. M. and Tang L. (2019). Coupling relationship between urban expansion and lake change – a case study of Wuhan. *Water*, **11**(6), 1215.

Yousefi S., Moradi H. R., Keesstra S., Pourghasemi H. R., Navratil O. and Hooke J. (2019). Effects of urbanization on river morphology of the Talar River, Mazandarn Province, Iran. *Geocarto International*, **34**(3), 276–292.

Zhang S., Meng Y., Pan J. and Chen J. (2017). Pollutant reduction effectiveness of low-impact development drainage system in a campus. *Frontiers of Environmental Science & Engineering in China*, **11**(4), 14.

Zhou Q. (2014). A review of sustainable urban drainage systems considering the climate change and urbanization impacts. *Water*, **6**(4), 976–992.

Zhou J., Zhang Y. T. and Yang X. Z. (2001). Oxygen restoration treatment of black and stench river by artificial aeration. *China Water & Wastewater*, **2017**(4), 47–49.

Zhou J., Falconer R. A. and Lin B. (2014). Refinements to the EFDC model for predicting the hydro-environmental impacts of a barrage across the Severn Estuary. *Renewable Energy*, **62**, 490–505.

Zhou J., Liu J., Shao W., Yu Y., Zhang K., Wang Y. and Mei C. (2018). Effective evaluation of infiltration and storage measures in Sponge City construction: a case study of Fenghuang City. *Water*, **10**(7), 937.

Chapter 14

Constructed wetlands for urban water ecological improvement

Mawuli Dzakpasu, Yucong Zheng and Xiaochang C. Wang

International Science and Technology Cooperation Centre for Urban Alternative Water Resources Development, Xian University of Architecture and Technology, No. 13 Yanta Road, Xian 710055, China

14.1 INTRODUCTION

The world today faces pressing issues due to rapid urbanisation and the continuing expansion of modern cities. The urban environment faces increasing challenges of urban runoff and stormwater management. Current problems are related to the ageing of existing infrastructure, changing precipitation patterns with more frequent and intense storm events, watershed deforestation, degradation of natural wetlands, and extensive use of impervious surfaces (e.g. roads, sidewalks, driveways, and parking lots) that all result in urban flooding and pollution of water sources. As climate change is projected to intensify these phenomena further, the adoption of urban wetlands could make a more beneficial and targeted use of the multiple ecosystem services to mitigate these impacts of climate change in urban areas (Lennon, 2015). In addition, as cities expand worldwide and become densely populated, there is a respective growing demand for improved sanitary and ecological conditions and a more sustainable way to exploit urban spaces. Moreover, the increasing water demand in urban areas and freshwater withdrawal calls for more efficient water utilization. This chapter

Editors: Xiaochang C. Wang and Guangtao Fu
doi: 10.2166/9781789060768_0383

explores the use of constructed wetlands for improving urban water ecological environments and, consequently, the sustainability, resilience, and liveability of cities. A case study of a small urban retrofit project located near Tianjin, China (Tianjin Harbor Eco-Wetland Park), is presented. The Eco-Wetland Park provides an example of the technology that contributes to building a sustainable urban environment. It was constructed to improve the water quality and management issues in the Tianjin Lingang Industrial Zone of China, as part of an integrated approach aimed at achieving efficient utilisation of urban water resources. Effluent from a wastewater treatment plant was used as a water source for the Eco-Wetland Park. In addition to water quality improvement, the wetland provided many other benefits to the environment and the local community. These include improving the habitat, biodiversity, and local climate of the coastal industrial zone. The Eco-Wetland Park is now a core ecological landscape feature of the area.

14.2 WETLANDS AS IMPORTANT URBAN ECOLOGICAL ELEMENTS

Wetland systems provide a sustainable solution to issues that mainly arise in urban environments. They offer an alternative to the traditional, mostly concrete-based 'grey' infrastructure systems. They also provide an approach to the utilisation of open areas in the urban environment to deliver various services or/and create new ecosystems (Mell, 2008). Applications of urban wetlands are varied and typically include urban runoff and stormwater management, reducing the effects of urban heat islands, and air quality improvement. Applications of urban wetlands can be further expanded to include other approaches, viz., provision of ecosystem services, biodiversity enhancement, reduction of greenhouse gas emissions, among others (Haines-Young & Potschin, 2012). Thus, urban wetlands represent an extensive set of solutions for increasing the resilience of urban environments.

The fact that modern cities face increasing risks due to the increase in frequency and extent of extreme events, such as urban flooding or extended droughts, provides additional impetus for the application of urban wetlands. Evidently, the goods and services that nature provides to humans (ecosystem services) possess immense value, which should not be neglected (Potschin & Haines-Young, 2011). Thus, the strong relation of urban wetlands to sustainability cannot be overemphasised.

14.2.1 Natural wetlands in urban environments

As transitional landscapes between land and water, the most important defining factor of wetlands is its hydrology. Hydrology creates conditions for all other wetland systems (Figure 14.1). Hydrology begins with climate and geomorphology, which dictate where wetlands occur and their water depth, flow

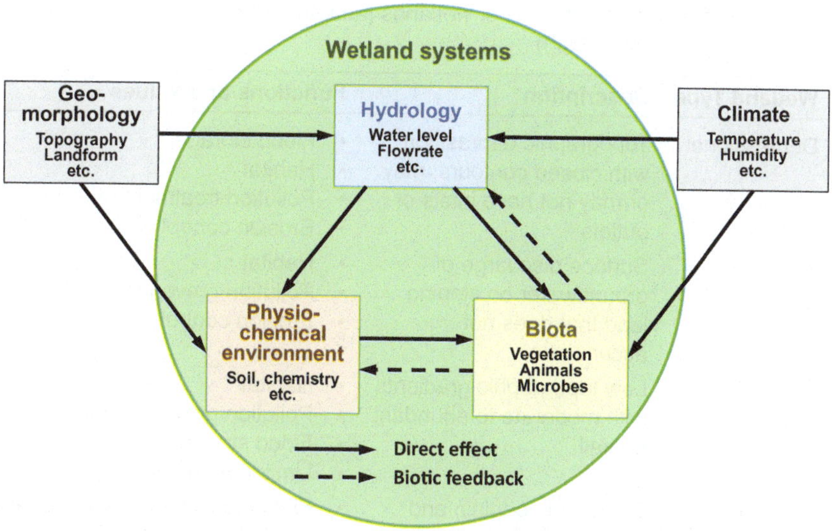

Figure 14.1 Basic components of wetlands (adapted from Mitsch & Gosselink, 2007).

patterns, duration and frequency of flooding. The major flows of water into a wetland are precipitation, surface inflows and outflows, groundwater, evapotranspiration, and tides. These water flows convey energy, sediments, nutrients, and organic material to the wetland. As a result, these flows shape the physicochemical conditions of the wetland, namely oxygen availability, nutrient availability, and pH. In turn, physicochemical conditions directly impact the biota in the wetland, affecting species diversity and ecosystem productivity (Mitsch & Gosselink, 2007). Wetland biota is, to a certain degree, adapted to the inherently dynamic environment of a wetland. However, the conditions of hydrologic flow, whether flowing or stagnant, can accelerate or slow the productivity of the wetland.

Wetlands perform a variety of functions in a landscape related to the type of wetland (Table 14.1). Wetlands offer flood protection by storing and slowly releasing floodwaters. The flood retention capacity of a wetland depends on its size, shape, location, depth to the water table, soil permeability, and slope. The ability to retain floodwaters not only protects downstream areas from flooding, but it also helps those areas avoid erosion and delay peak discharges. Wetland flood storage is most valuable in urban watersheds, where development has resulted in the production of high runoff rates. However, wetland flood storage capacity is also the most compromised in urban areas. Consequently, expensive engineered infrastructure, such as dams, channels, and levees, are typically constructed for flood drainage purposes.

Table 14.1 Watershed functions of wetlands (adapted from Wright *et al.* (2006) and Mitsch and Gosselink (2007)).

Wetland Type	Description	Functions and Values
Depressional	Topographic depression with closed contours, may or may not have inlets or outlets	• Flood storage • Habitat • Pollution treatment • Erosion control
Slope	Surface discharge of groundwater on sloping land that does not accumulate	• Habitat • Pollution prevention • Erosion control
Flat	Low topographic gradients with moderate to abundant rainfall	• Habitat • Pollution prevention • Flood storage • Limited recreation
Riverine	Occur in floodplain and riparian corridor of larger streams and rivers	• Flood conveyance and storage • Shoreline protection • Erosion control • Pollution treatment • Fish and waterfowl habitat • Recreation
Fringe	Adjacent to lakes or estuaries	• Habitat • Pollution treatment • Water supply protection • Shoreline protection • Erosion control • Recreation

Depending on the type of wetland, some wetlands recharge groundwater levels, while others are discharge points for groundwater, providing water for nearby wetlands and streams. In urban areas that rely on groundwater as drinking water supply, these functions of wetlands are useful. However, the relationship between a wetland and groundwater is complex and depends on the physical factors such as the soil permeability, vegetation density, and groundwater levels. Wetlands are connected to and help maintain the health of the larger hydrologic network, by reducing erosion, allocating water, stabilising and dispersing species communities, etc.

Fringe wetlands, or wetlands adjacent to lakes or estuaries, can provide shoreline erosion control. The vegetation and roots of fringe wetlands retain soil, absorb wave energy, and dissipate surface flows. Because wetlands retain soil, they can nourish stream banks and shorelines over time. The loss of wetlands can expose an area to unprecedented erosion, which can threaten shoreline properties and infrastructure.

Wetlands are well known for their capacity to provide habitat for a diverse number of aquatic, terrestrial, and avian species. Wetlands, noted as one of the most ecologically productive ecosystems in the world per unit area, can support high biodiversity because they produce high amounts of biomass and are composed of a patchwork of micro-habitats. Many bird species, especially migrating species, depend on wetlands for feeding, breeding, and nesting. Coastal wetlands are often nursery grounds for recreationally and commercially important fish and shellfish.

14.2.2 Constructed wetlands in urban environments

Wetland systems can transform or/and remove various pollutants (organics, nutrients, trace elements, etc.) through a series of biological, chemical, and physical processes to improve water quality (Stefanakis *et al.*, 2014). The wide range of economic and ecological benefits of wetlands stimulated the interest to exploit their natural water purification capacity for different applications, particularly for wastewater treatment. Constructed wetland ecosystems exploit many of these purification functions of natural wetlands, which have been used for the disposal and treatment of secondary and tertiary wastewater effluents for many years (Mander & Jenssen, 2002).

Nowadays, natural wetlands are occasionally used for the polishing of light-contaminated effluents in some areas. However, generally, their use for wastewater treatment purposes is mostly avoided around the world, because this could cause irreversible damage to their ecosystems. Therefore, constructed wetland systems seek to replicate the various processes occurring naturally in natural wetlands under controlled conditions for beneficial purposes. Thus, constructed wetlands are designed to mimic and enhance the functions of natural wetlands. Although the hydrology in constructed wetlands is largely artificial, they are designed with specific water flow rates in mind, and water flows are typically controlled by outlet and inlet structures.

Nonetheless, the ecology of constructed wetlands has been found to closely resemble that of natural wetlands, and offer the same general values and functions as natural wetlands such as the provision of a wide range of ecosystem services (Table 14.1). For example, constructed wetlands are capable of having as high or even higher population sizes and biodiversity as natural wetlands (Knight *et al.*, 2001). Free water surface flow constructed wetlands for various water treatment purposes are reported to support circa 361 bird species, 342 aquatic invertebrate species, 78 fish species (only one species, mosquito fish, being deliberately stocked), 22 mammal species, 10 amphibian species, and seven reptile species. In fact, constructed wetlands can reproduce all major animal groups and trophic levels that exist in natural wetlands. Also, it has been shown that constructed wetlands possess a higher value in terms of flood and stormwater control, water quality improvement and biodiversity restoration (Ghermandi *et al.*, 2010; Knight *et al.*, 2001; Masi *et al.*, 2016, 2018; Moore and Hunt, 2013; Stefanakis *et al.*, 2014). Their primary design characteristics make them more easily adopted and integrated into the urban environment.

14.3 TYPES OF WETLANDS FOR URBAN WATER ECOLOGICAL IMPROVEMENT

Constructed wetlands can be classified based on either the vegetation type or the water flow path through the system (Stefanakis *et al.*, 2014; Vymazal, 2007), as shown in Figure 14.2. Based on the flow path, two main types can be distinguished, namely, free water surface (FWS) constructed wetlands, and subsurface flow (SSF) constructed wetlands.

14.3.1 FWS wetlands

The design of FWS wetlands includes a water column of 10–50 cm above a substrate layer, usually soil. FWS wetlands most closely resemble natural wetlands and have the most ecological potential of the two types. This type of wetland is used for stormwater and other non-point source treatment purposes (Vymazal, 2013). Due to a relatively low cost per unit area, they generally find their greatest application in high flow volume, low pollutant concentration situations. In domestic and municipal wastewater treatment applications, they are usually found downstream of other treatment units and are often considered a tertiary or polishing step. Aesthetic and habitat values are often as important to the design as water quality improvement. The physical structure of an FWS wetland is as diverse as its potential application. They may be lined or unlined,

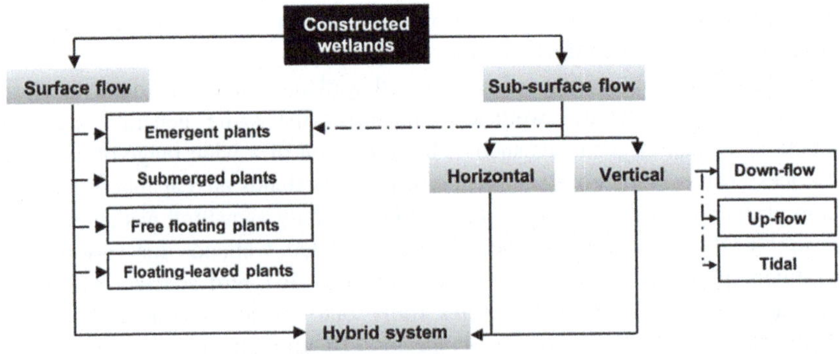

Figure 14.2 Classification of the various types of constructed wetlands (adapted from Vymazal and Kröpfelová, 2008).

constant or variable depth, completely or partially vegetated, the vegetation can be emergent, submerged or floating and they can vary in size from a few square meters to several square kilometres.

Nevertheless, there are several essential defining features. The water level is maintained above a rooting matrix of soil, sand or gravel that supports the growth of wetland plants that can survive continuously flooded conditions. Flow is horizontal but may take a circuitous path from inlet to an outlet at a very slow velocity (Figure 14.3).

FWS wetlands may comprise a forebay, marsh zone, and deep zones (Figure 14.3). The forebay is a deep pool, which receives water from an inlet structure before that water enters the main wetland. The depth of the forebay allows it to capture sediments, which are abundant in stormwater. Because of its function, it is usually constructed with hard edges that facilitate cleaning. The forebay is typically sized to store circa 10% of the total wetland's water volume. The main wetland is composed of alternating marsh and deep zones. Alternating these zones helps ensure that water is maintained at plug-flow conditions and that water is exposed to both shallow wetland areas and deep areas where different water treatment processes occur. Other applications of FWS wetlands include floating wetlands, stormwater wetlands, and permanently flooded bio-swales. FWS wetlands depend on a diverse set of pollutant removal mechanisms,

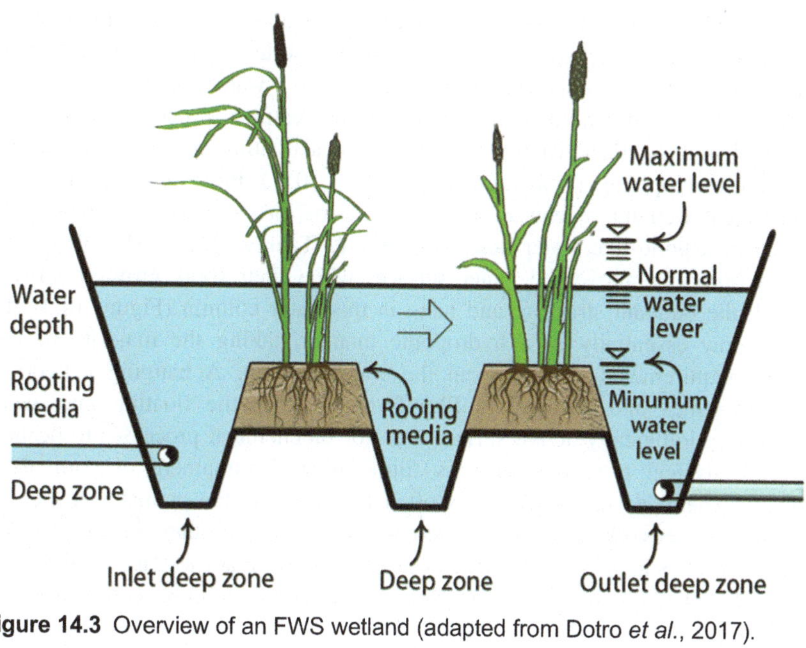

Figure 14.3 Overview of an FWS wetland (adapted from Dotro *et al.*, 2017).

including physical sedimentation and pollutant degradation by chemical, microbial and photo pathways. More so than other wetland variants, FWS wetlands simultaneously promote both aerobic and anaerobic processes and organic matter loading rates often determine which dominates. The rooting layer is largely anaerobic, especially after the system matures and a layer of detritus consisting of dead vegetation and incoming sediment establishes over it. The lower levels of the water column can range from anaerobic to aerobic depending on pollutant loading rates, depth of water column and distance from the flow entry point. FWS wetlands are designed so that the upper layers of the water column are always aerobic to prevent odour releases and promote the death of pathogenic organisms. Virtually all redox-dependent reactions, including nitrification and denitrification, are possible in the FWS wetlands due to this array of redox conditions. Open water areas allow sunlight to penetrate and enhance photodegradation. Plant uptake in an FWS system plays a more significant role in nutrient removal than in other wetland types. Plants also release small amounts of oxygen and organic carbon compounds into the rooting matrix, fuelling both aerobic and anoxic microbial degradation processes.

14.3.2 Floating wetlands

Floating wetlands represent a group of wetland technologies where a buoyant structure is used to grow emergent macrophytes on a pond, lake, river or similar water body. Floating wetlands offer opportunities to providing ancillary benefits, such as enhancement of habitat and aesthetic values of urban environments, applications which include stormwater, polluted water canals, combined sewer overflows (CSO), domestic wastewater, and water supply reservoirs. Furthermore, floating wetlands mimic the water treatment processes occurring in natural floating wetland islands and consist of a floating structure planted with emergent macrophytes (Schwammberger *et al.*, 2017). Floating wetlands consist of emergent wetland vegetation growing on a mat or structure that floats on the surface of a pond-like water body (Headley & Tanner, 2012). The plant stems remain primarily above the water surface, while their roots grow downwards through the buoyant structure and hang in the water column (Figure 14.4). The plants grow essentially in a hydroponic manner, taking the majority of their nutrient requirements directly from the water column. A hanging network of roots, rhizomes and attached biofilm forms beneath the floating mat, which provides a biologically active surface area for biochemical processes to occur as well as physical processes such as filtration and entrapment of particulates. Thus, a general design objective is often to maximise the contact between the root-biofilm network and the polluted water passing through the system. The depth of root penetration will depend largely on the plant species used and the physicochemical conditions that develop in the water column beneath the floating plants.

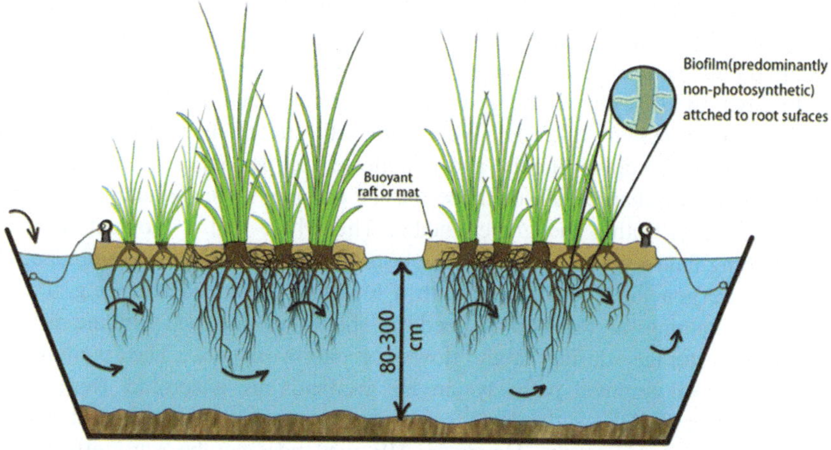

Figure 14.4 Schematic of a typical floating treatment wetland system (adapted from Headley and Tanner, 2012).

14.3.3 SSF wetlands

Subsurface flow wetlands are typically gravel beds and can be of either vertical flow (VF) or horizontal flow (HF). Figure 14.5(a) shows a schematic of a typical VF wetland. The sand and/or gravel bed is planted with emergent macrophytes. Influent water is loaded intermittently to the filter surface, and the large amount of water from a single loading causes good distribution of inflow water on the surface. The water percolates through the substrate, which gradually drains and is collected by a drainage network at the base of the filter. Between loadings, oxygen re-enters the pore space of the media, transporting oxygen into the filter bed in order to sustain aerobic microbial processes. The whole bed is isolated from the surrounding land by a combination of plastic liner and geotextile membrane. Due to the highly oxidising conditions in the filter bed, VF wetlands

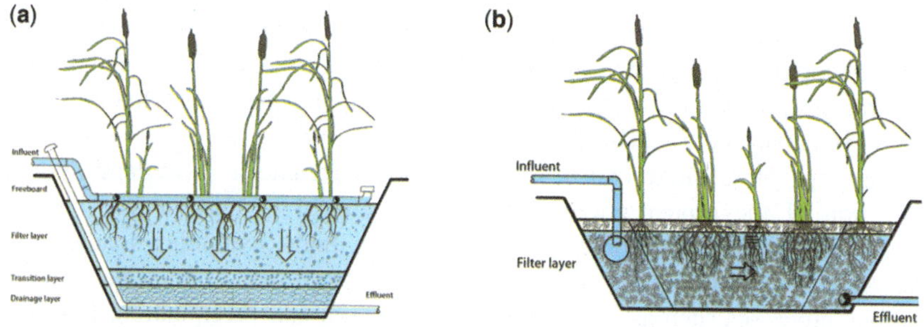

Figure 14.5 Schematic of subsurface flow wetlands: (a) VF wetland (b) HF wetland (adapted from Dotro *et al.*, 2017).

with intermittent loading are extremely efficient for the removal of organic matter. They are also suitable when strictly aerobic processes such as nitrification are required (Garcia *et al.*, 2010; Langergraber & Haberl, 2001; Dotro *et al.*, 2017).

In a typical HF wetland, the gravel bed is saturated and planted with emergent wetland plants (Figure 14.5(b)). Water enters the treatment system at one end, flows through the gravel media, and is collected on the opposite end of the bed prior to being discharged. A standpipe located outside of the wetland bed controls the water level within the gravel media. The whole bed is isolated from the surrounding land by a combination of plastic liner and geotextile membrane. HF systems are generally constructed with a longitudinal sloped base (e.g. 1%) to facilitate draining of the bed if needed. The remaining bed volume is used for water storage during high flows or storm events. The predominant microbiological removal pathways in HF wetlands are anaerobic. Removal of total nitrogen in HF wetlands is somewhat restricted due to limited aerobic conditions for nitrification. However, HF wetlands can be very effective at denitrification provided that there are sufficient nitrate and carbon present in the water column (Dotro *et al.*, 2017).

According to the vegetation type, wetlands may be further classified into those with emergent macrophytes or submerged macrophytes. The most common systems are those with rooted emergent macrophytes (Stefanakis *et al.*, 2014). When more than one wetland type is combined in one facility, this is called a hybrid wetland system.

The two key components of a wetland are the plant species and the substrate media. The most widely used emergent plant species are common reeds (*Phragmites australis*), cattails (*Typha latifolia*), and *Scirpus* spp. These species are found in most regions around the world (Stefanakis *et al.*, 2014), albeit other locally available species may also be used, for example, bamboo in tropical regions (Masi, 2009; Yu *et al.*, 2012). The primary consideration is that the selected species should be native, i.e. already adapted to the local climate and also tolerant against the pollutant loads. Indigenous species are always preferred in wetlands and not exotic ones, to prevent potential risks such as invasion of the exotic species and/or diseases (Stefanakis *et al.*, 2014). The role of wetlands plants in water treatment is mostly indirect, whereby they promote and support the development and growth of microbial communities along their roots, through the transfer and release of oxygen in microsites along the roots (Ramírez *et al.*, 2019; Stefanakis *et al.*, 2014, 2016).

The type of substrate matrix is also a critical parameter in wetland systems. Selection of the proper gravel grain size is crucial to prevent any clogging issues due to inappropriate porosity and/or high hydraulic loads. Also, substrates are capable of removing some constituents from water through various processes such as ion exchange, adsorption, and precipitation. Plants are established in the substrate matrix, which also provides filtration effects and, together with the plants, supports the various transformation and removal processes (Stefanakis *et al.*, 2014; Vymazal, 2007) that occur in wetlands. Common substrate materials used in wetland systems include natural materials (e.g. minerals, rocks and soils), synthetic materials (e.g. synthetic zeolites, activated carbon) and industrial by-products (e.g. steel slags) (Stefanakis *et al.*, 2014; Vohla *et al.*, 2011).

Wetlands can also be classified, according to their function and purpose, into three main application areas, namely, habitat creation, stormwater control, and wastewater treatment (Stefanakis *et al.*, 2014). Wetlands offer various opportunities to create new habitats to support wildlife and enhance biodiversity. The main goal is to exploit the ecological benefits of wetlands and not only their function as a treatment system (Knight, 1997). The main components of a wetland (i.e. presence of water and vegetation) make them suitable for the creation of new ecological habitats or the restoration of degraded ecosystems, by attracting wildlife species, especially birds, and establishing a green area. These systems can also be utilised as a source of food and fibre, and as public recreation and education sites (Knight, 1997; Sundaravadivel & Vigneswaran, 2001). By contrast, wetlands for stormwater control are systems used to receive urban runoff during flood events (Tsihrintzis & Hamid, 1997). Their implementation increases the stormwater storage capacity and infiltration volumes while reducing the volume of water reaching sewer systems and eventually, wastewater treatment plants. Within the urban hydrologic cycle, these wetland types may significantly contribute to the integrated management of urban water and also provide the ability to recycle stored water volumes (Shutes *et al.*, 2010; Sundaravadivel & Vigneswaran, 2001). The wetlands for wastewater treatment include systems designed to exploit the naturally occurring treatment processes occurring in wetlands to purify wastewater from various sources (Kadlec & Wallace, 2009; Stefanakis *et al.*, 2014; Vymazal, 2007).

14.4 URBAN WETLAND DESIGN, OPERATION AND MANAGEMENT

Wetland design approaches range from simple rules-of-thumb, regression-based approaches, and loadings charts to more advanced calculations that take into account factors such as hydraulic loading rates (HLR), non-ideal flow, background concentration, and pollutant weathering (Brix & Johansen, 2004; Kadlec & Wallace, 2009; Rousseau *et al.* 2004; Wallace & Knight, 2006; Dotro *et al.*, 2017). The design of physical and operational conditions varies depending on factors such as wetland type, influent and effluent concentrations, hydraulic and mass loadings, size, aspect ratio and depth, climate and associated water gains and/or losses (e.g. rainfall, evapotranspiration, etc.), ecology and plant community, and open water fraction of FWS wetlands. Except for the rules-of-thumb approach, all others consider a specific pollutant (e.g. BOD) to be removed for a particular water quality target. In practice, most treatment wetlands

Table 14.2 Processes required to reach specific treatment targets (compiled by authors).

Contaminant Group	Physical	Chemical	Biological
Suspended solids	Sedimentation Filtration Emulsification Flocculation	Adsorption Oxidation	Biodegradation Predation
Oxygen demand • Biochemical oxygen demand • Chemical oxygen demand	Sedimentation Filtration Flocculation	Oxidation	Biodegradation
Nitrogenous compounds • Organic N, NH_3, NH_4, NO_3^-, NO_2^-	Sedimentation Volatilization Absorption		Nitrification Biodenitrification Microbial uptake Plant uptake
Phosphoric compounds • Organic P, PO_4^{-3}	Sedimentation Filtration	Precipitation Adsorption Chelation	Microbial uptake Plant uptake
Metals • Al, As, Cd, Cr, Cu, Fe, Pb, Mn, Ni, Se, Ag, Zn	Sedimentation Filtration Flocculation Emulsification	Oxidation Precipitation Adsorption Ion exchange Chelation	Biodegradation Phytodegradation Phytovolatilization
Pathogens • Bacteria • Virus	Sedimentation Filtration Flocculation	Oxidation UV radiation Adsorption	Antibiosis Predation Die-off

are designed to remove multiple pollutants, and design calculations need to be carried out for all pollutants of interest to select the resulting design that enables all the target pollutants to be removed.

Once specific design objectives are defined, the processes that are required to deliver them need to be identified. Table 14.2 summarises the processes required to reach typical treatment targets. As the main treatment objective is improving water quality, most processes are related to this aspect. Table 14.3 summarises the major processes occurring in the various wetland types.

14.4.1 Design of wetlands for stormwater control

The urban water cycle is notably different from that of undisturbed natural land. Urban areas have lower rates of infiltration (both shallow and deep) and relatively

Table 14.3 Processes occurring in various wetland types (compiled by authors).

Treatment Process	VF Wetland	HF Wetland	FWS Wetland
Sedimentation			+
Filtration	+	+	o
Aerobic degradation	+	o	o
Anaerobic degradation		+	o
Nitrification	+		o
Denitrification		o	o
Sorption	o	o	
Precipitation		o	o
Plant uptake			o
Habitat creation	o	o	+

+ indicates primary processes occurring a wetland type.
o indicates processes occurring to some extent, but that the wetland type is not primarily designed for this process.

limited evapotranspiration due to extended coverage of land with building infrastructure, roads, and pavements, which considerably limits green spaces. The result is significantly higher volumes of runoff occurring in urban areas, which can exceed that of a natural land by about 50% (McGrane, 2016; Zölch *et al.*, 2017). For example, stormwater volumes generated from a unit area of parking lots have been estimated at 16 times higher than that generated from a meadow (Schueler, 1994). Public and private urban infrastructure today face frequent damages due to uncontrolled stormwater runoff.

Stormwater runoff from urban areas is extremely variable in both pollutants loads and water volumes. It is characterised by a significant microbial contamination, e.g. caused by animal excreta, and sometimes high organic loads due to littering and cleaning of roads. Combined sewer overflows and illicit connections in separate sewer systems contribute to nutrient and additional microbial loads in stormwater. Stormwater has a high load of very fine particles which do not necessarily settle, flocculate or precipitate even over extended periods due to their size, electrostatic charge and low organic load. A majority of the pollutant loads are attached to these fine particles (Boogaard *et al.*, 2014). In addition, the organic pollutants in stormwater runoff, especially those originating from highway runoff, are not readily biodegradable.

Wetlands treating stormwater or CSO are mainly targeted to hold and retain peak flows, to reduce the suspended solids load by filtration and to reduce the soluble and particulate pollutants through adsorption and biodegradation. Thus, wetlands for stormwater control may be designed to perform dual functions, namely, storage and treatment. Stormwater to be treated must be stored on the wetland surface, which requires an adequate storage volume and a throttled outflow. This is necessary to provide retention times (in the case of FWS wetlands) or filtration velocities (in case of SSF wetlands) compatible with good treatment efficiency. However, the storage function can also be a target by itself in order to assure flood protection of downstream areas. In some cases, an additional storage volume can be provided for water which does not have to undergo full treatment. In all cases, at some point excess water has to be evacuated by overflows. In some cases, legal limitations of the outflow can exceed the technical requirements for treatment and can thus become the key parameter for dimensioning.

In contrast, the treatment function is driven by the primary targets of solids, especially fine suspended solids, and, to a lesser extent, dissolved substances. Wetlands can also be designed to allow for the biodegradation and oxidation of dissolved organics during dry weather periods, where dissolved pollutants are retained by sorption on plants and sediments (in FWS wetlands) or the filter matrix (in SSF wetlands) during storm events. The treatment efficiency is, thus, at its best if the wetland works in two phases: a first phase, during the storm event, where the pollutants are retained by filtration or sorption, and a second phase of varying duration, during the resting period for biodegradation of the organic pollutants.

These types of wetlands are typically sized by one of two methods: design storm or percentage of the contributing watershed (Kadlec & Wallace, 2009). Both methods calculate a resulting wetland size and hydraulic loading rate (HLR), two variables that define the pollution reduction performance of a wetland. The storage capacity of the wetland can also be determined by hydraulic modelling, based on the maximum tolerable overflows in a given time. This gives the storm event to be stored and treated, e.g. the monthly or annual event, the stochastic occurrence of events and their intensity, the runoff patterns generated by these events and the throttled outflow of the wetlands. The treatment capacity needs to be adapted to the pollution loads and runoff patterns specific to the catchment, considering first flush effects.

14.4.2 Design of wetlands for wastewater treatment

The predominant microbiological removal pathways in HF wetlands are anaerobic. When used for secondary treatment of domestic wastewater, HF wetlands are generally capable of removing BOD and TSS to a reasonable extent, albeit the performance of individual systems varies considerably depending on influent concentrations and HLRs. Removal of total nitrogen in HF wetlands is somewhat restricted due to limited aerobic conditions for nitrification. However, HF wetlands can be very effective at denitrification provided that sufficient nitrate-nitrogen and carbon are present in the water column. Removal of phosphorus in HF wetlands is not sustainable over the long-term unless special reactive media is used.

Consequently, guidance for HF wetlands design varies greatly. They can be sized using simple, specific surface area requirements (m^2/PE), maximum areal loading rates (e.g., g BOD/m^2·d), or more sophisticated methods such as loading charts or the P-k-C* approach (Kadlec & Wallace, 2009; Dotro *et al.*, 2017). Examples of such design guidance values are 5–10 m^2/PE for wetlands used as a secondary treatment step or 4–8 g BOD/m^2 d at HLRs of 20–40 mm/d (Vymazal & Kröpfelová, 2008; Vymazal *et al.*, 1998; Wallace & Knight, 2006; Dotro *et al.*, 2017).

VF wetlands with intermittent loading are especially effective in removing contaminants that are degraded aerobically. For domestic and municipal wastewater, bulk organic contaminants (BOD or COD) and ammonia-nitrogen are removed mainly through aerobic microbial processes. In contrast, suspended solids, including pathogenic organisms, are removed via physical filtration. The treatment efficiency of VF wetlands is directly related to the type of filter material used. Where fine materials are used, the retention time of the water in the filter is more prolonged, often enabling higher removal efficiencies. However, the HLRs are limited, as it takes longer for water to infiltrate and the potential for clogging increases. Coarser filter materials enable higher HLRs and less clogging potential, albeit removal efficiencies are lower. This can be partially overcome in some cases by increasing the depth of the main layer. Available design guidelines for VF wetlands are based on empirical rules-of-thumb, such as those using specific surface area requirements of 3–4 m^2/PE or maximum organic loading rates of 20–27 g COD/m^2 d (Brix & Johansen, 2004; DWA, 2017; ÖNORM, 2009; Dotro *et al.*, 2017).

14.4.3 P-k-C* approach

The most recent kinetic equation for representing pollutant degradation in wetlands is the modified first-order equation with a non-zero background concentration (Dotro *et al.*, 2017). The performance of wetlands is demonstrated to be well represented by the P-k-C* model (Equation (14.1)) (Kadlec & Wallace, 2009; Dotro *et al.*, 2017):

$$\frac{C_o - C^*}{C_i - C^*} = \frac{1}{(1 + k/Pq)^P} \tag{14.1}$$

where C_o = effluent concentration, mg/L; C_i = influent concentration, mg/L; C^* = background concentration, mg/L; k = first-order reaction coefficient, m/d; q = hydraulic loading rate, m/d; P = apparent number of tanks-in-series (TIS), dimensionless.

The P-k-C* equation models the reduction of a specific pollutant in a wetland. If the area of the wetland is varied, the P-k-C* model yields a curve of the pollutant removal. The gains in reducing the effluent concentration of a pollutant decrease as the wetland size increases. Thus, after a certain point, reductions in pollution concentrations may not be worth increasing the wetland size. Consequently, the pollution removal curves can be used to choose an optimal wetland size for the reduction of a specific pollutant. Therefore, the P-k-C* model offers the opportunity to experiment with wetland shapes and configurations, which are not a factor in the model. In theory, so long as plug-flow conditions are maintained, there is the freedom to be creative with the wetland design to fulfil other goals, such as provide amenity values.

Information required for calculating the rate coefficient using the P-k-C* approach includes the physical attributes of the system such as length, width, and effective depth of the treatment cell, as well as the porosity of the porous medium, operational data such as flow rate(s), effluent water temperature, influent and effluent pollutant concentrations, as well as estimated parameters (for systems providing secondary treatment of domestic wastewater, P and C^* are often estimated) (Kadlec & Wallace, 2009; Dotro *et al.*, 2017).

14.4.4 Hydrologic budget

For wetlands used to treat stormwater or the excess flow of domestic wastewater during rain events in CSOs, the importance of a hydrologic budget to proper design cannot be overemphasised. This is because the majority of the water in the wetland is responding to rainfall from areas outside of the wetland. Nonetheless, hydrology must be considered even in cases where the wetland is designed exclusively to treat domestic wastewater (Dotro *et al.*, 2017). Several water fluxes that must be considered in addition to the influent and effluent include stream inflow and outflow, precipitation, catchment runoff, snowmelt, groundwater discharge and recharge, bank loss, and evapotranspiration (Kadlec & Wallace, 2009; Dotro *et al.*, 2017). A hydrologic budget is represented by Equation (14.2) (Kadlec & Wallace, 2009; Dotro *et al.*, 2017):

$$Q_i - Q_o + Q_c - Q_b - Q_{gw} + Q_{sm} + (P \times A) - (ET \times A) = \frac{dV}{dt} \tag{14.2}$$

where A = wetland surface area, m^2; ET = evapotranspiration rate, m^3/d; P = precipitation rate, m/d; Q_b bank loss rate, m^3/d; Q_c = catchment runoff, m^3/d; Q_{gw} = infiltration rate to groundwater, m^3/d; Q_i = influent flow rate, m^3/d; Q_o = effluent flow rate, m^3/d; Q_{sm} = snowmelt rate, m^3/d; t = time, d; V = water volume in wetland, m^3.

Once the fluxes are estimated, water flow through the wetland is determined by the application of two fundamental concepts, namely, conservation of mass (continuity) and conservation of momentum. Application of the momentum equation is different for FWS, HF, and unsaturated VF wetlands. A power law similar to Manning's equation is usually applied to FWS wetlands, and Darcy's Law is usually applied to HF systems (Kadlec & Wallace, 2009; Dotro *et al.*, 2017).

14.4.5 Operation and maintenance

The most critical operational issue for subsurface wetlands is clogging. Clogging occurs when the pore spaces in the filter media are filled with either organic or inorganic solids, thus limiting the contact area and time between the biofilm and the water to be treated. Clogging can occur in any (biological) filter and has been reported for both HF and VF wetlands (Knowles *et al.*, 2011). For HF wetlands providing treatment of domestic wastewater, clogging is most commonly caused by excessive organic and/or solids loading onto the gravel bed. This is often due to improper maintenance of the pre-treatment systems, where the HF wetland is used for secondary treatment or final settling tanks in the case of tertiary HF wetlands, or poor dimensioning of the wetland itself. Hydraulic and solids loading rates that are at the top end of recommended values have been suggested as the main factors resulting in the reported clogging of HF systems. This can be a result of poor design or deliberate use of HF beds for solids storage rather than treatment (Dotro & Chazarenc, 2014). In either case, it is the net accumulation of solids in the pore spaces that results in overland flow and a clogged system. Clogging can thus be minimised and the bed lifespan extended by selecting appropriate media (e.g. gravel vs. sand) and both hydraulic and mass pollutant loading rates. Also, upstream processes should be correctly maintained to enable the bed to operate within the range of its intended design.

Clogging in VF wetlands may also result from the insufficient removal of sludge from the primary treatment step (e.g. septic tank). Unremoved sludge, when transported to the filter surface, will clog the filter. Several other operational problems can result from poor design and/or problems during the construction phase. Problems during design and/or construction that should be avoided include insufficient protection of VF wetland surface from surface water and superficial runoff, unsuitable filter media, uneven slope of the filter surface, and uneven distribution of wastewater in intermittently loaded systems (Mitterer-Reichmann, 2012). Routine checks for proper O&M of wetlands include the following (Dotro *et al.*, 2017):

- Upstream treatment tanks (secondary treatment HF) and final settling tanks (tertiary treatment HF) must be emptied regularly to prevent solids carryover to the HF wetland. The sludge from the primary treatment unit must be removed in order to prevent sludge drift to VF wetland beds. In addition, if pumping is required, the equipment must be maintained according to the manufacturer's specifications (e.g. lubrication).
- Uneven influent distribution can result in solids or organic loading over a small portion of the intended influent area, and result in clogging. For surface-loaded systems, it is important to ensure that influent water is evenly delivered across the width of the wetland bed. For HF wetlands that have subsurface loading, the distribution pipes must be adequately designed and should contain inspection ports so that the influent header can be periodically washed out and/or cleaned. Intermittent loading to VF wetlands should be checked frequently by measuring the height difference in the well before and after a loading event.

- Outlet level control structures should be checked on a routine basis. The water level should be maintained 5–10 cm below the surface of the gravel. If a decrease in the height of the outlet control structure does not result in a decrease in the water level within the gravel bed, further investigations may be necessary to assess the extent of clogging in the gravel bed.
- Surface-loaded systems should be monitored for sludge accumulation. Sludge accumulation in the inlet zone of the bed should be measured at least once a year. Action limits should be set to trigger intervention (refurbishment) actions based on the sludge build-up rate (cm/yr) and the available storage capacity within the freeboard of the wetland cells.
- Wetland vegetation should be monitored to ensure that unwanted plant species (weeds) do not overtake the intended plant community. In the first two full growing seasons, weeds should be removed as needed. In temperate climates, the plant litter provides extra insulation during the winter. In hot and arid climates, thatch may accumulate indefinitely, and plant harvesting may be necessary.

For wetlands treating stormwater pollution, regular maintenance will be required to maintain high treatment performance. One of the most important maintenance tasks is clearing the forebay of sediments, approximately every 2–5 years, depending on sediment levels.

14.5 TECHNOLOGY ADVANCES FOR INTENSIFYING WETLAND PERFORMANCE

14.5.1 Hybrid wetlands

Due to the limited treatment performance and different pollutant removal mechanisms of all types of wetlands, hybrid wetlands have been developed, which include VF, HF or FWS wetlands. Hybrid wetland systems are combinations of various wetland types, mainly VF wetland and HF wetlands, aiming to improve the overall treatment efficiency (Cooper, 1999; Cooper *et al.* 1999; Vymazal, 2011; Vymazal *et al.* 1998, 2006). The concept is to exploit the advantages of one wetland type to offset the disadvantages of the other. For example, hybrid systems combining VF and HF wetlands in sequence was developed mainly for nitrification-denitrification treatment trains to produce good quality effluents (Cooper *et al.*, 1999; Kadlec & Wallace, 2009; Vymazal, 2013).

Thus, the fact that HF wetlands have lower nitrification capacity due to limited oxygen transfer capacity (OTC) can be offset with VF wetlands which are more effective in nitrification (higher OTC) (Vymazal *et al.*, 1998). On the other hand, HF wetlands provide suitable conditions for denitrification, in contrast to VF wetlands. The first attempt to combine various wetland types was the design of a two-stage system: parallel VF wetlands followed by HF wetlands in series (Seidel, 1965). Generally, there are two common types of hybrid systems, a stage with VF wetland units followed by HF wetland units in series and an HF wetland stage followed by VF units (Cooper, 1999, 2001). Today, the first type is the most widely used hybrid system (Kadlec & Wallace, 2009; Vymazal, 2005).

However, to achieve higher removals of total nitrogen, other types of hybrid wetlands, including FWS and multistage wetlands, have also been established (Vymazal, 2013). Thus, the need for aerobic and anaerobic environments to reach specific water quality targets highlights the potential interest in coupling reductive and oxidative processes in wetlands (Vymazal, 2005). Hence, research into combinations of the different types of wetlands for the enhanced removal of conventional and emerging contaminants is burgeoning (e.g. Ávila *et al.*, 2014; Nivala *et al.*, 2019; Sgroi *et al.*, 2018).

The available evidence indicates that specific processes are involved in the removal of particular types of pollutants in wetlands, albeit complex biological and physicochemical processes may simultaneously occur, including photodegradation, volatilisation, sorption, plant uptake and accumulation, as well as biodegradation (aerobic and anaerobic), mainly depending on the design of the wetlands.

14.5.2 Aerated wetlands

The use of artificial aeration in wetlands has been proposed to improve oxygen transfers. The idea is to use an air blower or injector connected to a subsurface network of air distribution pipes to provide compressed air to the wetland bed. This concept has been implemented in HF wetlands, where oxygen availability is highly limited (Stefanakis *et al.*, 2014). Nonetheless, artificial aeration of VF wetlands, usually saturated VF wetland systems, have also been reported.

In FWS wetlands and ponds, aeration creates air bubbles, causing a hydrodynamic mixing effect, which results in a more uniform distribution of the dissolved oxygen at a considerable distance from the air diffuser (Nivala *et al.*, 2013a). By contrast, in SSF wetlands containing gravel matrix, the hydrodynamic mixing is limited, as is the distance an air bubble can reach, usually not exceeding 30 cm. Therefore, adequate aeration in SSF wetlands could be implemented via the uniform distribution of small air quantities across the bottom of the wetland bed (Wallace, 2001). Supplemental aeration of VF wetlands in this way combines the action of the vertical downward drainage of the wastewater with the up-flow movement of air bubbles, resulting in a perfect water mixture within the wetland bed. Supplemental aeration of wetlands has, thus, been promoted as a positive alternative to enhance total nitrogen removal.

Improved results are reported for aerated saturated VF wetlands receiving primarily treated wastewater (Nivala *et al.*, 2013b) over non-aerated ones. Intermittent aeration (multiple on-off aeration cycles per day) favours total nitrogen removal through denitrification processes facilitated by alternating aerobic and anoxic conditions in the wetland bed (Dong *et al.*, 2012; Pan *et al.*,

2012). Also, artificial aeration enhances the organic matter decomposition and the nitrification process, compared to non-aerated systems (Pan *et al.*, 2012). Aerated wetlands are reported to have organic matter removal rates 10- to 100-fold higher than conventional wetlands (Nivala, 2012). Also, aerated VF wetlands can intensify the ammonia-nitrogen removal via intense nitrification, up to 10 times higher than non-aerated wetlands (Wallace *et al.*, 2006). Moreover, this efficiency could also be sustained even at low temperatures. This means that the footprint is also significantly reduced, making aerated wetlands a somewhat compact system.

Aerated wetlands are generally designed with a coarse gravel media and a saturated depth of at least 100 cm. In most cases, the water level in the wetland is kept 5–10 cm below the surface of the matrix layer (Dotro *et al.*, 2017). Overall, artificial aeration appears as a promising alternative when additional oxygen amount is desirable because such wetland systems are capable of treating strong wastewater at high loading rates, with considerably lower footprints. Nevertheless, the use of aeration devices increases the total cost of the facility as well as its energy consumption. Therefore, the use of renewable energy sources such as wind or solar power has been suggested to reduce operation costs (Nivala *et al.* 2013a).

14.5.3 Effluent recirculation

Recirculation involves returning and mixing a portion of the wetland effluent with the influent. Effluent recirculation has been proposed as an operational modification to improve organic matter and nitrogen removal, especially in highly aerobic VF wetlands. Removal of total nitrogen is enhanced because effluents with appreciable nitrate concentrations but limited organic matter is mixed with influents low in nitrate but high in organic carbon, allowing denitrification to take place. This modification increases the wastewater retention time within the bed, reduces the washout of necessary nitrifying microorganisms, and provides additional oxygen input by convection which is directly related to the ratio of recirculated water (Platzer, 1999; Sun *et al.*, 1999). The end result is improved performance, especially concerning the nitrification of the ammonia load.

Where recirculation sends nitrified effluents back to the primary stage (e.g. sedimentation tank), the carbon content of the raw wastewater coupled with the anoxic conditions in the tank could also favour denitrification (Arias *et al.*, 2005). For this pre-denitrification step, Platzer (1999) reported that recirculation rates of up to 200% could be used, but not for very high nitrogen loads (in this case, an HF wetland should be used for post-denitrification after the VF wetland). Recirculation was reported to increase mean oxygen saturation in wetlands from 56 to 67% (Bahlo, 2000). Sun *et al.* (1999) reported significantly enhanced nitrification.

Also, effluent recirculation seems to enhance the oxygen consumption by microorganisms for organic matter degradation and nitrification. Effluent recirculation ratios of 100, 200, and 300% reportedly resulted in fully nitrified effluents with increases in total nitrogen removals of 52, 66, and 68%, respectively, over one with no recirculation (Arias *et al.*, 2005). The findings indicated that recycling the effluent back to the sedimentation tank improves denitrification in the tank and nitrification in the VF wetlands. However, optimum recirculation ratios between 50 and 200% were suggested to avoid possible negative impacts of the higher loading rates on the performance. Higher ratios return more nitrate for additional denitrification but simultaneously increase the hydraulic loading and, therefore, decrease the HRT of the first pass influent. Thus, the proper recirculation ratio is specific to the hydraulic and nutrient loading rates of the system.

14.5.4 Flow reciprocation

Sequential filling and draining of water have been proposed to increase subsurface oxygen availability, and thus removal of oxygen demanding compounds such as organic matter and ammonia-nitrogen. These wetlands are commonly known as tidal flow, fill-and-drain, or reciprocating wetlands (Stefanakis *et al.*, 2014; Dotro *et al.*, 2017). Frequent water level fluctuation or operation in fill-and-drain mode has been shown to increase treatment performance compared to wetlands with a static water level. Reciprocation refers to the alternating filling and draining of pairs of wetland cells. In contrast, tidal flow or fill-and-drain wetland cells can either be configured pairwise or in series (Dotro *et al.*, 2017).

Consecutive fill and drain cycles (up- and down-flow movement of water) promote passive aeration of the wetland. The rate of oxygen transfer into reciprocating wetlands is related to the frequency of the water level fluctuation. First, in the filling cycle, water is introduced at the bottom of the wetland bed, which gradually fills up until water reaches and floods the wetland surface. The water is left to maintain contact with the wetland components (media, rootzone, biofilm) for a defined period. In the drain cycle, air is introduced within the bed and diffuses in the void space of the substrate media to gradually restore aerobic conditions. During the subsequent fill cycle, the thin water film on the media surface is surrounded by anaerobic or anoxic water, and reducing conditions prevail. The alternating oxic-anoxic sequence is repeated multiple times per day (6–24 cycles), which creates unique conditions that develop a microbial community that is diverse and robust. As a result, reciprocating wetlands are particularly well suited for the removal of pollutants from complex wastewaters and have shown high removal rates, especially for total nitrogen.

Such an operational regime considerably limits the clogging problems of the wetland surface compared to traditional VF wetlands fed on top of the surface. Due to their operational regime, these wetlands possess a higher treatment capacity than traditional VF wetlands, especially concerning nitrification and organic matter decomposition. This enables the use of such wetlands for the treatment of various wastewater types and of higher loading rates and pollutant concentrations.

14.6 CASE STUDY: TIANJIN HARBOR ECO-WETLAND PARK

14.6.1 Background

The Tianjin Lingang Industrial Zone, with a total area of 200 km^2, was founded in June 2003. The industrial zone is located at the central area of Binhai New Area, China, with Bohai Bay in the east, Haihe River Estuary in the north, Haibin Avenue in the west and Jinjin Expressway in the south. It is 46 km away from downtown Tianjin and 15 km away from downtown Tanggu. The Tianjin Harbor Eco-Wetland Park is located in this Industrial Zone, a bird's eye view of which is shown in Figure 14.6. The park covers a total area of about 630,000 m^2, of which the water surface covers about 170,000 m^2. The construction of Tianjin Harbor Eco-Wetland Park was completed in August 2011.

The park is composed of a regulating pond, constructed wetlands area and landscape lake area (Figure 14.7). The total volume of water is 200,000 m^3. With a designed HRT of 20 d, the replenishment demand for water is about 10,000 m^3/d. However, the park is located in an area with high salinity and severe water shortage. There is no available natural water source within 5 km of the park, and its supplementation with only a small amount of rainfall is far from meeting the

Figure 14.6 Bird's eye view of the Tianjin Harbor Eco-Wetland Park (photo source: Management Committee of Tianjin Lingang Industrial Zone).

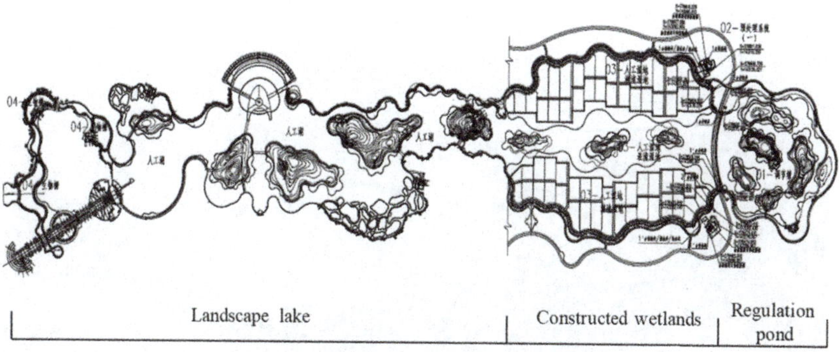

Figure 14.7 Schematic diagram of the Tianjin Harbor Eco-Wetland Park (figure by authors).

basic water demand of the park. Therefore, the effluent from a municipal sewage treatment plant located about 4 km from the park is the only available water source that can meet the water demand of the park. The treatment capacity of the sewage treatment plant is 20,000 m^3/d. Part of the effluent water (about 5000 m^3/d) is fed to the park through a 3.2 km long pipe. The rest is discharged into the Dagua River. However, the concentrations of total nitrogen (TN), ammonium-nitrogen (NH_4-N) and total phosphorus (TP) in the water source are high and fall within the range for algae propagation. Therefore, there is a high potential risk of algae overgrowth and eutrophication by use of this water as a replenishment source for the park. Thus, ecological purification was considered for its further purification.

14.6.2 Components of the Eco-Wetland Park

For the efficient utilization of urban water resources, a partial bypass circulation purification facility was set up at the front of the ecological park, which was connected with the regulating pond through pipelines. Moreover, constructed wetlands were built in the lower reaches of the regulating pond, mainly comprising of subsurface flow and free water surface flow constructed wetlands, to improve the water quality further. The effluent of the wetlands enters the central water body and the ecological lake.

14.6.2.1 Bypass treatment facilities

The regulating pond is the injection point of incoming water from the sewage plant. It has a water surface area of 12,000 m^2, water depth of 1.5–2 m and a storage volume of 21,000 m^3. As shown in Figure 14.8, the effluent of the wastewater treatment plant is discharged to the front of the regulating pond and then pumped

Figure 14.8 Aerial view of the regulation pond (photo source: Management Committee of Tianjin Lingang Industrial Zone).

to the bypass circulating treatment facility from the downstream of the regulating pond via a lifting pump. After being treated in the bypass circulating treatment facility, the water is pumped back to the upstream of the regulating pond. Therefore, this circulation flow of the regulating pond significantly reduces the nitrogen and phosphorus loading, laying a foundation for the stability of the water quality.

The bypass circulating treatment facility adopts the cross-flow biological filter technology. The treatment capacity of the biological filter is 2000 m^3/d, with an area of 1670 m^2, and the filter bed depth is 1 m. The lower part comprises 0.2 m of coarse gravel with a particle size of 60–90 mm, and the upper part is composed of 0.8 m of volcanic rock with a particle size of 20–50 mm. As shown in Figure 14.8, the two ends of the biological filter are symmetrical, and the heights of the inlet and outlet water weirs can be adjusted to obtain an alternate operation of the filter in both directions. In the forward influent condition, the front part of the filter is aerobic and the rear, anoxic. In this process, biofilm is gradually formed on the surface of the substrates in the front to accumulate biomass. After a period of operation, the flow direction is changed, and the original aerobic section becomes anoxic. The accumulated biomass provides a carbon source for denitrification. The cross-flow biological filter could remove excess suspended matter, organic matter and nutrients in the influent. Thus, the impact of excessive nitrogen and phosphorus loading was reduced. Moreover, the construction of the facility shortens the HRT of the regulating pond to 1.5 days.

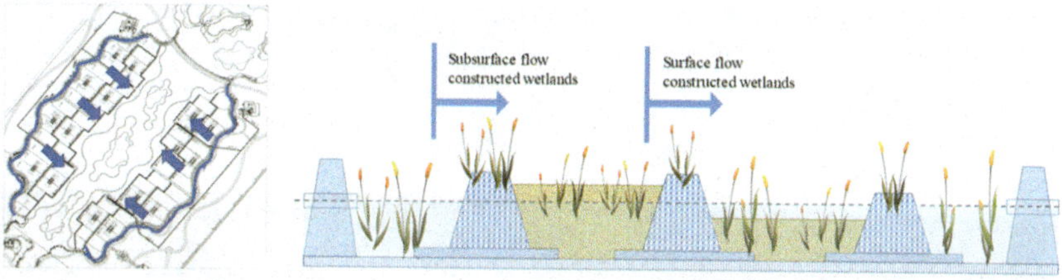

Figure 14.9 Schematic diagrams of the constructed wetlands (figure by authors).

14.6.2.2 Hybrid constructed wetlands

As shown in Figure 14.9, the hybrid constructed wetlands comprised of subsurface flow and free water surface flow constructed wetlands. The water flowing from the regulating pond was first distributed to the subsurface flow constructed wetlands through the distribution channel, which then flowed into the free water surface flow constructed wetlands. The subsurface flow wetlands cover an area of 46,800 m^2 and are divided into 36 groups. The free water surface flow wetlands cover an area of 9500 m^2. Gravel layers were laid at distances of 1.5 m before and after the hybrid constructed wetland, which were used as the distribution zones of the subsurface flow constructed wetlands and the collection zones of the free water surface flow constructed wetlands respectively. The main substrate of the wetlands was gravel, which was laid in two layers. The lower layer was a 0.6 m with coarse gravel within the diameter range of 60–90 mm, and the upper layer was 0.4 m with gravel at the diameter range of 20–50 mm. The bottom slope of the wetlands was 0.5%. The wetlands were planted mainly with reeds, cattails, scallions and calamus. The removal of COD, TN and TP in the hybrid constructed wetland was about 30, 60 and 60%, respectively.

14.6.2.3 Landscape lake district

In the landscape lake area, the self-purification capacity of the water body was enhanced by combining flow pattern optimization, aeration and oxygen enrichment, and ecological revetment (Figure 14.10).

14.6.3 Performance of the eco-wetland park

The effluent of the wastewater treatment plant met the basic water demand for the operation of the eco-wetland park. The HRT of the regulating pond, constructed wetland and landscape lake area are 1.5, 3.5 and 8 d respectively. The bypass circulation system and constructed wetlands significantly reduced the nutrients loading and improved the sensory properties. The subsequent ecological protection measures helped create a good waterscape. Moreover, the lake

Figure 14.10 Overview of the ecological intensification treatment system (photo source: Management Committee of Tianjin Lingang Industrial Zone).

supplied water for the greening of the surrounding area directly, and all the flowers, grasses and tree species in the park grow well, creating rich biodiversity. After treatment by the bypass circulation and constructed wetlands system, the removal efficiencies of TN, DTN, TP and DTP all exceeded 60%. The concentrations of these pollutants in the wetland area decreased to 2.46, 1.01, 0.5 and 0.12 mg/L, respectively (Figure 14.11).

Moreover, there was also a significant improvement in the value of the sensory trait index represented by turbidity and chroma (Figure 14.12). The chroma of the water in the lake was very low and, therefore, the lake appeared very clear. However, turbidity increased in the lake because the area of the ecological lake area was vast, and atmospheric dust, to a certain extent, increased the concentration of suspended matter in the water. Nonetheless, the turbidity of the water was still less than 20 NTU, which did not affect the sensory effect of the lake.

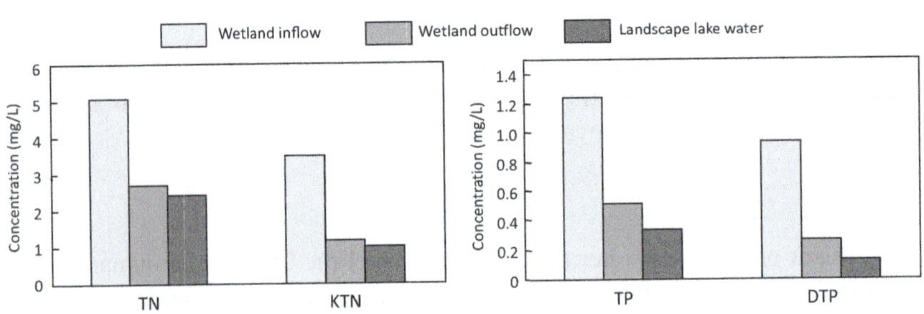

Figure 14.11 Concentrations of nitrogen and phosphorus in the three water areas (figure by authors).

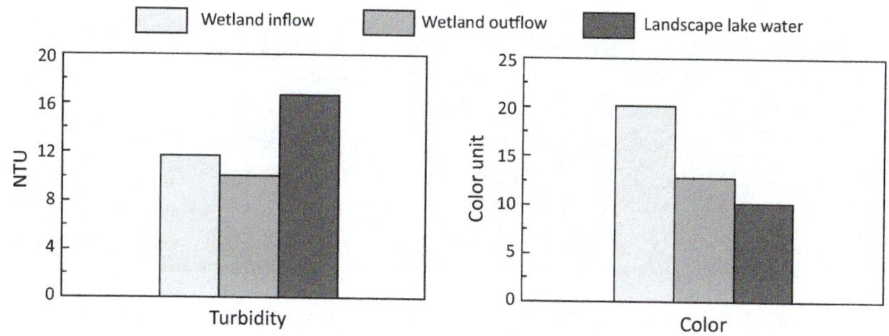

Figure 14.12 Fluctuations in sensory trait indexes in the three water areas (figure by authors).

14.6.4 Outcomes of the Eco-Wetland park

The Eco-Wetland park has created a unique habitat that has improved biodiversity. The planted vegetation has thrived since construction and is now self-seeding. Anecdotal and photographic evidence suggests the wetland is well-populated with native birds, macroinvertebrates, and other aquatic biota. It is envisaged that the vegetation has helped to reduce urban heat in the area.

Urban wetlands help to provide ecosystem services to cities and diversify water management, thereby improving a city's resilience to climate change. Tianjin Harbor Eco-Wetland Park is an example of such a technology that contributes to building a sustainable urban environment. It was constructed to improve the water quality and management issues in the Tianjin Lingang Industrial Zone of China. The pollution reduction efficiency of the wetland was determined mainly by the size of the space available. However, it was part of an integrated approach aimed at achieving efficient utilisation of urban water resources. Although the wetland was not designed to attain substantial pollution reduction, other benefits have been evidenced. These include improving the habitat, biodiversity and local climate of the coastal industrial zone.

Effluent from a wastewater treatment plant was used as a water source for the Eco-Wetland Park. Through a two-stage ecological purification treatment (bypass circulation and constructed wetlands) and subsequent ecological measures, the concentrations of nutrients were reduced significantly, and the excessive growth of algae was effectively controlled. The original site of the Eco-Wetland Park was saline and alkaline. Following the construction of the Eco-Wetland Park, it has become a core ecological landscape feature of the industrial zone. In addition, reuse of the wastewater treatment plant effluent significantly reduced the pollution loading to the coastal waters, which helped in alleviating the pollution problems of the Haihe estuary. An overview of the park is shown in Figure 14.13.

Figure 14.13 Overview of the Tianjin Harbor Eco-Wetland Park (photo source: Management Committee of Tianjin Lingang Industrial Zone).

REFERENCES

Arias C. A., Brix H. and Marti E. (2005). Recycling of treated effluents enhances removal of total nitrogen in vertical flow constructed wetlands. *Journal of Environmental Science and Health*, **40**, 1431–1443.

Ávila C., Matamoros V., Reyes-Contreras C., Piña B., Casado M., Mita L., Rivetti C., Barata C., García J. and Bayona J. M. (2014). Attenuation of emerging contaminants in a hybrid constructed wetland system under different hydraulic loading rates and their associated toxicological effects in wastewater. *Science of the Total Environment*, **470–471**, 1272–1280.

Bahlo K. (2000). Treatment efficiency of a vertical-flow reed bed with recirculation. *Journal of Environmental Science and Health A*, **35**(8), 1403–1413.

Boogaard F. C., van de Ven F., Langeveld J. G. and van de Giesen N. (2014). Stormwater quality characteristics in Dutch (urban) areas and performance of settlement basins. *Challenges*, **5**, 112–122.

Brix H. and Johansen N. H. (2004). Guidelines for the Establishment of Reed Bed Systems up to 30 PE (in Danish: Retningslinier for Etablering af beplantede filteranlæg op til 30 PE). Århus, Denmark.

Cooper P. (1999). A review of the design and performance of vertical-flow and hybrid reed bed treatment systems. *Water Science and Technology*, **40**(3), 1–9.

Cooper P. (2001). Nitrification and denitrification in hybrid constructed wetlands systems. In: Transformation of nutrients in Natural and Constructed Wetlands, J. Vymazal (ed.), Backhuys Publishers, Leiden, The Netherlands, pp. 256–270.

Cooper P., Griffin P., Humphries S. and Pound A. (1999). Design of a hybrid reed bed system to achieve complete nitrification and denitrification of domestic sewage. *Water Science and Technology*, **40**(3), 283–289.

Dong H., Qiang Z., Li T., Jin H. and Chen W. (2012). Effect of artificial aeration on the performance of vertical-flow constructed wetland treating heavily polluted river water. *Journal of Environmental Sciences*, **24**(4), 596–601.

Dotro G. and Chazarenc F. (2014). Solids accumulation and clogging. *Sustainable Sanitation Practice Journal*, **18**, 8–14.

Dotro G., Langergraber G., Molle P., Nivala J., Puigagut J., Stein O. and von Sperling M. (2017). Treatment Wetlands. Biological Wastewater Treatment Series, Volume 7. IWA Publishing, London, UK.

DWA (2017). Grundsätze für Bemessung, Bau und Betrieb von Kläranlagen mit bepflanzten und unbepflanzten Filtern zur Reinigung häuslichen und kommunalen Abwassers, in German. (Principles of design, construction and operation of planted and unplanted filters for treatment of domestic wastewater). Deutsche Vereinigung für Wasserwirtschaft, Abwasser und Abfall e.V. (DWA), Hennef, Germany.

Garcia J., Rousseau D., Morato J., Lesage E. L. S., Matamoros V. and Bayona J. (2010). Contaminant removal processes in subsurface-flow constructed wetlands: A Review. *Critical Reviews in Environmental Science and Technology*, **40**(7), 561–661.

Ghermandi A., van den Bergh J. C. J. M., Brander L., De Groot H. L. F. and Nunes P. A. L. D. (2010). Values of natural and human-made wetlands: A meta-analysis. *Water Resources Research*, **46**, 1–12.

Haines-Young R. and Potschin M. (2012). Common International Classification of Ecosystem Services (CICES): Consultation on Version 4, August–December 2012. EEA Framework Contract No EEA/IEA/09/003. European Environment Agency, Denmark.

Headley T. R. and Tanner C. C. (2012). Constructed wetlands with floating emergent macrophytes: an innovative stormwater treatment technology. *Critical Reviews in Environmental Science and Technology*, **42**, 2261–2310.

Kadlec R. H. and Wallace S. D. (2009) Treatment Wetlands, 2nd edn. CRC Press, Boca Raton, FL, USA.

Knight R. L. (1997). Wildlife habitat and public use benefits of treatment wetlands. *Water Science and Technology*, **35**, 35–43.

Knight R. L., Clarke R. A. and Bastian R. K. (2001). Surface flow (SF) treatment wetlands as a habitat for wildlife and humans. *Water Science and Technology*, **44**(11–12), 27–37.

Knowles P., Dotro G., Nivala J. and García J. (2011). Clogging in subsurface-flow treatment wetlands: Occurrence and contributing factors. *Ecological Engineering*, **37** (2), 99–112.

Langergraber G. and Haberl R. (2001). Constructed wetlands for water treatment. *Minerva Biotecnologica*, **13**(2), 123–134.

Lennon M. (2015). Green infrastructure and planning policy: A critical assessment. *Local Environment*, **20**, 957–980

Mander Ü. and Jenssen P. (2002). Natural Wetlands for Wastewater Treatment in Cold Climates. Advances in Ecological Sciences. WIT Press, Southampton, UK.

Masi F. (2009). Water reuse and resources recovery: The role of constructed wetlands in the Ecosan approach. *Desalination*, **247**, 28–35.

Masi F., Rizzo A., Bresciani R. and Conte G. (2016). Constructed wetlands for combined sewer overflow treatment: Ecosystem services at Gorla Maggiore, Italy. *Ecological Engineering*, **98**, 427–438.

Masi F., Rizzo A. and Regelsberger M. (2018). The role of constructed wetlands in a new circular economy, resource oriented, and ecosystem services paradigm. *Journal of Environmental Management*, **216**, 275–284.

McGrane S. J. (2016). Impacts of urbanisation on hydrological and water quality dynamics, and urban water management: A review. *Hydrological Sciences Journal*, **61**, 2295–2311.

Mell I. C. (2008). Green infrastructure: Concepts and planning. *FORUM Ejournal*, **8**, 69–80.

Mitsch W. J. and Gosselink J. G. (2007). Wetlands, 4th edn. John Wiley and Sons, Inc., Hoboken, New Jersey.

Mitterer-Reichmann G. M. (2012). Treatment wetlands in Austria: Practical experiences in planning, construction and maintenance. *Sustainable Sanitation Practice Journal*, **12**, 4–8.

Moore T. L. C. and Hunt W. F. (2013). Predicting the carbon footprint of urban stormwater infrastructure. *Ecological Engineering*, **58**, 44–51.

Nivala J. (2012). Effect of design on treatment performance, plant nutrition and clogging in subsurface flow treatment wetlands. PhD Thesis. Aarhus University, Department of Bioscience, Aarhus, Denmark.

Nivala J., Wallace S., Headley T., Kassa K. and Brix H. (2013a). Oxygen transfer and consumption in subsurface flow treatment wetlands. *Ecological Engineering*, **61**(B), 544–554.

Nivala J., Headley T., Wallace S., Bernhard K., Brix H., van Afferden M. and Müller R. A. (2013b). Comparative analysis of constructed wetlands: The design and construction of the ecotechnology research facility in Langenreichenbach, Germany. *Ecological Engineering*, **61**(B), 527–543.

Nivala J., Kahl S., Boog J., Afferden M., Reemtsma T. and Müller R. A. (2019). Dynamics of emerging organic contaminant removal in conventional and intensified subsurface flow treatment wetlands. *Science of the Total Environment*, **649**, 1144–1156.

ÖNORM (2009). ÖNORM B 2505: Bepflanzte Bodenfilter (Pflanzenkläranlagen) – Anwendung, Bemessung, Bau und Betrieb (Subsurface flow constructed wetlands – Application, dimensioning, installation, and operation) [in German]. Österreichisches Normungsinstitut, Vienna, Austria.

Pan J., Zhang H., Li W. and Ke F. (2012). Full-scale experiment on domestic wastewater treatment by combining artificial aeration vertical- and horizontal-flow constructed wetlands system. Water, Air, and Soil Pollution, **223**, 5673–5683.

Platzer C. (1999). Design recommendations for subsurface flow constructed wetlands for nitrification and denitrification. *Water Science and Technology*, **40**(3), 257–264.

Potschin M. B. and Haines-Young R. H. (2011). Ecosystem services Exploring a geographical perspective. *Progress in Physical Geography*, **35**, 575–594.

Ramírez S., Torrealba G., Lameda-Cuicas E., Molina-Quintero L., Stefanakis A. I. and Pire-Sierra M. C. (2019). Investigation of pilot-scale Constructed Wetlands treating simulated pre-treated tannery wastewater under tropical climate. *Chemosphere* **234**, 496–504.

Rousseau D. P. L., Vanrolleghem P. A. and De Pauw N. (2004). Model-based design of horizontal subsurface flow constructed treatment wetlands: a review. *Water Research*, **38**, 1484–1493.

Schueler T. R. (1994). Performance of grassed swales along east coast highways. *Watershed Protection Techniques*, **1**, 122–123.

Schwammberger P., Walker C. and Lucke T. (2017). Using floating wetland treatment systems to reduce stormwater pollution from urban developments. *International Journal of GEOMATE*, **12**, 45–50.

Seidel K. (1965). Neue Wege zur Grundwasseranreicherung in Krefeld, Teil. II. Hydrobotanische Reinigungsmethode (New methods for groundwater recharge in Krefeld – Part 2: Hydrobotanical treatment method). *GWF Wasser/Abwasser*, **30**, 831–833.

Sgroi M., Pelissari C., Roccaro P., Sezerino P. H., García J., Vagliasindi F. G. A. and Ávila C. (2018). Removal of organic carbon, nitrogen, emerging contaminants and fluorescing organic matter in different constructed wetland configurations. *Chemical Engineering Journal*, **332**, 619–627.

Shutes B., Revitt M. and Scholes L. (2010). Constructed wetlands for flood prevention and water reuse. Proceedings of the 12th International Conference on Wetland Systems for Water Pollution Control, Venice, Italy. International Water Association, Italy.

Stefanakis A. I., Akratos C. S. and Tsihrintzis V. A. (2014). Vertical Flow Constructed Wet-lands: Eco-engineering Systems for Wastewater and Sludge Treatment, 1st edn. Elsevier Science, Waltham, MA, USA.

Stefanakis A. I., Seeger E., Dorer C., Sinke A. and Thullner M. (2016). Performance of pilot-scale horizontal subsurface flow constructed wetlands treating groundwater contaminated with phenols and petroleum derivatives. *Ecological Engineering*, **95**, 514–526.

Sun G., Gray K. R., Biddlestone A. J. and Cooper D. J. (1999). Treatment of agricultural wastewater in a combined tidal flow: Downflow reed bed system. *Water Science and Technology*, **40**(3), 139–146.

Sundaravadivel M. and Vigneswaran S. (2001). Constructed wetlands for wastewater treatment. *Critical Reviews in Environmental Science and Technology*, **31**, 351–409.

Tsihrintzis V. A. and Hamid R. (1997). Modelling and management of urban stormwater runoff quality: A review. *Water Resources Management*, **11**, 137–164.

Vohla C., Kõiv M., Bavor H. J., Chazarenc F. and Mander Ü. (2011). Filter materials for phosphorus removal from wastewater in treatment wetlands – A review. *Ecological Engineering*, **37**, 70–89.

Vymazal J. (2005). Horizontal sub-surface flow and hybrid constructed wetlands systems for wastewater treatment. *Ecological Engineering*, **25**, 478–490.

Vymazal J. (2007). Removal of nutrients in various types of constructed wetlands. *Science of the Total Environment*, **380**, 48–65.

Vymazal J. (2011). Constructed wetlands for wastewater treatment: five decades of experience. *Environmental Science & Technology*, **45**, 61–69.

Vymazal J. (2013). Emergent plants used in free water surface constructed wetlands: A review. *Ecological Engineering*, **61**(B), 582–592.

Vymazal J. and Kröpfelová L. (2008). Wastewater Treatment in Constructed Wetlands with Horizontal Sub-surface Flow. Springer, the Netherlands.

Vymazal J., Brix H., Cooper P. F., Green M. B. and Haberl R. (1998). Constructed Wetlands for Wastewater Treatment in Europe. Backhuys Publishers, Leiden, The Netherlands.

Vymazal J., Greenway M., Tonderski K., Brix H. and Mander Ü. (2006). Constructed wetlands for wastewater treatment. In: Wetlands and Natural Resource Management. Ecological Studies. J. T. A. Verhoeven, B. Beltman, R. Bobbink and D. F. Whigham (eds), Springer-Verlag, Berlin, Germany, pp. 69–94.

Wallace S. D. (2001). Patent: system for removing pollutants from water. United States: US 6,200,469 B1. United States Patent and Trademark Office, USA.

Wallace S. D and Knight R. L. (2006). Small-scale Constructed Wetland Treatment Systems: Feasibility, Design Criteria, and O&M Requirements. Water Environment Research Foundation (WERF), Virginia, USA.

Wallace S., Higgins J., Crolla A., Kinsley C., Bachand A. and Verkuijl S. (2006) High-rate ammonia removal in aerated Engineered Wetlands. Proceedings of the 10th International Conference on Wetland Systems for Water Pollution Control, Lisbon, Portugal. International Water Association, Portugal.

Wright T., Tomlinson J., Schueler T., Cappiella K., Kitchell A. and Hirschman D. (2006). Direct and indirect impacts of urbanisation on wetland quality. Wetlands and Watersheds Article Series, Office of Wetlands, Oceans and Watersheds U.S. Environmental Protection Agency, Washington, DC.

Yu R., Wu Q. and Lu X. (2012). Constructed wetland in a compact rural domestic wastewater treatment system for nutrient removal. *Environmental Engineering Science*, **29**, 751–757.

Zölch T., Henze L., Keilholz P. and Pauleit S. (2017). Regulating urban surface runoff through nature-based solutions – an assessment at the micro-scale. *Environmental Research*, **157**, 135–144.

Chapter 15

Boscastle case of flash flood modelling and hazards reduction

Ioanna Stamataki

Department of Architecture & Civil Engineering, University of Bath, UK

15.1 INTRODUCTION

An increase in the frequency and magnitude of flooding is one expected consequence of climate change. The UN Office for Disaster Risk Reduction (UNISDR) and the Belgian-based Centre for Research on the Epidemiology of Disasters (CRED) in their 2015 report 'The Human Cost of Weather Related Disasters' associated 157 000 deaths with flooding since 1995. In the last 20 years, floods accounted for 47% of all other weather disasters with 3062 individual events resulting in 2.3 billion people being affected by floods, an alarming number (UNISDR, 2015).

The event occurrence of different geophysical, meteorological, hydrological and climatological events from 1970 to 2017 were assembled by the International Disaster Database in CRED and are presented in Figure 15.1. A significant increase is seen in extreme events and especially the number of floods (EMDAT, 2017).

In line with the above-mentioned predictions, a considerable increase is also expected in the intensity and frequency of extreme precipitation events. This chapter's focus will be on flash floods which are a destructive natural hazard with one of the highest mortalities. They are short duration floods associated with excessive amounts of rainfall and their different causes include a short duration

Editors: Xiaochang C. Wang and Guangtao Fu
doi: 10.2166/9781789060768_0415

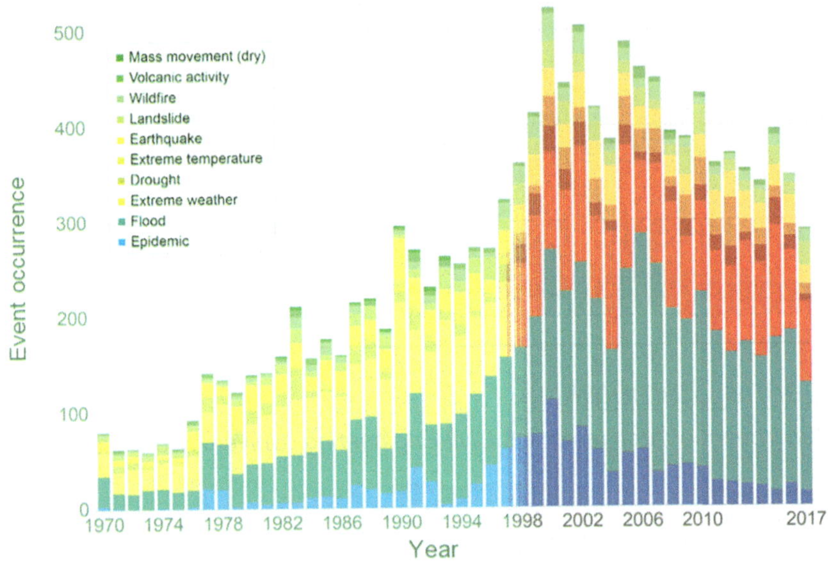

Figure 15.1 Number of disasters from 1970 to 2017 looking at geophysical, meteorological, hydrological and climatological events (EMDAT, 2017).

intense rainfall event, snow melt events, hydraulic structure failures or glacier lake outbursts (Archer & Fowler, 2015; World Meteorological Organisation, 2012).

The current state of climate change and its impact is regularly assessed, and Synthesis Reports are published every few years. All reports target the increase in frequency and intensity of extreme events such as flash floods. The 4th Synthesis Report (AR4) specifically discusses the possibility of an accelerated water cycle. This in turn would lead to an increased storage capacity of water in the atmosphere which would result in higher frequency and intensity storms (IPCC, 2007). The 5th Synthesis Report (AR5) also mentions the probable increase in intensity and frequency of extreme precipitation events (IPCC, 2014) and based on a scoping session in 2017 it is an issue that will also be included in the 6th Synthesis Report (AR6) which will be published in 2022 (IPCC, 2017). This increase in the intensity and frequency of extreme precipitation events will lead to an increase in flash flood events and therefore it is important that the scientific community works to develop new and improved tools to enhance the resilience of urban areas to the threat of extreme flooding through prediction, preparedness strategies and accurate modelling.

In 2018, the European Severe Storms Laboratory (ESSL), taking also into account parts of northern Africa and the Middle East, accounted for 152 fatalities due to flash floods. They plotted all heavy rain events and flash floods events associated with fatalities on the map shown in Figure 15.2. The major highlighted events in Europe were on October 15th in Trebes, France with 13 casualties, on

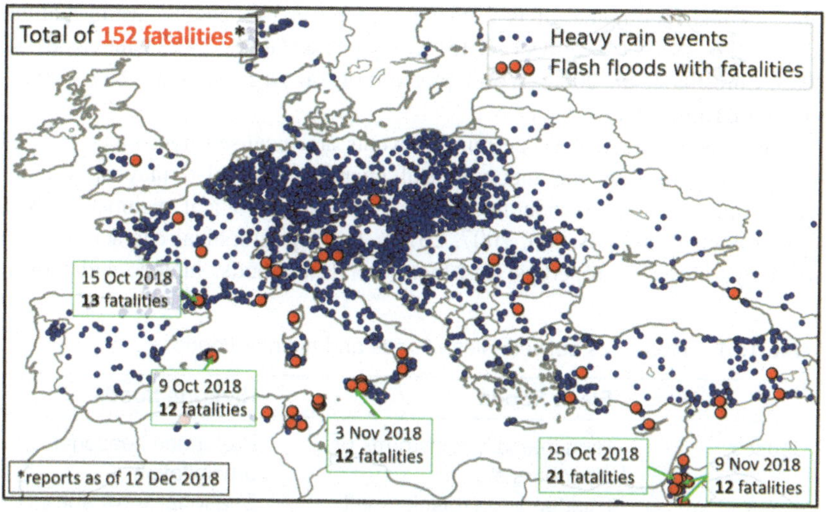

Figure 15.2 Map of deadly flash floods in 2018 produced by the European Severe Storms Laboratory (ESSL, 2018).

October 9th in Mallorca, Spain with 12 casualties and on November 3rd in Sicily, Italy with 12 casualties (ESSL, 2018). Furthermore, there were more flash floods recorded with no casualties in Montenegro, Netherlands, Sweden, Belgium, Poland, Luxembourg, Slovakia, Slovenia, Bulgaria, Macedonia (FYROM), Greece, Ireland, Switzerland, Ukraine and Norway.

It has been recognised that traditional flood management approaches for flooding are not necessarily applicable to flash floods (Kobiyama & Goerl, 2007; World Meteorological Organisation, 2012) and in order to create a more appropriate framework, the differences between these types of events needs to be understood. In this research it is therefore essential to firstly define what a flash flood is, and then to clarify the difference between a large riverine flood and a flash flood. Flash floods have been defined in many different ways but the World Meteorological Organisation (WMO), provides a very descriptive definition of a flash flood as follows (World Meteorological Organisation, 2012):

> 'A flash flood is a short and sudden local flood with great volume. It has a limited duration which follows within few (usually less than six) hours of heavy or excessive rainfall, rapid snow melt caused by sudden increases in temperature or rain on snow, or after a sudden release of water from a dam or levee failure, or the break-up of an ice jam'.

Discussing the main differences between riverine floods and flash floods was first attempted by Xu *et al.* (2006) who, considering the management of flash floods and sustainable development in the Himalayas, created a table including the main differences between riverine floods and flash floods. Nonetheless, as the table seemed incomplete, it is further improved in this chapter using additional sources (i.e., Archer & Fowler, 2015; Kobiyama & Goerl, 2007; Merz & Bloschl, 2003; Shrestha *et al.* 2008; World Meteorological Organisation, 2012, 2017), as shown in Table 15.1.

Flash floods due to extreme rainfall events are localised hydro-meteorological phenomena and thus the topographical characteristics also play an important role as they have a considerable effect on all hydrological parameters (World Meteorological Organisation, 2007, 2012). The topographical characteristics that affect hydrological properties, and therefore flash floods, include soil moisture,

Table 15.1 Differences between flash floods and riverine floods.

	Flash floods	Riverine floods
Causes	High intensity rainstorms or cloudbursts Sudden snow/glacier melt Dam breaks Levee breaches Wet/dry catchment	Prolonged seasonal precipitation Seasonal snow and glacial melt Saturated catchment
Characteristic features	Quick onset Short storm/flood duration Quick water level rise Peak flow in minutes/few hours Quick recession Not related to base flow Rapid response to rainfall, short lag time Limited spatial extent (<30 km^2) Steep slope catchments	Slow onset Long storm/flood duration Slow water level rise Peak flow in hours/days Slow recession High base flow Slow response to rainfall, medium long lag time Regional to large spatial extent All catchments
Associated problems	Large amount of debris High hydraulic force associated with erosion and structural damage	Inundation/flooding
Frequency	All year	Rainy season
Affected areas	River plains, valleys Local extent Small to medium areas	River plains, valleys Local to regional extent Large areas
Forecasting	Forecasting difficult Local information essential Hydro-meteorological problem Coordination for flood response in real-time difficult	Forecasting possible Local information not essential Hydrological problem Coordination for flood response in real-time possible

soil depth, soil permeability, land use, catchment size and the catchment slope (World Meteorological Organisation, 2012). Thus, it has been concluded that topography is an important characteristic in an area's predisposition to flash floods (World Meteorological Organisation, 2012). Thus, small steep upland catchments often have a naturally 'flashy' response to intense rainfall (meaning an almost immediate response to rainfall) resulting in severe damage from small and localised events (Werner & Cranston, 2009).

In the last 20 years, there have been several major flash flood events in the UK, including the 2004 flash flood in Boscastle, Cornwall, where 200 mm of rain fell in 5 h, equivalent to 20% of the annual average rainfall. During this event, 100 people were evacuated, 60 buildings were flooded/damaged and 116 vehicles were carried by the flow (Bettess, 2005; Xia *et al.* 2011a). Second, the 2007 large flood in Hull, Yorkshire, where 135 mm of rain was measured in 24 h, equivalent to 20% of the annual average rainfall, and 8657 houses and 600 streets were flooded/damaged (Coulthard *et al.* 2007; Marsh & Hannaford, 2007). Then, in 2011 the Bournemouth, Dorset, event where 40.6 mm of rain was recorded in 1 h, equivalent to 78% of the monthly average rainfall and 270 houses were flooded and/or damaged (Ambrose, 2011). In 2012, a flash flood in Honister Pass, Cumbria, flooded 100 houses and 71 mm of rain fell in 24 h, equivalent to 40% of the monthly average rainfall (Met Office, 2011b, 2013). Also in 2012, in Aberystwyth, Wales, 125 mm of rain fell in 24 h, equivalent to twice the monthly average rainfall and 150 people had to be evacuated (Climate Data, 2018; Webb, 2013). Finally, in 2018 in Birmingham, West Midlands, 81 mm of rain fell in 1 h, equivalent to 1.3 times the monthly average rainfall, resulting in one casualty (Met Office, 2011a; Muchan *et al.*, 2018).

Flash floods remain a global problem and due to their dynamic nature combined with their limited spatial and temporal scales and short lead times, observation, modelling and forecasting of these events continues to be a challenge (World Meteorological Organisation, 2012). However, even though the accuracy of flood estimation for extreme events and flash floods has been identified to be a common problem, shared databases or common guidelines do not exist, and each individual country is focusing their efforts primarily on national and localised projects. In China, for example, since 2003 (Figure 15.3) the number of flash floods that have resulted in casualties has been decreasing and this can be attributed to China's national flash flood prevention projects, especially the flash flood early-warning systems (Liu *et al.*, 2018).

This restricted and site-specific approach has led to often simplistic and rarely generalised approaches and strategies resulting in further uncertainty in the reliability of flash flood prediction, estimation and mitigation (Kjeldsen *et al.*, 2014). As very limited field data exist from flash floods, a practical approach to generate flash floods, both numerically and experimentally, is through a dam break. This guarantees the main characteristic features of flash flood events including rapid onset and the rate of rise in water level (Archer & Fowler, 2015).

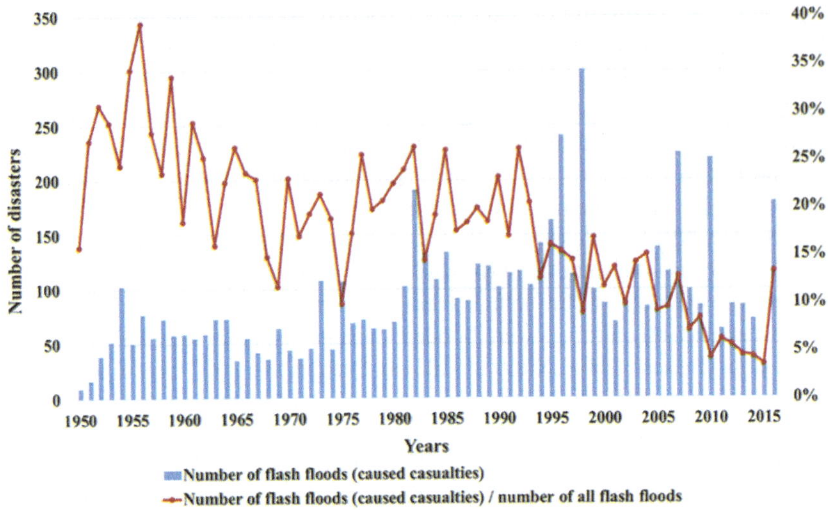

Figure 15.3 Number of flash floods and related casualties in China from 1950 to 2015 (Liu *et al.*, 2018).

15.2 BACKGROUND

15.2.1 Numerical methods used to model extreme events

Hydrodynamic modelling of flood events is usually considered through the use of mathematical models of varying complexity (Xia *et al.*, 2011a). Regardless of a model's complexity, all numerical models make approximations and thus present limitations that can easily lead to inaccurate predictions (Rowiński & Radecki-Pawlik, 2015; Toombes & Chanson, 2011). The main problems presented in regard to hydraulic modelling of floods are (Liang & Borthwick, 2009; Néelz & Pender, 2010; Zech, *et al.*, 2015):

(1) the numerical instabilities present in high-resolution grids,
(2) the computational time,
(3) the modelling of the moving wet-dry interface, specifically the arrival time of the wave front in fluvial floods,
(4) the maximum water depth and, finally
(5) the representation of complex boundaries.

All previously mentioned issues remain challenging limitations and emphasise the need for further advancement in numerical hydrodynamic modelling techniques.

There are several verified 2D hydraulic models commonly used to predict flood inundation extents, but their performance in extreme events such as flash floods, where the flows are fast-transient, remains an active area of research (Huang *et al.*, 2015). Flash flood characteristics, especially their limited spatial and temporal scales, make modelling of these events challenging and complicated.

They are rarely captured in the field and the data associated with such events is very limited. Specific flow features are difficult to model accurately and thus several researchers when modelling flash floods have tried to find a balance between model complexity and computational time, taking into account specific physical mechanisms such as infiltration for example (Huang *et al.*, 2015). They are also localised impact events and therefore local knowledge is important for their modelling (World Meteorological Organisation, 2012).

When predicting the hydrodynamic behaviour of flash floods, a common problem is that the performance of most hydraulic models is not consistent across event magnitudes (Horritt & Bates, 2002). The majority of numerical models are calibrated using a limited number of historical events and thus, assessing a model's ability to predict the flash flood dynamics of the most extreme events (i.e., model validation) is an essential task to ensure the model's credibility (Horritt & Bates, 2002). Considering this in addition to all the previously mentioned challenges (i.e., numerical instabilities in high-resolution grids on complex topographies, computational time, modelling the wet-dry interface dynamics, and the sharp flood wave front), further research through both experimental and numerical modelling is needed.

15.2.2 Flash flood models

There have been a limited number of publications on models specifically designed for flash flood modelling and as support tools for flash flood warning systems. In a laboratory setting the most prominent flash flood experiment is the Testa *et al.* (2007) experiment which was part of the IMPACT project; a project that assessed the risks from extreme flooding. Another large-scale experiment, part of the CADAM Project, was the Chatelet experiment which assessed the effect of a dam break on a triangular bottom sill, in a 38 m long channel (Ferreira *et al.*, 2006). The same experiment was later replicated as part of the IMPACT project on a smaller scale (Soares-Frazão, 2007). Other experiments include Chanson's flash flood surges (Chanson, 2004).

As very limited field data exist from flash floods, a practical approach to generate flash floods, both numerically and experimentally, is through a dam break as this guarantees the main characteristic features of flash flood events including the rapidity of onset and the rate of rise in water level (Archer & Fowler, 2015). The dam break problem has become a widely researched problem and it has been modelled both experimentally and numerically. Research started as early as 1960 with the US Army Engineers Waterways Experiment Station publishing a report on experimental cases on floods resulting from suddenly breached dams (Corps of Engineers, 1960). The research continued from simple experimental studies such as the initial stages of a dam-break (Stansby *et al.*, 1998) to more complicated problems such as dam-break induced mudflows (Peng and Chen, 2006). Numerically, the dam-break problem has been modelled in 1D, 2D and

3D (Marsooli and Wu, 2014; Zhainakov & Kurbanaliev, 2013) and experimental and numerical results have been compared by several researchers (Aureli *et al.*, 2015; Peng & Chen, 2006).

When fluids interact with structures the complexity of the numerical simulation increases exponentially and requires considerations of the structural dynamics which are not simulated accurately by any numerical scheme. Wave structure interaction is mainly investigated in the design of coastal and offshore structures as they are exposed to extreme situations with breaking waves that can result in very high impact forces on small temporal scales (Chella *et al.*, 2012). Thus, many experimental and numerical studies have been used to examine wave loading, run-up and scattering around such structures (Chen *et al.*, 2014b). Nevertheless, the majority of applications are for offshore applications and in dam break flows there is only very limited research describing the dynamics of these events and studying flood wave structure interaction, such as the work of Trivellato (2004), Kleefsman *et al.* (2005), Bukreev & Zykov (2008), Bukreev (2009), Chen *et al.* (2014a) and Lobovský *et al.* (2014).

15.3 BOSCASTLE, UK

The August 2004 Boscastle flash flood is one of the most known flash flood events in the UK. In 2004 a severe flash flood took place in the Boscastle village in Cornwall where 200 mm of rain was recorded in 5 hours (London's yearly average precipitation is 583.6 mm and Beijing's is 610 mm) and it only took 25 minutes from the moment the river breached its banks to the moment cars and vans were swept by the flow. There were no casualties from the event, but the property damage was extensive, leading to an estimated cost of damage of £15 million. The Boscastle flash flood is a common event in flood risk modelling and has already been modelled by several researchers both numerically and experimentally and both from a hydrological and a meteorological perspective.

15.3.1 Catchment description

To understand the background and the extremity of the 2004 flash flood, the catchment area will first be described in terms of geographical location, geology, catchment description and climate before outlining the 2004 event. Boscastle is a village located in North Cornwall on the southwest coast of the UK and has an annual rainfall total of 961 mm (Met Office, 2010). Figure 15.4 shows the average monthly rainfall in Boscastle where November is typically the wettest month and April the driest. For comparison, Figure 15.5 shows the average rainfall around the UK from 1821 to 2010 for: (a) annual average, (b) November (Boscastle's wettest month) and (c) April (Boscastle's driest month) (Met Office, 2010; World Weather & Climate, 2016). Thus, when considering the presented rainfall profiles for Boscastle (Figures 15.4 and 15.5) it is apparent that the highest monthly average is 100 mm of precipitation in November which is a medium average for the UK.

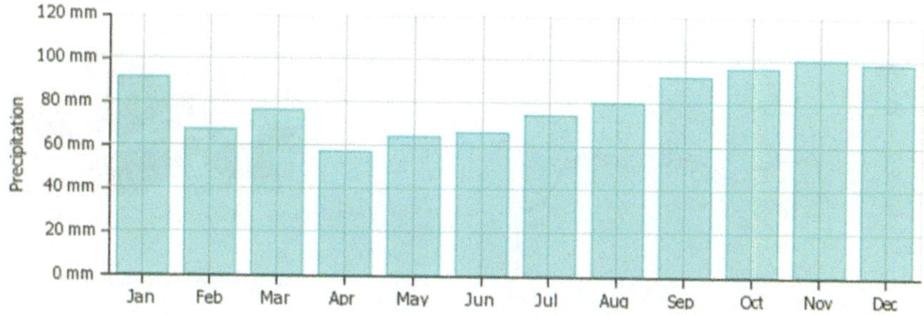

Figure 15.4 Average monthly rainfall in Boscastle (World Weather & Climate, 2016).

Boscastle is positioned in the Valency catchment at the bottom of the valley where two rivers, the Valency River and the Jordan River, meet (Into Cornwall, 2015). The catchment that drains into Boscastle is the Valency catchment which has a round shape (Figure 15.6a), is 8.04 km in length with an area of 20.4 km^2 (Environment Agency, 2016). It is mainly rural and areas of woodland surround the main river (Xia *et al.*, 2011a).

The bedrock geology of the area is a Yeolmbridge formation which contains slate (Figure 15.6b) but there is also sedimentary bedrock and pelagite deposits, due to past domination of sea water (British Geological Survey, 2016). Slate has a hydraulic conductivity of 5×10^{-9} to 5×10^{-6} m/s and a low conductivity value, resulting in relatively slow infiltration through the strata (British Geological Survey, 2006). This, in combination with the small steep rocky catchment, results in increased runoff potential and a steep rising limb in the flood hydrographs. Thus, such a catchment which is characterised by an almost instant response to intense rainfall falls into the category of 'flashy catchments'.

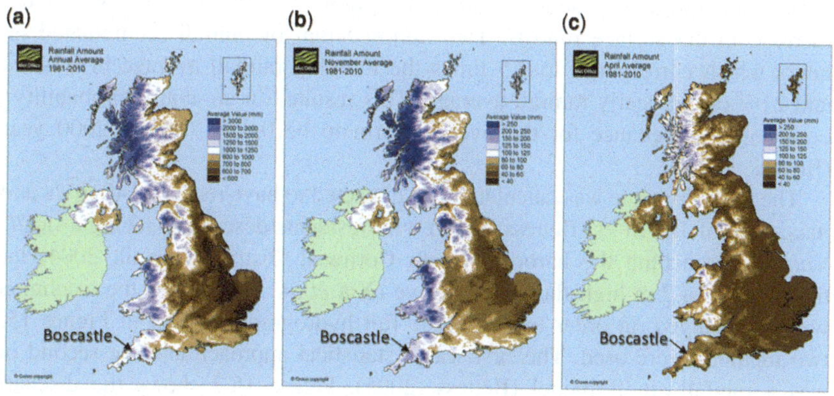

Figure 15.5 Average rainfall 1821–2010: (a) annually; (b) November; and (c) April (Met Office, 2010).

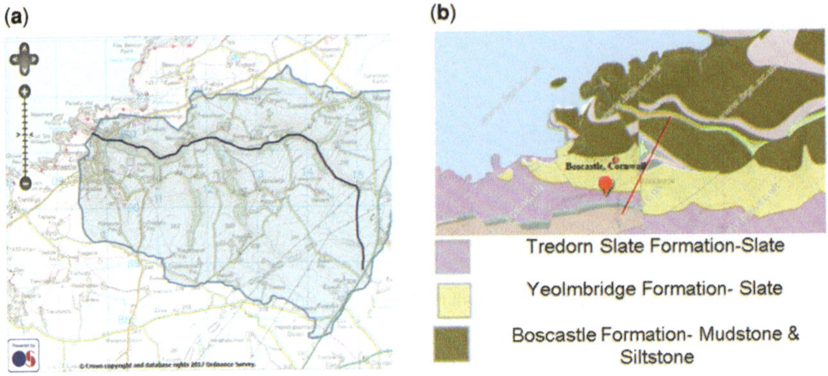

Figure 15.6 (a) Valency catchment (Environment Agency, 2016), (b) Boscastle and surrounding area's geology (British Geological Survey, 2016).

15.3.2 The 2004 flash flood event

Heavy rainfall on the 16th August 2004 caused severe flooding in the Valency catchment and the River Jordan. This resulted in a flash flood in Boscastle which caused severe damage. Even though there were no casualties, at least 100 people had to be evacuated, 60 buildings were flooded, with some of them completely wrecked, and 116 vehicles were carried by the flow (Xia *et al.*, 2011a). From a meteorological point of view, a cyclonic scale in the Atlantic Ocean resulted in a humid and unstable environment over the region of Cornwall. Due to the lack of wind, clouds assimilated and moved north-east resulting in very concentrated rainfall on the Valency catchment with peak rates of precipitation of up to 400 mm/h (Golding *et al.*, 2005). On that day, the soil was already saturated at the onset of the heavy rainfall from previous rainfall. This combined with the overlay of impermeable rock (easily saturated), the steepness of the slopes (1/20) and the static cumulonimbus clouds, resulted in a rapid saturation of the soil and an increase in the surface run-off. Two hundred mm of rainfall accumulated in 5 hours which corresponds to 2.5 times the monthly rainfall average in Boscastle and 21% of the yearly rainfall average. This resulted in an annual probability of occurrence exceedance for the overall storm to be 0.05% (one in 2000 years) (Bettess, 2005).

The peak flow rate was calculated and expected to have reached 140 m^3/s and a maximum of 180 m^3/s (Bettess, 2005) with residents describing seeing a '*wall of water*' approaching the harbour (North Cornwall District Council, 2004) later translated to a 2 m high flash flood wave (Xia *et al.*, 2011a). As the catchments were not gauged, in order to derive the full hydrographs shown in Figure 15.7, two methods were used. The first was a statistical approach and the second one was a rainfall-runoff model (Bettess, 2005). Figure 15.7 shows the discharge hydrographs for different locations along the River Valency. Velocities were not

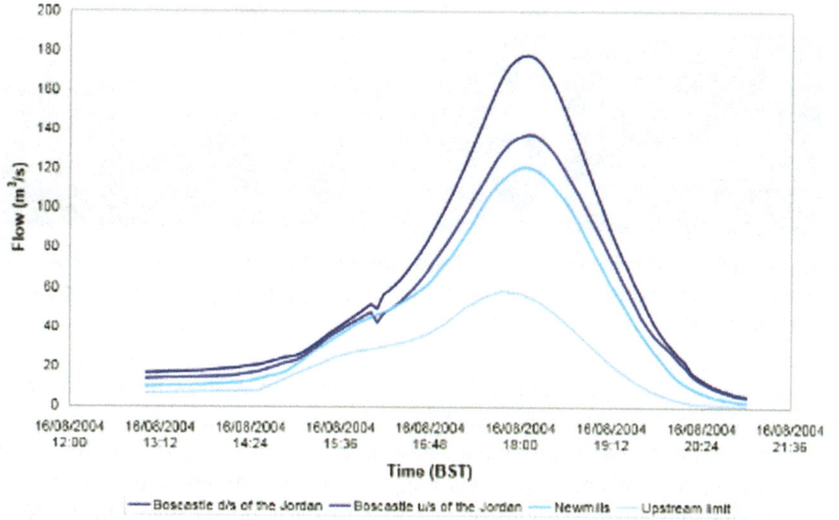

Figure 15.7 Discharge hydrographs for different locations on the River Valency (Bettess, 2005).

measured during the flash flood but they were computed based on the acquired data and measurements (Bettess, 2005). Figure 15.8 shows the maximum velocities modelled for the flood which were at the Valency River, reaching a maximum value of 10 m/s.

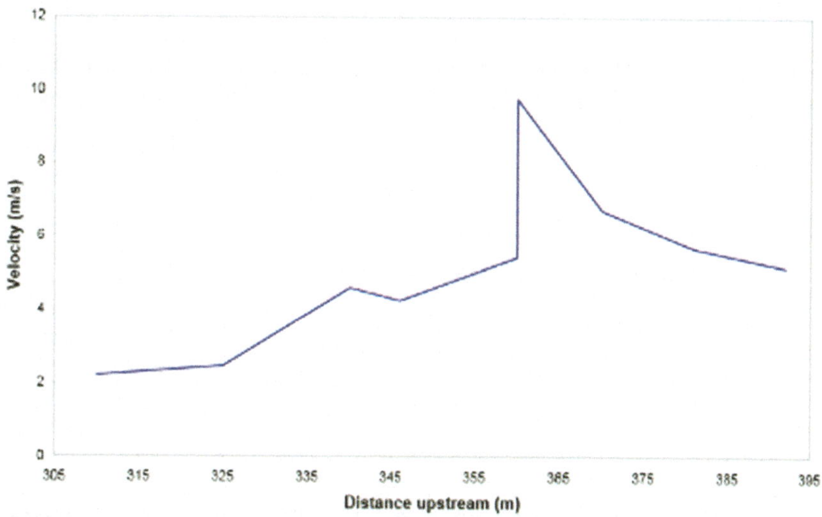

Figure 15.8 Maximum calculated velocities in the Valency River (Bettess, 2005).

(a) (b) (c)

Figure 15.9 Boscastle 2004 flash flood: (a) Flooding of the Valency River and blockage of the bridge; (b) Flooding in the town; (c) Flooding of the Valency River (Bettess, 2005).

Some photos from the flood can be seen in Figure 15.9, showing the extent and damage. Despite the 2004 flash flood being a very famous and catastrophic event, it was not the first recorded case of a notably large flood or flash flood in the village. The most important events since 1827 are listed below, supporting the concept that some catchments may be more predisposed to flash floods than others (North Cornwall District Council, 2004):

- 28th October 1827 – No recorded rainfall
- 16th July 1847 – No recorded rainfall
- 6th September 1950 – No recorded rainfall
- 8th June 1957–140 mm in 2.5 h
- 3rd June 1958 – River rose 4.5 m in 20 min
- 6th February 1963 – No recorded rainfall

15.3.3 Mitigation solutions

Following the 2004 event, mitigation solutions were implemented by the Environmental Agency in the village (Figure 15.10) including a £4.2 m project in

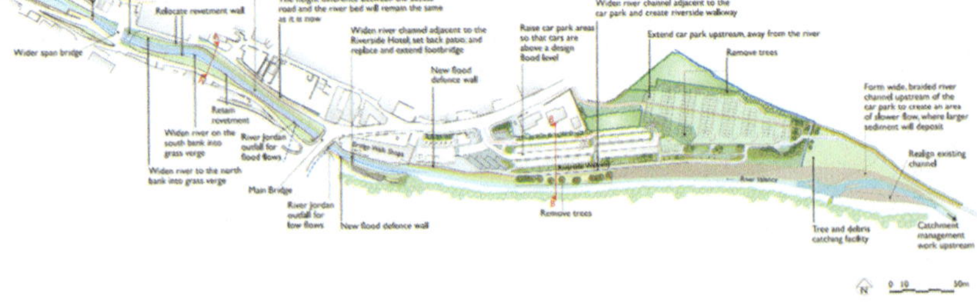

Figure 15.10 Flood defence mitigation Boscastle (Nicholas Pearson Associates, 2012).

October 2006. The most important mitigation solutions were (Halcrow Group Ltd, 2017; Nicholas Pearson Associates, 2012):

(1) erosion control mats which could accommodate 5 m/s flows,
(2) raising the car park level which previously flooded,
(3) installation of SUDS and permeable paving, river dredging, widening and realignment to avoid blockage from fallen trees and slow down its flow,
(4) installation of a flood overflow culvert for the River Jordan,
(5) installation of concrete toe-rail at the foot of the embankment, and
(6) new flood defence walls and new wider span bridge downstream with a one in 100 year flood design life designed to fail in case of a similar event.

The Boscastle event was selected as an inspiration to conduct further laboratory experiments and was simplified and scaled down for experimental purposes. The configuration consisted of an elevated reservoir, followed by a 1/20 slope (slope of Penally Hill, the hill leading to Boscastle harbour), followed by a flat area where different combinations of buildings were positioned, the urban settlement.

15.3.4 Research

Boscastle is a common case in flood risk modelling and the characteristics surrounding the event have been analysed from many different perspectives.

A detailed study by HR Wallingford (2005) described the meteorological, hydrological and hydraulic aspects of the flash flood event. The event was reconstructed numerically and propagation mechanisms, peak flows and peak water levels were presented (Bettess, 2005; HR Wallingford, 2005). Next, Roca and Davison (2010) analysed main flash flood processes using a 2D numerical model and investigated specifically the flow regime changes, the blockage of structures, changes in flow paths and the effect of the geomorphology on flow characteristics. Xia *et al.* (2011a, 2011b, 2014, 2018) conducted extensive research, both experimentally and numerically, looking at submerged vehicles during a flash flood and used the Boscastle flash flood as a case study for their analysis. The Boscastle event has also been modelled extensively hydraulically. Important work was presented by Lhomme *et al.* (2010) who looked at flood extents and forces on buildings using a 2D model, Falconer (2012) who looked at flow interactions of supercritical flow with buildings and Xia *et al.* (2011a) who modelled flash flood risk in urban areas, taking into account not only the flood extent but also the risk to people and properties.

Research has also been conducted on the Boscastle flash flood from a meteorological perspective. The forecasting department of the Met Office analysed the meteorological conditions before the flash flood both from observations and also by using output from a high-resolution land surface model (Golding *et al.*, 2005). Burt (2005) specifically discussed the rainfall observations recorded during the event and compared the Boscastle flash flood to other historical storms in the UK, concluding that even though it is considered as a very extreme event, the historical perspective is important as it showed that there have been many other severe events in the area. Murray *et al.* (2012) modified a flash flood severity assessment, previously created by Collier and Fox (2003), and determined from a hydrometeorological point of view and using a scoring system, the flood susceptibility and severity of a catchment to extreme events. Finally, Warren *et al.* (2014) discussed the similarity of another quasi-convective stationary system in 2010 in the southwest of England which had many similar characteristics to the Boscastle event.

15.4 FLASH FLOOD EXPERIMENT: BOSCASTLE

As part of a PhD research at the University of Bath, flash floods were generated in a controlled laboratory environment for the validation of numerical hydrodynamic models and the investigation of the effect of land use and intensity on flash flood propagation (Stamataki *et al.*, 2018). A new experimental dataset for flash floods in a controlled environment was developed and the impacts of water levels and loads on downstream urban settlements were investigated. The importance of the experimental study lay within the fact that it is important for flash flood experiments to obtain an impact stage from a flash flood wave in an urban settlement which would not have been possible without a dam break experiment. Thus, this allowed for the effect of land use (vegetated, non-vegetated slope) and the intensity of flash flood characteristics (different initial water depths) to be investigated in a controlled environment.

15.4.1 Description of experiments

The experiments were conducted in a flume located in the Department of Mechanical Engineering in University College London (UCL). The flume is 20 m long and 1.2 m wide, and wave gauges and ultrasonic sensors were installed along its length. An elevated reservoir was built in the upstream part of the experimental apparatus separated by a gate and containing a controlled volume of water allowed to be released instantly upon the opening of the gate. The water was then discharged onto a 6 m long slope with 1/20 gradient followed by a horizontal floodplain area, where buildings were installed, the urban settlement (Figures 15.11 and 15.12).

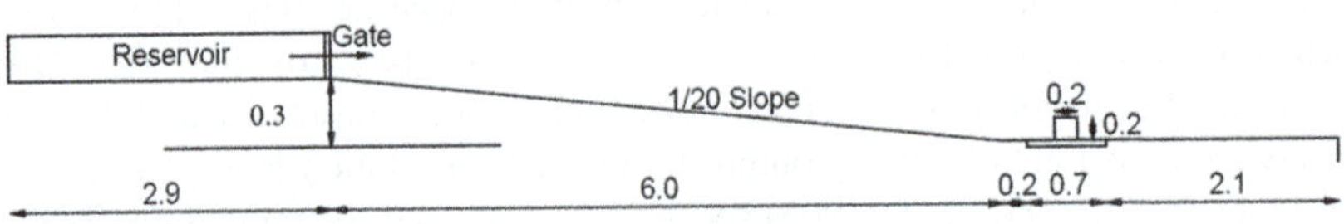

Figure 15.11 Dimensions of experimental set up.

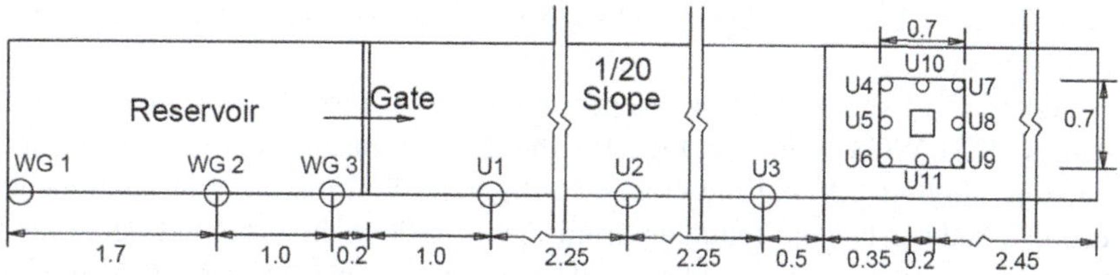

Figure 15.12 Wave gauges (WG1-WG3) and ultrasonic sensor positions (U1-U11) on the experimental setup.

For the scope of this chapter, six main experimental test cases will be discussed in this section:

(1) B1_H100: Single building case with 0.1 m initial water level in the reservoir, no roughness layer on the slope (unvegetated slope).
(2) B1_H200: Single building case with 0.2 m initial water level in the reservoir, no roughness layer on the slope (unvegetated slope).
(3) B1_H100G: Single building case with 0.1 m initial water level in the reservoir, roughness layer on the slope (vegetated slope).
(4) B1_H200G: Single building case with 0.2 m initial water level in the reservoir, roughness layer on the slope (vegetated slope).
(5) B0_H100: No building case with 0.1 m initial water level in the reservoir, no roughness layer on the slope (unvegetated slope).
(6) B0_H200: No building case with 0.2 m initial water level in the reservoir, no roughness layer on the slope (unvegetated slope).

15.4.2 Results

The case B1_H100, which as previously described has an initial water depth in the reservoir of 0.1 m, no roughness layer on the slope and a single building in the urban settlement, will be analytically presented below. Figure 15.13 presents a schematic representation of the case showing the location of selected measurement points.

Once the gate was released, a dam break wave starts propagating downstream along the slope and a negative wave starts moving upstream within the reservoir. The first instruments to record a change were the wave gauges (WG1-WG3) in

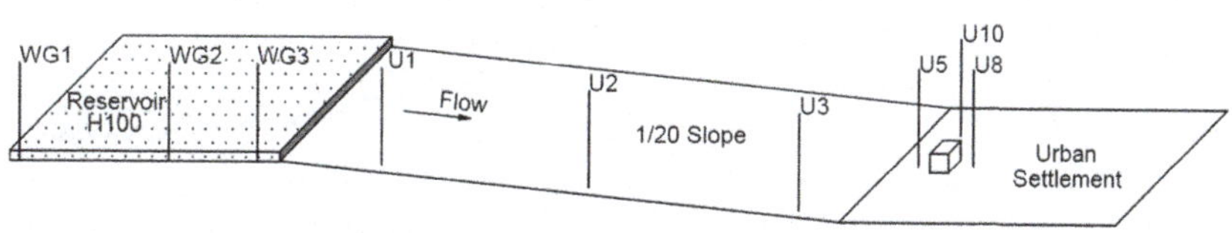

Figure 15.13 Schematic representation of B1_H100 case showing the location of some key instruments.

the reservoir which recorded the change in water depth during the emptying of the reservoir.

Figure 15.14 shows the water depth changes over time, first for the three wave gauges WG1-WG3 (top) and then for the three ultrasonic probes U1-U3 along the slope (bottom). Being the closest to the gate, WG3 is the first instrument to record a sudden change reducing from 0.1 to 0.06 m in 1.2 s and decreases to 0.055 m, where it reaches a plateau. The negative wave reaches WG2 after just under 1 s, which decreases to 0.055 m less suddenly. WG1 has the most delayed response when the negative wave reaches it 2.5 s later and also reaches 0.055 m.

The water depth evolution along the slope is visible from the ultrasonic probes U1-U3 where the arrival of the dam break wave to different positions along the slope is recorded by the probes. The flow on the slope is supercritical and characterised by two components, the propagation of the dam break wave and the presentation of a uniform flow between U2 and U3 from $t = 4$–7 s. The increase in velocity is apparent from the arrival of the dam break wave to the ultrasonic sensors as it travels a distance of 2.25 m from U1 to U2 in 1.6 s (1.4 m/s) and the same distance from U2 to U3 in 1.35 s (1.6 m/s). After $t = 8$ s, it presents the same exponential decay which is evident for all three sensors, with a very small difference in water depth between them. It is important to note the first peak noticeable in U1 at $t = 0.6$ s can be attributed to splashing from the gate opening. The opening is not completely instantaneous and the flow, due to the friction and the gate's sealing,

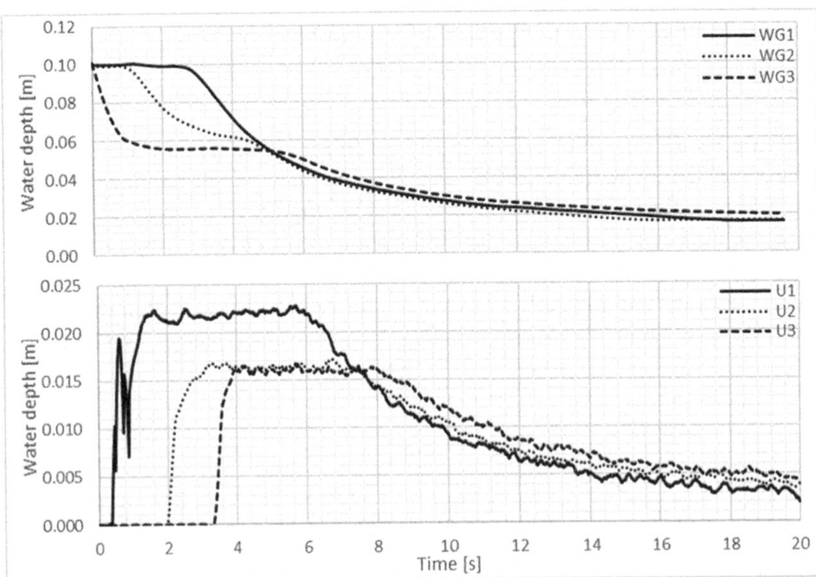

Figure 15.14 Water depth evolution in the reservoir and along the slope for H100.

takes some time to restructure after the gate opening, thus creating the different shapes in water depth evolution at U1 than at U2 and U3 (Figure 15.14, bottom.

Looking further at the other test cases, Figure 15.15 compares the water depth evolution for B0_H100 and B1_H100, thus comparing the changes in water depth in the urban settlement due to the blockage created by the single building.

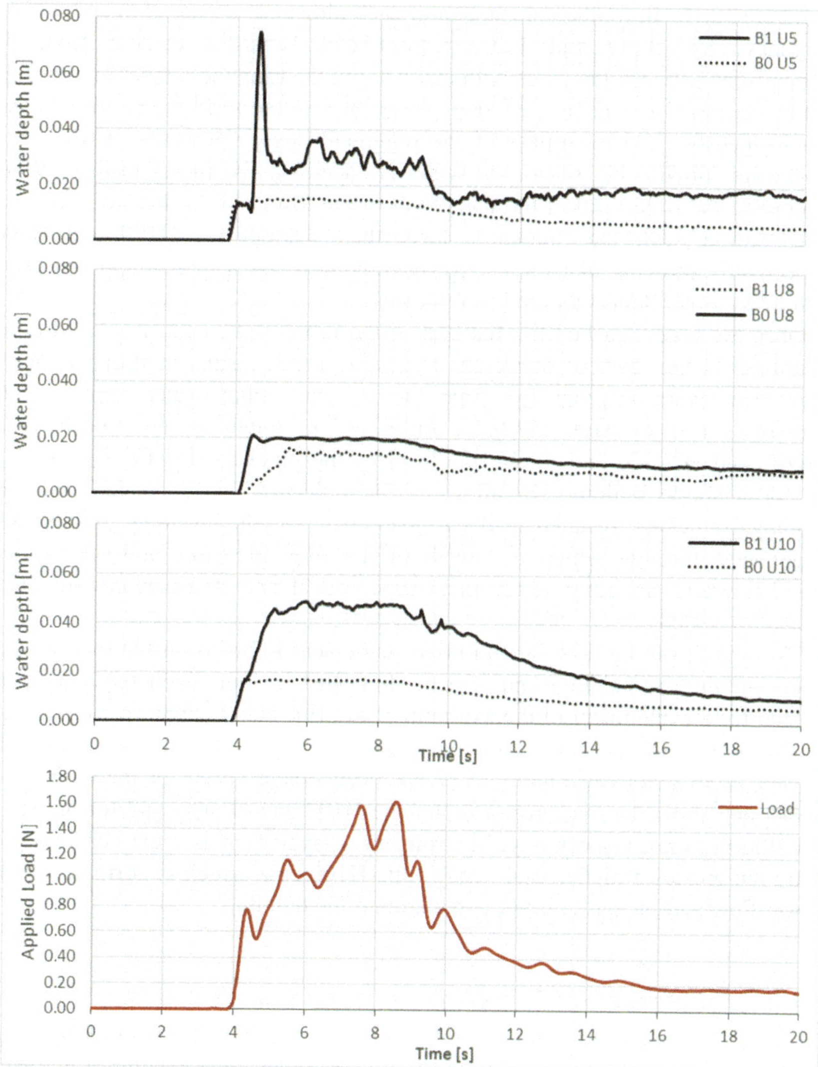

Figure 15.15 Water depth evolution around the building (U5, U8, U10) for B0_H100 and B1_H100 and load acting on the structure for B1_H100.

The top graph shows U5, the second corresponds to U8, followed by U10 and finally the load acting on the building is presented. The highest water level depth is in front of the building as would be expected due to the blockage and the creation of the hydraulic jump while U8 is less affected by the blockage, resulting in similar water depths to those observed for the B0_H100 case. U10 shows an increase that is, attributed to the reflection from the blockage thus increasing the water depth in the B1_H100 case.

Figure 15.16 shows the comparison of each ultrasonic sensor with and without roughness for H100 and H200 respectively. What is evident from the comparisons between the cases with and without the roughness layer on the slope is the decrease in velocity. In both cases, the water reaches the first sensor simultaneously (U1) regardless of the roughness layer. However, as the water propagates down the slope the velocity decrease is more visible. When comparing the time it takes the dam break wave to reach U3, it is 1.37 and 1.2 times slower with the roughness layer than without for H100 and H200, respectively, showing that the change in roughness has a more important effect with lower water depths and lower velocities.

Once the water reached the flat part of the urban settlement, it slowed down regardless of the level of blockage. The water depth results in this part of the experiment were affected by three factors: the initial water depth in the reservoir, 0.1 m for H100 and 0.2 m for H200; the roughness layer in the cases H100G and H200G, and finally the level of the blockage B0 for no building, B1 for a single building. In all cases the reflection wave created from the buildings' blockage resulted in the formation of a hydraulic jump, a stationary surge wave through which the depth of the flow increases and occurs in a situation where the flow upstream is supercritical and downstream subcritical (Chaudhry, 2008).

The impact on the downstream urban settlement is based on the theory of an object in supercritical flow and can be described in four distinctive stages: (i) Impact, (ii) Development of the hydraulic jump, (iii) Steady high Fr flow (around an obstacle) and (iv) Decaying quasi-steady flow with decreasing high Fr number. Both graphs in Figure 15.17 has been synchronised for the moment of impact and show the load over time for the H100 and H100G and for the H200 and H200G cases, respectively. The roughness layer decreases the peak load for H200 but creates a higher peak load in the H100 case which is attributed to the slower flow and increased water depth around the building.

15.4.3 Discussion

Once the flow reached the urban settlement in the experiment, the reflection and blockage from the building made the flow subcritical, thus creating a hydraulic jump in front of the building. The different configurations affected the flow in a

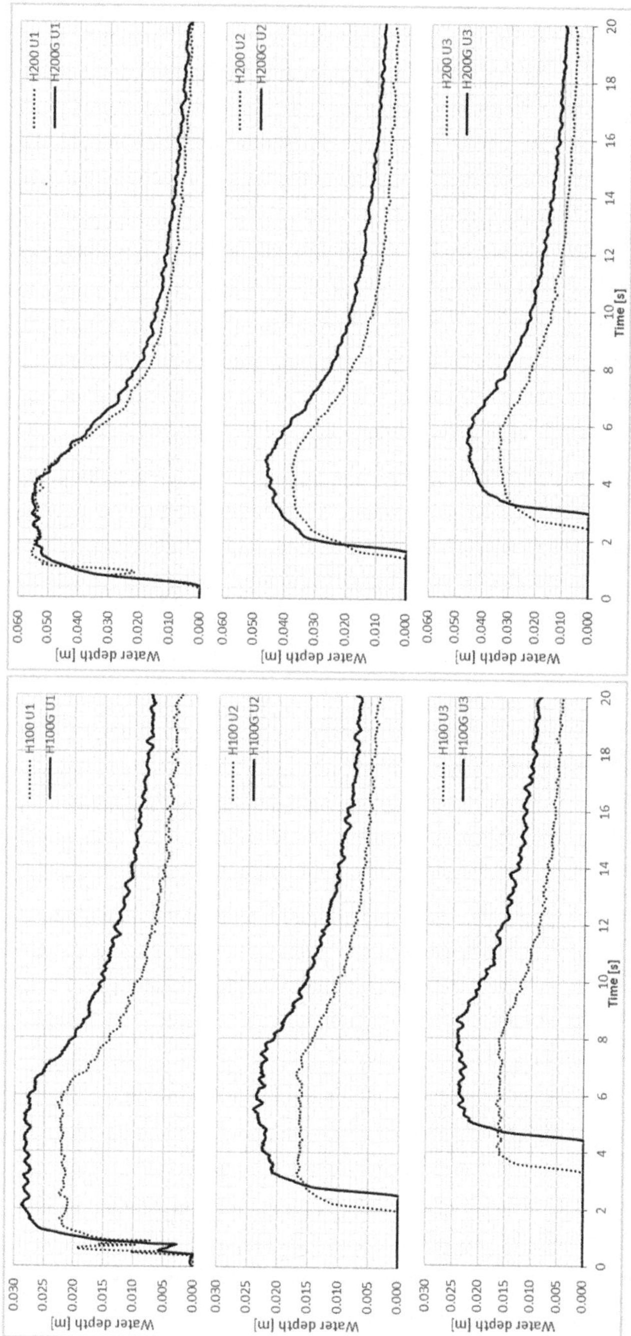

Figure 15.16 Comparison of water depths at U1, U2 and U3 for (left) H100 and H100G (with roughness) and (right) H200 and H200G (with roughness).

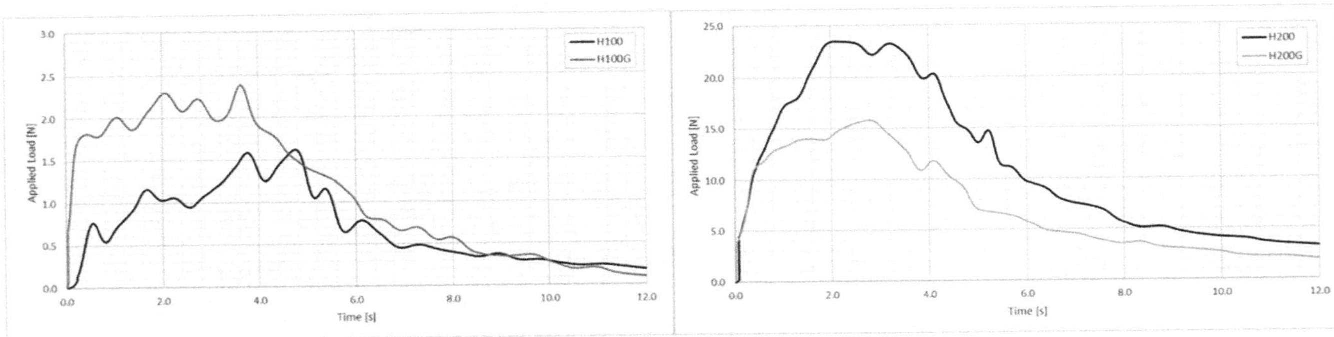

Figure 15.17 Applied load over time for (left) H100 and H100G and (right) H200 and H200G.

three-dimensional way creating different cross-waves and flow patterns depending on the blockage investigated. As expected, the vegetated slope increased the friction, thus slowing down the flow and reducing its Froude number considerably. This translated to a decrease in applied load on the buildings in the higher water depth cases. In terms of applied load on the urban settlements, the level of blockage had no effect on the higher water depth cases while it aggravated the applied load in the lower water depths. This was attributed to the hydraulic jump created in the lower water depths resulting in the buildings being submerged for longer.

15.5 NUMERICAL MODELLING OF FLASH FLOODS

The Boscastle inspired laboratory experiment was also used as a validation case to investigate flash floods numerically and to develop a methodology and an optimal parametrisation for the hydraulic modelling of these types of events. Two- and three-dimensional OpenFOAM models were used to further investigate the interaction of the flood wave with the urban settlement of the experiment.

15.5.1 OpenFOAM software

OpenFOAM is a C++ toolbox used for the solving of computational fluid dynamic problems (Damián, 2012), developed in the 1980s and finally released as an open-source software in 2004 (Damián, 2012). The multiphase solver *interFoam* (part of OpenFOAM's CFD solver library) which models the interface between the water and the air was used here to provide further understanding of the physical processes of flash floods. *InterFoam* solves the Navier–Stokes equations and records the position of the water/air interface, using the VoF method (Volume of Fluid).

15.5.2 Slope

The dam break itself and the accelerated supercritical flow on the slope were 2D in nature, and thus a two-dimensional OpenFOAM model was used to represent the flow propagation. A parametric analysis was undertaken to best represent the flash flood event. First, following a sensitivity analysis for different mesh sizes, a mesh of 0.00025 m was selected and tested for different Courant numbers. Following that, a Courant number of 0.2 was selected as the most converged solution as it seemed to best capture the highest water depth. This agrees with Berberović *et al.* (2009) who states that for these types of open channel simulations, the Courant number criteria should be always set to less than 0.2. Then, different turbulence parameters were tested and $k = 0.2$ and $\epsilon = 0.2$ were selected, resulting in an eddy viscosity of $\mu_t = 18\ \mathrm{m/s^2}$. Finally, the results presented in Figure 15.18 are a combined turbulent and laminar model where a

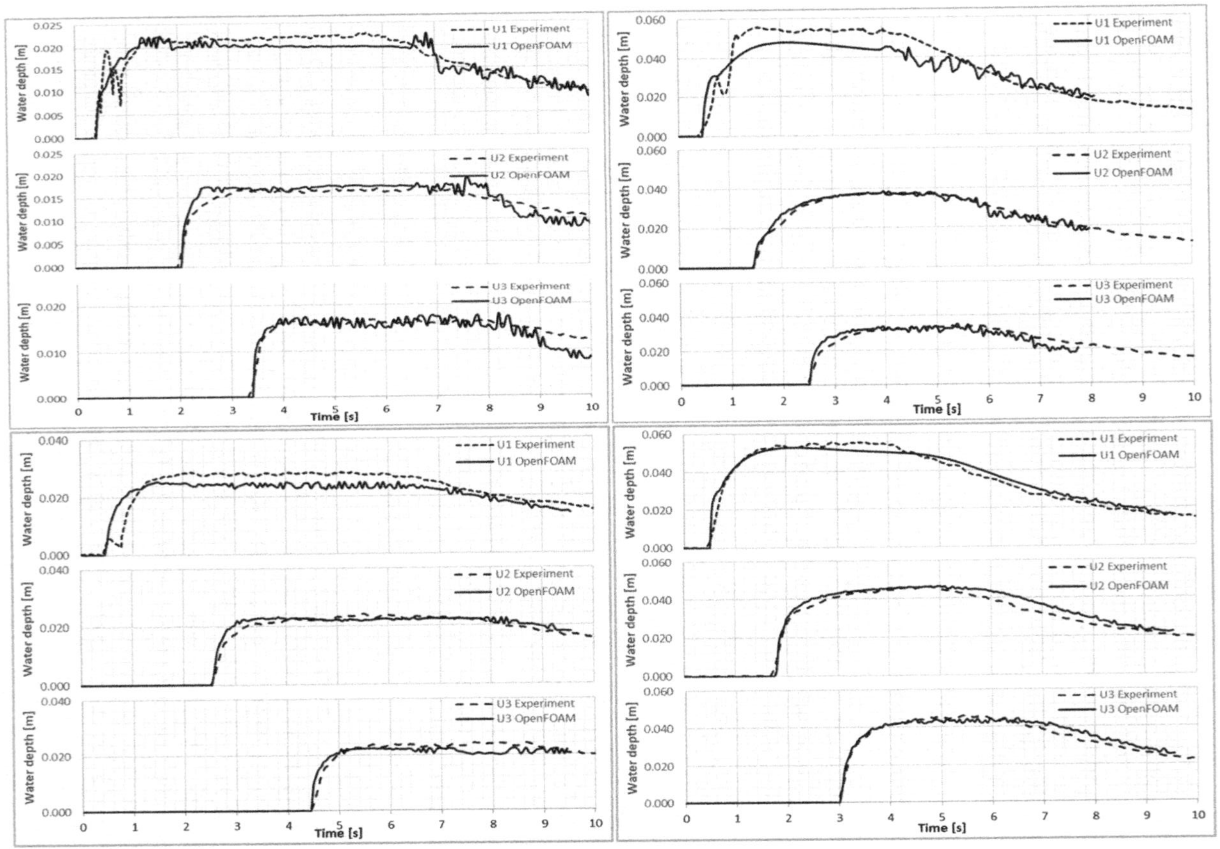

Figure 15.18 Comparison of experimental (dashed line) and numerical water depth (solid line) evolution along the slope for H100, H200, H100G and H200G at locations U1, U2 and U3.

turbulent model has been used to represent the initial stages of the water depth propagation and a laminar model to represent the stabilisation and decrease in the water depth elevation.

Figure 15.18 shows the comparison between experimental and simulated water depth results over time for the three ultrasonic probes, U1-U3, along the slope for H100, H200, H100G and H200G. A very good agreement is achieved between the model simulations and the experimental data at locations U2 and U3 and disparities found in location U1 are attributed to splashing from the gate opening and is an acceptable error for the numerical model.

In the 2D numerical simulations, using the combination of turbulent and laminar flow the model outputs were found to provide a very good fit to the experimental results in all four cases. In terms of the numerical simulation, the application of a 2D OpenFOAM model has highlighted the sensitivity of the flow to the model's parametrisation, the two-dimensionality of the flow in this part of the experiment has also shown that OpenFOAM is capable of simulating the supercritical flow on the slope very accurately.

15.5.3 Urban settlement

While the dam break itself and the accelerated supercritical flow on the slope described in the previous section were 2D in nature, and could be accurately approximated in a two-dimensional plane, modelling the interactions between the urban settlement and the dam break wave requires three dimensions.

Experimental photos of B1_H100 and B1_H200 are compared in Figures 15.19 and 15.20, with the relevant modelled snapshots in the 3D OpenFOAM simulations. In the OpenFOAM modelled images (A_3, A_4, B_3, B_4, C_3, C_4) arrows show the velocity direction and the different colours represent the range of the velocity magnitude from blue to red, representing a range of 0.00021 to 2.1 m/s.

The 3D OpenFOAM model proved an appropriate tool for modelling dam break events and wave structure interactions and accurately captured the hydraulic features of the flow in the urban settlement that were not modelled with the 2D model. The model is capable of reproducing the different flow characteristics, hydraulic jumps and wake zones and match substantially the arrival time of reflected waves. However, the parametrisation of such events is complex due to the range of specifications and variable factors that include mesh accuracy, refinement and alignment, the choice of the order of accuracy of the numerical model and the selection of the eddy viscosity and roughness coefficients. Simulation run speed in 2D and coarse meshes remain practical for consultants, engineers and designers but the additional detail provided in a 3D model leads to a deeper understanding of the fluid dynamics of the events, confirming that the use of 3D models can have positive effects on flood risk management decision making.

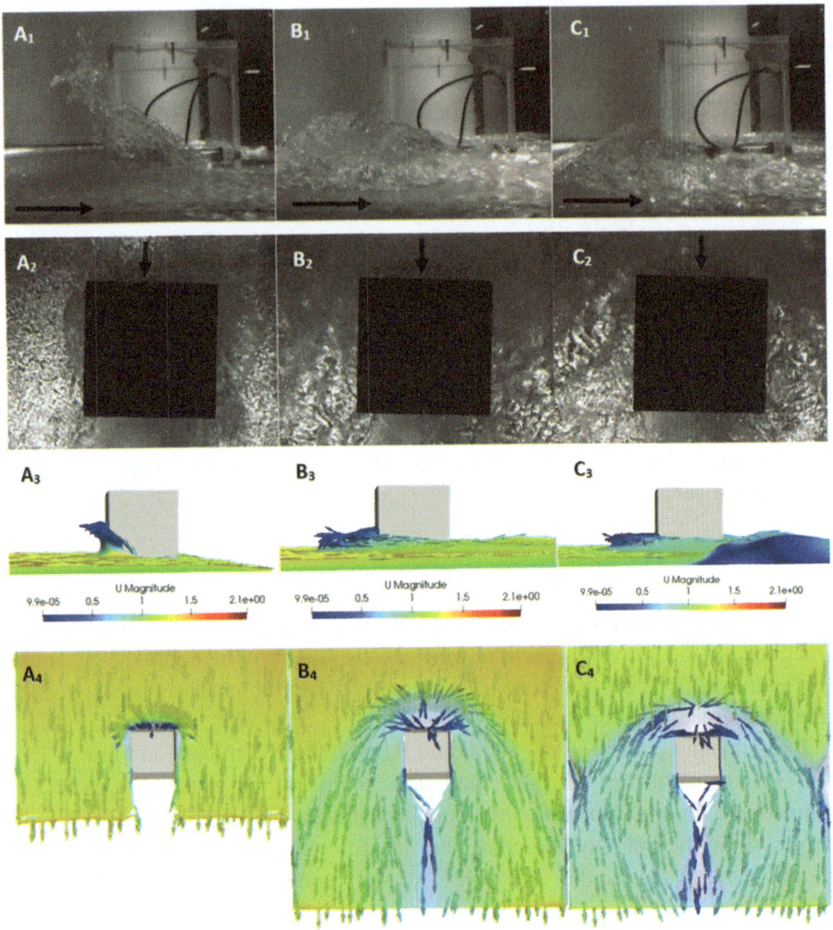

Figure 15.19 Comparison of photos and numerical snapshots of water impact on building for B1_H100 case at times $t = 4.18$ s (A_1, A_2, A_3, A_4), $t = 5.16$ s (B_1, B_2, B_3, B_4) and $t = 10$ s (C_1, C_2, C_3, C_4) from side (first and third row) and top view (second and fourth row).

15.6 FLASH FLOOD MODELLING FOR FLOOD RISK ANALYSIS

An increase in the frequency and magnitude of flooding is one of the severe expected consequences of climate change. Flash floods are of a challenging nature and as they are expected to be exacerbated by climate change understanding flash flood dynamics and the effect different drivers have on the influence of flood propagation is essential. It is therefore crucial to know how to accurately predict flood propagation and inundation extents in order to contribute

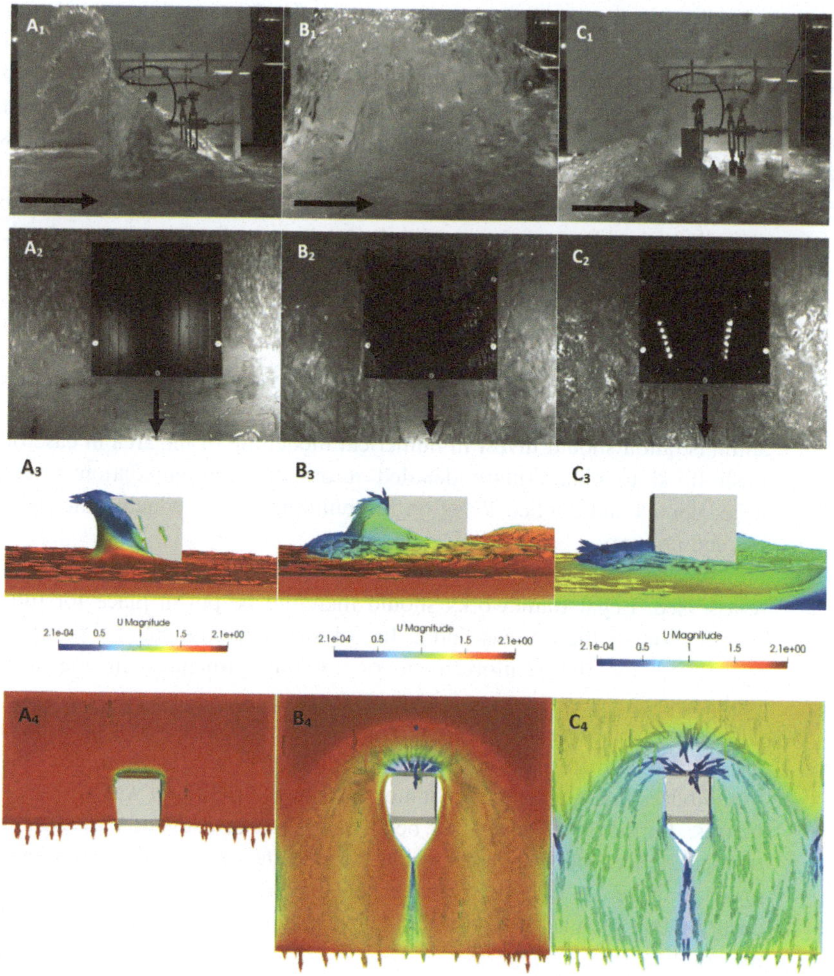

Figure 15.20 Comparison of photos and numerical snapshots of water impact on building for B1_H100 case at times $t = 2.96$ s (A_1, A_2, A_3, A_4), $t = 3.45$ s (B_1, B_2, B_3, B_4) and $t = 10$ s (C_1, C_2, C_3, C_4) from side (first and third row) and top view (second and fourth row).

to the development of better adaptation and preparedness strategies in flash flood prone areas.

15.6.1 Mitigations

Flash floods can be unpredictable but a factor that worsens them is land use due to human activities, projects and river interventions which strengthens the deterioration of the eco-geological systems (Arlikatti *et al.*, 2018). In terms of mitigation strategies, research has shown that in flood prone areas, the lack of preparation planning for recovery and mitigation strategies inevitably results in higher susceptibility and a deficient approach (Arlikatti *et al.*, 2018). Mitigation strategies that can be considered in flash flood management can be separated into two categories: (i) structural mitigations and (ii) non-structural mitigations. These include but are not limited to: erosion control mats, sustainable drainage systems SuDS, permeable paving, river dredging and realignment, overflow culverts, defence walls, rebuilding of bridges and flood protection structures, books, leaflets and documentaries.

Some further mitigation initiatives, in addition to the mitigations already in place, mainly applicable to the UK are presented below:

- Flash flood prone catchment areas should be identified, especially for catchments with historical flash floods (e.g., Boscastle). The local administration should invest in numerical modelling of the area in case of a flash flood to obtain more detailed information on inundation extents, water depths and applied loads on the buildings, thus reducing the risk to life and property.
- Policies and building guidelines in flash flood prone areas should then be re-assessed. Legal frameworks should therefore be put in place for future construction in these areas and a list of structural mitigations should be considered for the reinforcement of existing structures in the urban settlements.
- Blockage level: Depending on the blockage level of an urban settlement the residents should invest collectively on reinforcement, for example, of the front houses that would be the first exposed to a flash flood wave.
- Fences: Further work needs to be done in the investigation of different types of fences (ideal heights, widths and distance from the building) in order to have an effective breakwater for the first impact wave while ensuring that such a fence would not create additional water submersion for the buildings.
- Wall reinforcement: Reinforcement of existing walls should be considered for the buildings that would be strongly impacted by the flood waves.
- Low-vegetated slopes and higher roughness roads should be incorporated in catchment management plans as it has been shown that they can lead to a considerable reduction in the applied loads on the buildings.
- Outreach programs with educational material (e.g., videos) are necessary to emphasise the dangers and raise awareness for flash floods in flash flood prone areas. Residents need to be aware of potential solutions that can even be applied on a resident level and visual aids are a strong persuasion tool.

With the threat of an increase in the intensity, frequency and magnitude of extreme events, today more than ever we should continue to find the most suitable ways to manage flash flood prone catchments so that extreme events do not overwhelm and overthrow existing mitigation strategies.

REFERENCES

Ambrose P. (2011). Flooding in Bournemouth: 18th August 2011. 2011 Flooding Task and Finish Group, Bournemouth Bournemouth Borough Council, UK. Available from: www.bournemouth.gov.uk/environment-and-sustainability/Documents/flooding-in-bournemouth-18-august-2011.pdf, (accessed 15 November 2020).

Archer D. R. and Fowler H. J. (2015). Characterising flash flood response to intense rainfall and impacts using historical information and gauged data in Britain. *Journal of Flood Risk Management*, **11**(S1), S121–S133.

Arlikatti S., Maghelal P., Agnimitra N. and Chatterjee V. (2018). Should I stay or should I go? Mitigation strategies for flash flooding in India. *International Journal of Disaster Risk Reduction*, **27**, 48–56.

Aureli F., Dazzi S., Maranzoni A., Mignosa P. and Vacondio R. (2015). Experimental and numerical evaluation of the force due to the impact of a dam-break wave on a structure. *Advances in Water Resources*, **76**, 29–42.

Berberović E., van Hinsberg N. P., Jakirlić S., Roisman I. and Tropea C. (2009). Drop impact onto a liquid layer of finite thickness: Dynamics of the cavity evolution. *Physical Review. E*, **79**(3), 1–15.

Bettess R. (2005). Flooding in Boscastle and North Cornwall, August 2004. Project Report. HR Wallingford Ltd, UK.

British Geological Survey. (2006). Guide to Permeability Indices. Keyworth, Nottingham.

British Geological Survey. (2016). Geology of Britain Viewer [Online]. Available from: http://mapapps.bgs.ac.uk/geologyofbritain/home.html, (accessed 15 March).

Bukreev V. I. (2009). Force action of discontinuous waves on a vertical wall. *Journal of Applied Mechanics and Technical Physics*, **50**(2), 278–283.

Bukreev V. and Zykov V. (2008). Bore impact on a vertical plate. *Journal of Applied Mechanics and Technical Physics*, **49**, 926–933.

Burt S. (2005). Cloudburst upon Hendraburnick Down: The Boscastle storm of 16 August 2004. *Weather*, **60**(8), 219–227.

Chanson H. (2004). Experimental study of flash flood surges down a rough sloping channel. *Water Resources Research*, **40**(3), W03301.

Chaudhry M. H. (2008). Open-Channel Flow. Springer, New York.

Chella M. A., Tørum A. and Myrhaug D. (2012). An overview of wave impact forces on offshore wind turbine substructures. *Energy Procedia*, **20**, 217–226.

Chen H. Y., Xu W. L., Deng J., Xue Y. and Li J. (2014a). Experimental investigation of pressure load exerted on a downstream dam by dam-break flow. *Journal of Hydraulic Engineering – ASCE*, **140**, 199–207.

Chen L. F., Zang J., Hillis A. J., Morgan G. C. J. and Plummer A. R. (2014b). Numerical investigation of wave–structure interaction using OpenFOAM. *Ocean Engineering*, **88**, 91–109.

Climate Data. (2018). Climate: Aberystwyth [Online]. Available from: https://en.climate-data.org/location/6805/, (accessed 28 June).

Collier C. G. and Fox N. I. (2003). Assessing the flooding susceptibility of river catchments to extreme rainfall in the United Kingdom. *International Journal of River Basin Management*, **1**(3), 225–235.

Corps of Engineers. (1960). Flood Resulting from Suddenly Breached Dams. Mississippi. U. S. Army Engineer Waterways Experiment Station, Vicksburg, Mississippi.

Coulthard T., Frostick L., Hardcastle H., Jones K., Rogers D., Scott M. and Bankoff G. (2007). The June 2007 Floods in Hull. Hull City Council, UK. Available from: www.coulthard.org.uk/downloads/floodsinhull1.pdf, (accessed 15 November 2020).

Damián S. M. (2012). Description and Utilization of InterFoam Multiphase Solver. Available from: http://infofich.unl.edu.ar/upload/3be0e16065026527477b4b948c4caa7523c8ea52.pdf, (accessed 15 November 2016).

EMDAT. (2017). Number of Disasters [Online]. Université catholique de Louvain, Brussels, Belgium. Available from: www.emdat.be/, (accessed 23 January 2019).

Environment Agency. (2016). Catchment Data Explorer [Online]. Available from: http://environment.data.gov.uk/catchment-planning/WaterBody/GB108049007170, (accessed 8 February).

ESSL. (2018). Deadly Flash Floods in 2018 [Online]. Available from: www.essl.org/cms/deadly-flash-floods-in-2018/, (accessed 14 December).

Falconer R. A. (2012). Modelling extreme flood events and associated proxesses in rivers, estuaries and coastal environments. Proceedings of 10th International Conference on Hydroscience and Engineering (ICHE 2012), 4–8 November, Orlando, USA.

Ferreira R. M. L., Alves E. C. T. L., Leal J. G. A. B. and Cardos A. H. (2006). River Flow 2006, two volume set. Proceedings of International Conference on Fluvial Hydraulics, 6–8 September 2006, Lisbon, Portugal. Taylor and Francis Group, Florida.

Golding B., Clark P. and May B. (2005). The Boscastle flood: Meteorological analysis of the conditions leading to flooding on 16 August 2004. *Weather*, **60**(8), 230–235.

Halcrow Group Ltd. (2017). Boscastle Flood Defences. Available from: https://cms.esi.info/Media/documents/37236_1400680833562.pdf, (accessed 25 January 2017).

Horritt M. S. and Bates P. D. (2002). Evaluation of 1D and 2D numerical models for predicting river flood inundation. *Journal of Hydrology*, **268**(1–4), 87–99.

HR Wallingford. (2005). The Boscastle flood of 16 August 2004: Characteristics, causes and consequences. 40th Defra Flood and Coastal Management Conference, 5–7 July, York, UK. Available from: https://eprints.hrwallingford.com/546/, (accessed 15 November 2020).

Huang W., Cao Z. X., Qi W. J., Pender G. and Zhao K. (2015). Full 2D hydrodynamic modelling of rainfall-induced flash floods. *Journal of Mountain Science*, **12**(5), 1203–1218.

Into Cornwall. (2015). Boscastle [Online]. Available from: www.intocornwall.com/engine/azabout.asp?guide=Boscastle, (accessed 15 March).

IPCC. (2007). Climate Change 2007: Synthesis Report. IPCC, Geneva, Switzerland.

IPCC. (2014). Climate Change 2014: Synthesis Report. IPCC, Geneva, Switzerland.

IPCC. (2017). Scoping of the IPCC Sixth Assessment Report (AR6). Forty-sixth session of the IPCC, 6–10 September Montreal, Canada. IPCC, Geneva, Switzerland.

Kjeldsen T. R., Macdonald N., Lang M., Mediero L., Albuquerque T., Bogdanowicz E., Brázdil R., Castellarin A., David V., Fleig A., Gül G. O., Kriauciuniene J., Kohnová S., Merz B., Nicholson O., Roald L. A., Salinas J. L., Sarauskiene D., Šraj M.,

Strupczewski W., Szolgay J., Toumazis A., Vanneuville W., Veijalainen N. and Wilson D. (2014). Documentary evidence of past floods in Europe and their utility in flood frequency estimation. *Journal of Hydrology*, **517**, 963–973.

Kleefsman K. M. T., Fekken G., Veldman A. E. P., Iwanowski B. and Buchner B. (2005). A volume-of-fluid based simulation method for wave impact problems. *Journal of Computational Physics*, **206**(1), 363–393.

Kobiyama M. and Goerl R. F. (2007). Quantitative method to distinguish flood and flash flood as disasters. *SUISUI Hydrological Research Letters*, **1**, 11–14.

Lhomme J., Gutierrez-Andres J., Weisgerber A., Davison M., Mulet-Marti J., Cooper A. and Gouldby B. (2010). Testing a new two-dimensional flood modelling system: analytical tests and application to a flood event. *Journal of Flood Risk Management*, **3**(1), 33–51.

Liang Q. and Borthwick A. G. L. (2009). Adaptive quadtree simulation of shallow flows with wet–dry fronts over complex topography. *Computational Fluids*, **38**(2), 221–234.

Liu C., Guo L., Ye L., Zhang S., Zhao Y. and Song T. (2018). A review of advances in Chnia's flash flood early-warning system. *Natural Hazards*, **92**, 619–634. https://doi.org/10.1007/s11069-018-3173-7

Lobovský L., Botia-Vera E., Castellana F., Mas-Soler J. and Souto-Iglesias A. (2014). Experimental investigation of dynamic pressure loads during dam break. *Journal of Fluids and Structures*, **48**, 407–434.

Marsh T. J. and Hannaford J. (2007). The Summer 2007 Floods in England and Wales – a Hydrological Appraisal. NERC/Centre for Ecology and Hydology, Wallingford, UK. Available from: http://nora.nerc.ac.uk/id/eprint/2814/, (accessed 15 November 2020).

Marsooli R. and Wu W. (2014). 3-D finite-volume model of dam-break flow over uneven beds based on VOF method. *Advances in Water Resources*, **70**, 104–117.

Merz R. and Bloschl G. (2003). A process typology of regional floods. *Water Resources Research*, **39**(12), 5.1–5.20.

Met Office. (2010). UK Climate [Online]. Available from: www.metoffice.gov.uk/public/weather/climate/#?tab=climateMaps, (accessed 27 August).

Met Office. (2011a). Birmingham Climate [Online]. Available from: www.metoffice.gov.uk/public/weather/climate/gcqdt4b2x, (accessed 14 January).

Met Office. (2011b). Keswick Climate [Online]. Available from: www.metoffice.gov.uk/public/weather/climate/gcty8njjs, (accessed 14 January).

Met Office. (2013). Exceptionally Wet Weather – November 2012 [Online]. Available from: www.metoffice.gov.uk/climate/uk/interesting/november-2012, (accessed 28 June).

Muchan K. M., Nikos B. L., Turner S., Lewis M. and Clemas S. (2018). Hydrological Summary for the United Kingdom: May 2018. (CEH Project no. C04954). NERC/Centre for Ecology & Hydrology, Wallingford, UK.

Murray S. J., Smith A. D. and Phillips J. C. (2012). A modified flood severity assessment for enhanced decision support: Application to the Boscastle FLASH FLOOD of 2004. *Weather Forecast*, **27**(5), 1290–1297.

Néelz S. and Pender G. (2010). Benchmarking the Latest Generation of 2D Hydraulic Modelling Packages. Environment Agency, Bristol.

Nicholas Pearson Associates. (2012). Boscastle Flood Alleviation Scheme [Online]. Available from: www.npaconsult.co.uk/projects.asp?gid=99&pid=16&pkeyword=&prelated=&pproject=, (accessed 5 April).

North Cornwall District Council. (2004). North Cornwall District Council [Online]. Available from: www.ncdc.gov.uk, (accessed 2 April).

Peng S. H. and Chen S. C. (2006). Comparison of numerical and experimental study of dam-break induced mudflow. *Sediment Dynamics and the Hydromorphology of Fluvial Systems*, **306**, 548–555.

Roca M. and Davison M. (2010). Two dimensional model analysis of flash-flood processes: application to the Boscastle event. *Journal of Flood Risk Management*, **3**(1), 63–71.

Rowiński P. and Radecki-Pawlik A. (2015). Rivers – Physical, Fluvial and Environmental Processes, 1st edn. Springer, Switzerland.

Shrestha A. B., Shah S. H. and Karim R. (2008). Resource Manual on Flash Flood Risk Management. International Centre for Integrated Mountain Development (ICIMOD), Kathmandu.

Soares-Frazao S. (2007). Experiments of dam-break wave over a triangular bottom sill. *Journal of Hydraulic Research*, **45**, 19–26. https://doi.org/10.1080/00221686.2007.9521829

Stamataki I., Zang J., Buldakov E., Kjeldsen T. and Stagonas J. (2018). Study of dam break flow interaction with urban settlements over a sloping channel. *E3S Web of Conferences*, **40**, 06006. https://doi.org/10.1051/e3sconf/20184006006

Stansby P. K., Chegini A. and Barnes T. C. D. (1998). The initial stages of dam-break flow. *Journal of Fluid Mechanics*, **374**, 407–424.

Testa G., Zuccalà D., Alcrudo F., Mulet J. and Soares-Frazão S. (2007). Flash flood flow experiment in a simplified urban district. *Journal of Hydraulic Research*, **45**(Extra issue), 37–44.

Toombes L. and Chanson H. (2011). Numerical limitations of hydraulic models. Proceedings of 34th IAHR World Congress – Balance and Uncertainty, Brisbane, Australia. Engineers Australia, Brisbane, Australia, pp. 2322–2329.

Trivellato F. (2004). Experimental and numerical investigation of bore impact on a wall. *Transactions of the Built Environment*, **71**, 3–12.

UNISDR. (2015). The Human Cost of Weather Related Disasters – 1995–2015. Available from: www.unisdr.org/files/46796_cop21weatherdisastersreport2015.pdf, (accessed: 28 August 2020).

Warren R. A., Kirshbaum D. J., Plant R. S. and Lean H. W. (2014). A 'Boscastle-type' quasi-stationary convective system over the UK Southwest Peninsula. *Quarterly Journal of the Royal Meteorological Society*, **140**, 240–257.

Webb S. (2013). Heavy rain and flooding in and around Aberystwyth on 8–9 June 2012. *Weather*, **68**(6), 162–165.

Werner M. and Cranston M. (2009). Understanding the value of radar rainfall nowcasts in flood forecasting and warning in flashy catchments. *Meteorological Applications*, **16** (1), 41–55.

World Meteorological Organisation. (2007). Guidance on Flash Flood Management – Recent Experience from Central and Eastern Europe. WMO, Geneva, Switzerland.

World Meteorological Organisation. (2012). Management of Flash Floods. Available from: https://library.wmo.int/index.php?lvl=notice_display&id=16348#.X0tqMOhKjD4, (accessed 15 November 2020).

World Meteorological Organisation. (2017). Flash Flood Guidance Systems. WMO, Geneva, Switzerland.

World Weather & Climate. (2016). CLimate in Boscastle [Online]. Available from: https://weather-and-climate.com/average-monthly-precipitation-Rainfall-inches,boscastle-cornwall-gb,United-Kingdom, (accessed 28 August 2020).

Xia J., Falconer R. A. and Lin B. (2011a). Incipient velocity for partially submerged vehicles in floodwaters AU - Shu, Caiwen. *Journal of Hydraulic Research*, **49**(6), 709–717.

Xia J., Falconer R. A., Lin B. and Tan G. (2011b). Modelling flash flood risk in urban areas. *Proceedings of the Institution of Civil Engineers-Water Management*, **164**(6), 267–282.

Xia J., Falconer R., Xiao X. and Wang Y. (2014). Criterion of vehicle stability in floodwaters based on theoretical and experimental studies. *Natural Hazards*, **70**, 1619–1630.

Xia J., Teo F., Falconer R. A., Chen Q. and Deng S. (2018). Hydrodynamic experiments on the impacts of vehicle blockages at bridges. *Journal of Flood Risk Management*, **11**(S1), S395–S402.

Xu J., Eriksson M., Ferdinand J. and Merz J. (2006). Managing Flash Floods and Sustainable Development in the Himalayas. ICMOD, Kathmandu, Nepal.

Zech Y., Soares-Frazão S. and Van Emelen S. (2015). Modelling of fast hydraulic transients: issues, challenges, perspectives. *La Houille Blanche*, **5**, 5–15. https://doi.org/10.1051/lhb/20150049

Zhainakov A. Z. and Kurbanaliev A. Y. (2013). Verification of the open package OpenFOAM on dam break problems. *Thermophys Aeromech+*, **20**(4), 451–461.

Index

IWA Publishing's authorised EU representative for General Product Safety Regulations is Diane D'Arras, 15 rue Duret, 75116 Paris, France, e-mail: safety@iwap.co.uk.

Printed and bound by CPI Group (UK) Ltd, Croydon, CR0 4YY
17/09/2026
02218797-0003